EARTH
IN THE
BALANCE

EARTH
IN THE
BALANCE

EARTH
IN THE
BALANCE

ECOLOGY AND THE HUMAN SPIRIT

AL GORE

Author of *An Inconvenient Truth*

A *NEW YORK TIMES* BESTSELLER
WITH A NEW FOREWORD BY THE AUTHOR

RODALE

Rodale books may be purchased for business or promotional use or for special sales. For information, please write to: Special Markets Department, Rodale, Inc., 733 Third Avenue, New York, NY 10017

Rodale, Inc. makes every effort to use acid-free ∞, recycled paper ♲.

Library of Congress Cataloging-in-Publication data is available.

ISBN: 0-618-05664-5 hardcover
ISBN: 13: 978–1–59486–637–1 paperback
ISBN: 10: 1–59486–637–6

Printed in the United States of America

MP 10 9 8 7 6 5 4 3 2 1

Credits appear on page 408.

We inspire and enable people to improve their lives and the world around them
For more of our products visit **rodalestore.com** or call 800-848-4735

TO MY SISTER

Nancy LaFon Gore Hunger

January 23, 1938–July 11, 1984

Contents

PART III
STRIKING THE BALANCE

Foreword:

The Coming
"Environment Decade"

LOOKING BACK

I began writing *Earth in the Balance* in the spring of 1989 and finished it in the fall of 1991. At the time I was still in the U.S. Senate after running for president in 1987–88, in part to bring more visibility to the issue of the climate crisis. I was in the midst of rethinking all of my priorities, and one of the outcomes of that process was that I had already decided not to run for president again in 1992, although I was getting lots of encouragement to do so.

While writing this book, I started giving the slide show that would later become the basis for the documentary movie and book *An Inconvenient Truth*. As I was writing *Earth in the Balance* that fall, I was at the same time in the midst of planning a trip to the Earth Summit in Rio de Janeiro where I would lead the U.S. Senate delegation the following summer. As a matter of fact, it was while I was in Rio participating in the Summit that I received a telephone call from Warren Christopher—my first contact from him on the subject—asking if I would allow my name to be placed on the list of those being considered as a vice-presidential candidate on Bill Clinton's ticket. I didn't think being named to the ticket was a serious possibility at the time. Since Clinton was from Arkansas and I was from the neighboring state of Tennessee, this combination would not have made sense according to the traditional political calculus. But the old tradition, surprisingly, was upended. I was selected and became vice president. I had no idea what would follow: the successful 1992 campaign with Clinton, the vice presidency for eight years, the 2000 campaign, the thirty-six days that followed Election Day 2000, the Supreme Court decision in Bush v. Gore, and all that has happened since in the U.S. and the world.

As I look back over the past fifteen years since I finished writing this book, it is striking to recall what a central role *Earth in the Balance* has played in my life. It helped me to focus on the mission of solving the climate crisis—a mission I am still pursuing, with an ever-increasing sense of urgency and determination. It spawned the principal attacks on me in the 1992 election by then President George H. W. Bush, when he referred to me as "Ozone Man." And it formed my principal agenda for eight years in the White House.

The book is a snapshot in time that captures the evolution of my personal and intellectual involvement in the issue of global environment, and the 2000 update (also included in this volume) provides the other bookend for the work of the Clinton-Gore years on the environment. And now, this new edition of *Earth in the Balance* is offered to complement my latest book, *An Inconvenient Truth*.

When I first wrote *Earth in the Balance*, I could not have predicted the political upheavals to come. Nor could I have predicted how long it would take for the scientific consensus to solidify. The nature and severity of the climate crisis had seemed painfully obvious to me for quite a long time; but in retrospect, I wish that we could have had in the 1990s the deafening scientific consensus that has emerged in more recent years. It would have been much easier to galvanize the public and persuade the Congress to act.

THE GATHERING STORM

Since *Earth in the Balance* was released in January 1992, voluminous new scientific research has greatly strengthened our grasp of the basics: Global warming is real, it is getting worse rapidly, it is mainly caused by human beings, we need to act now to avoid the worst of its consequences—and it is not too late.

Most everyone, by now, understands that the burning of fossil fuels (like coal, oil, and gas) thickens the normally thin atmospheric blanket around the globe, and in the process, traps much more of the sun's heat close to the earth's surface. The unnatural levels of atmospheric heating that result radically destabilize the climatic balance that has existed for all of human history.

To put it another way, we have radically transformed the fundamental relationship between humankind and the earth. This is due to a combination of factors. First, we have quadrupled the human population of our planet in just the past hundred years.

It took 10,000 human generations to reach a population of 2 billion when my generation—the baby boomers—was born. Now, in the course of a single lifetime—ours—we are increasing in numbers from 2 billion to 9 billion (over the next forty-five years). We have already passed the 6.5 billion mark.

Second, the power of the new technologies now at our disposal has magnified by thousands of times the impact each individual can have on the natural world. Our old habits, once largely benign, are now pursued with such enhanced power that we have become like the proverbial "bull in a china shop."

Third, our bizarre focus on short-term thinking and instant gratification—not just as individuals, but more important in the behavior of markets, national economies, and political agendas—has led to a systematic exclusion of long-term consequences in our decisions and policies.

The results of this profoundly new relationship between humans and the earth are devastating. It is now not so much a relationship as a collision.

The scientific community has deluged us with evidence of the tremendous changes we are wreaking on the planet, their version of shouting from the rooftops.

In the original 1992 edition of this book, I depicted information from an ice core going back 160,000 years demonstrating that, at the time, carbon dioxide levels in the earth's atmosphere had never been higher. In the 2000 reissue of the book, I used a newer version that went back 420,000 years—and it pointed even more clearly to the same conclusion. Now, in these pages, you can see that current CO_2 levels are at the highest they have been in the past 650,000 years! As with the rest of the evidence, it still points to exactly the same conclusion that this book presented fifteen years ago; it's just that the evidence is even more overwhelming now.

Since 1992, four prestigious scientific panels have prepared new compendiums of studies that provide staggering amounts of data and have created the strongest consensus imaginable on this issue to advise policymakers:

• Two voluminous reports have been issued by the Intergovernmental Panel on Climate Change, a panel of more than 2,000 of the world's experts, which conclude that humans are having an impact on the earth's climate and detail many of the terrible consequences that are occurring now and the infinitely worse consequences that will happen in the future if nothing is done to address the climate crisis.

• The National Academy of Sciences, the gold standard of U.S. research, has issued several reports, including one in 2006, saying that we are "probably" in the warmest period in the past two millennia and has advised the Bush administration on key questions relating to the basics of the climate system.

• The U.S. Global Change Research Program released its 2000 National Assessment, detailing, for the first time, the regional impacts of climate change in terms of geography and on key sectors (like agriculture, human health, and forests) here in the United States.

• In 2004, the Arctic Climate Impacts Assessment was released. The assessment described arctic temperatures that are increasing at almost twice the rate as those of the rest of the world, largely because the highly reflective arctic snow and ice is melting and leaving behind darker land and ocean surfaces, which in turn absorb more of the sun's heat, further warming the region. It also concluded that reductions in sea ice will drastically shrink marine habitat for polar bears, ice-inhabiting seals, and some seabirds, pushing some species to extinction.

In addition to these major studies, other pieces of the puzzle have emerged including:

• A September 2005 study in *Nature* concluding that increasing atmospheric levels of carbon dioxide reduce levels of pH in the ocean, which makes its waters more acidic. If this cycle continues, key marine organisms, like corals and some plankton (which form the bottom of the oceanic food chain), will have trouble maintaining and building their calcium carbonate skeletons. The findings of this study indicate that these conditions could develop within decades, not centuries, as previously believed.

- Several studies in *Science* and a recent study in both *Nature* and *Geophysical Research Letters* confirm the emerging consensus in the climate community that the climate change is warming the ocean waters, exacerbating the destructive energy of hurricanes.

- A July 2006 study in *Geophysical Research Letters* reports that alpine glaciers in Europe could nearly disappear within this century.

- A July study in *Science* reports that higher temperatures have contributed to an increase in large costly wildfires that have hit the western United States. This phenomenon is the result of the fact that summer arrives earlier and lasts longer, creating drier and drier vegetation.

- And perhaps most alarming, NASA recorded new satellite data demonstrating that the Greenland ice sheet is more unstable than previously believed. Researchers at Harvard have evidence of glacial earthquakes registering between 4.0 and 5.0 on the Richter scale. This is in addition to the collapse of huge ice shelves the size of Rhode Island, which have broken off of the peninsula of Antarctica, and growing concerns over the stability of the Western Antarctic ice sheet. If we were to destabilize and melt the 10,000-foot-thick mound of ice on top of Greenland or the equally enormous mass of ice of Western Antarctica, either one would raise the sea level worldwide more than 20 feet.

Incredibly, we could be setting these changes into motion in the lifetime of our children and grandchildren. Unless we act boldly and quickly, some of the leading scientific experts are now telling us that without dramatic changes to cut the pollution that causes global warming, we are in grave danger of crossing a point of no return within the next ten years!

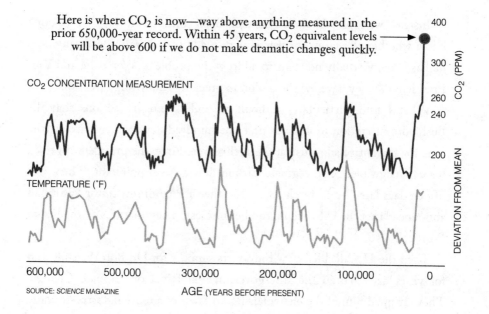

Here is where CO_2 is now—way above anything measured in the prior 650,000-year record. Within 45 years, CO_2 equivalent levels will be above 600 if we do not make dramatic changes quickly.

CO_2 CONCENTRATION MEASUREMENT

TEMPERATURE (°F)

CO₂ (PPM)

DEVIATION FROM MEAN

400
300
260
240
200

600,000   500,000   300,000   200,000   100,000   0

SOURCE: *SCIENCE* MAGAZINE   AGE (YEARS BEFORE PRESENT)

POLITICS AND THE PRIVATE INTEREST

In 1992 and in 2000, I was filled with a sense of hope, and the strong feeling that we were building toward addressing the climate crisis. I had helped to negotiate the Framework Convention on Climate Change and the Kyoto Protocol. We were developing new partnerships with the U.S. auto industry for more efficient automobiles and creating new standards for more efficient appliances. We had not only acknowledged the crisis but also had begun preparing to deal with its implications.

I knew how difficult it would be; the Congress had already refused to give serious consideration to the Kyoto Treaty and had rebuffed strong proposals for a carbon tax.

But I was confident that the "bully pulpit" I hoped to gain as president—and the new stronger scientific consensus I knew was about to emerge—would mean that we in the U.S. were about to shift into high gear in addressing the climate crisis.

Instead, we took a U-turn. President Bush continues to say that we don't know whether the climate crisis is man-made or naturally occurring, and he has taken virtually no steps to address the problem. Worse, he and Vice President Cheney have led the nation in precisely the wrong direction.

What I find particularly difficult to understand are actions that the Bush administration took in its first one hundred days, which have set the tone for the entire administration in this area. First, the president reversed his campaign pledge to regulate carbon dioxide as a pollutant. Then, just fifteen days later—on March 28, 2001—then EPA Administrator Whitman announced that the U.S. had no further interest in negotiations on the Kyoto Protocol.

Then the U.S. decided to withdraw its support for Dr. Bob Watson, who for years had chaired the Intergovernmental Panel on Climate Change. They stripped him of the chairmanship in favor of a more industry-friendly candidate. Their motivations for this dizzying reversal were not entirely clear until later, when under a Freedom of Information Act Request, the National Resources Defense Council obtained a faxed memo from ExxonMobil's Washington office to the Council on Environmental Quality at the White House dated February 6, 2001. The memo spelled out a plan.

ExxonMobil requested the removal of Dr. Watson stating, "Can Watson be replaced now at the request of the U.S.?" The memo also suggested removal of other officials who worked on the U.S. National Assessment on Climate Change, including Drs. Rosina Bierbaum and Mike MacCracken. The two subsequently left the panel. The same ExxonMobil memo also proposed the name of Dr. Harlan Watson to the post of lead climate negotiator at the State Department and this "suggestion" was also implemented by the administration. Dr. Watson was the person who, as the lead U.S. representative, torpedoed the latest international effort to strengthen the fight against global warming.

Other links between ExxonMobil and the Bush administration appeared as well. For example, Philip Cooney was appointed as chief of staff for the

White House Council on Environmental Quality. Prior to joining the Bush administration, Cooney was a lobbyist at the American Petroleum Institute (an oil-industry lobby group). During his tenure, Mr. Cooney censored a number of official scientific reports on climate change, despite the fact that he had no formal scientific training. In June 2005, after the embarrassing discovery of his sabotage of scientific integrity on behalf of the oil industry, he resigned his position with the Bush White House and was immediately hired by ExxonMobil.

Sadly, the trend illustrated by the Cooney appointment is pervasive in the current Administration. On February 18, 2004, more than sixty leading scientists—Nobel laureates, leading medical experts, former federal agency directors, university chairs and presidents—signed a statement voicing their concerns over the misuses of science by the Bush White House. Their concerns ranged across a wide array of issues from climate change to childhood lead poisoning and reproductive health. They cited incidences of outright censorships and political "oversight" of government scientists at a variety of agencies. The statement, organized by the Union of Concerned Scientists says in part, "The distortion of scientific knowledge for partisan political ends must cease if the public is to be properly informed about issues central to its well-being, and the nation is to benefit fully from its heavy investment in scientific research and education." The signatories' list had grown to include forty-nine Nobel laureates, sixty-three National Medal of Science recipients, and 175 members of the National Academies.

The Bible says, "Where there is no vision, the people perish." We must have the unfettered vision derived from the best science in order for our leaders to make the best decisions for the earth and its inhabitants.

GLOBAL WARMING HITS WALL STREET

One of the most heartening signs of progress since 1992 is the tremendous response from the private sector to the market signals sent by a plethora of new policy initiatives including the Regional Greenhouse Gas Initiative in the northeast, the Western Governors' climate agreement, the emissions trading market in the Kyoto Protocol, and many others. More than forty companies, including Boeing, Duke Energy, DuPont, Georgia-Pacific, IBM, Whirlpool, and Weyerhaeuser, now belong to the Business Environmental Leadership Council. They have committed to a range of proactive measures to address climate change, including setting targets for greenhouse gas emissions reductions, implementing new energy efficiency measures, participating in emissions trading, and investing in carbon sequestration research.

FedEx, Kinkos, General Motors, Johnson & Johnson, Pitney Bowes, Staples, Starbucks, Wal-Mart, and other top companies are now trading renewable-energy credits to purchase and use green power across the country.

And a large number of institutional investors, with trillions of dollars in assets, have called on the Securities and Exchange Commission to require corporate disclosure of the financial risks of climate change.

Wall Street knows we have a problem, and I believe that with the right signals and the right investments, we can create the kinds of public-private partnerships that will unleash the power of the markets to solve the climate crisis.

ECOLOGY AND THE HUMAN SPIRIT

Abraham Maslow, the psychologist, once said, "If the only tool you have is a hammer, you tend to see every problem as a nail." It is this myopic thinking that has kept so many from seeing the truth about the climate crisis.

But since 1992, we have seen increasing evidence of moral leadership emerging from a variety of quarters. Most recently, eighty-six evangelical leaders, including Rev. Dr. Rick Warren, joined in the call to act now to solve the climate change problem. They know what President Abraham Lincoln sensed at the time of our nation's greatest trial, "We must disenthrall ourselves, and then we shall save our country." America is beginning to awaken. If we do so, we will save our planet.

The Chinese expression for "crisis" consists of two characters. The first is a symbol for danger, the second is a symbol for opportunity. And this is our opportunity to rise to meet this crisis successfully, to see the truth of our circumstances, and to chart our own course for a better world.

The climate crisis brings us the opportunity to experience what few generations in history have had the privilege of knowing: a generational mission, the exhilaration of a compelling moral purpose, a shared and unifying cause, and the thrill of being forced by circumstances to put aside the pettiness and conflict that so often stifle the restless human need for transcendence.

Right now, it is difficult to imagine that we could cut global emissions of the pollutants that cause global warming by 70 to 80 percent, that the roadways could soon be filled with hybrid electric and plug-in hybrid vehicles, that the greenest buildings could be generating power that they are actually selling back to the utilities. It is perhaps hard now to fully see the potential of hydrogen power, to imagine a smart superelectric grid, to envision new biofuels running our vehicles or having access to the most energy efficient machines and appliances as well as the newest nanotechnologies and manufacturing techniques—all of this would be new to us. But examples of nearly all this technology really do exist today or will exist in the very near future. New market technologies are as difficult for us to see today as the Internet was for workers in the 1980s.

We have encountered and accepted other great challenges in the past. We declared our liberty and then won it, establishing a new country. We devised a new form of government. We freed the slaves. We gave women the right to vote. We took on Jim Crow and segregation. We cured polio and helped eradicate smallpox. We landed on the moon. We brought down communism and helped end apartheid.

We even solved a previous global environmental crisis—the hole in the stratospheric ozone layer—because Republicans and Democrats, rich nations and poor nations, businessmen and scientists, all came together to shape a solution.

We cannot wait any longer to solve this crisis.

We have nearly all the tools we need to solve this problem, perhaps with one exception: What we are missing is the political will that would be required to really affect change. Thankfully, in a democracy like ours, political will is a renewable resource.

Introduction

Writing this book is part of a personal journey that began more than twenty-five years ago, a journey in search of a true understanding of the global ecological crisis and how it can be resolved. It has led me to travel to the sites of some of the worst ecological catastrophes on the planet and to meet some of the extraordinary men and women throughout the world who are devoting their lives to the growing struggle to save the earth's environment. But it has also led me to undertake a deeper kind of inquiry, one that is ultimately an investigation of the very nature of our civilization and its relationship to the global environment.

The edifice of civilization has become astonishingly complex, but as it grows ever more elaborate, we feel increasingly distant from our roots in the earth. In one sense, civilization itself has been on a journey from its foundations in the world of nature to an ever more contrived, controlled, and manufactured world of our own imitative and sometimes arrogant design. And in my view, the price has been high. At some point during this journey we lost our feeling of connectedness to the rest of nature. We now dare to wonder: Are we so unique and powerful as to be essentially separate from the earth?

Many of us act — and think — as if the answer is yes. It is now all too easy to regard the earth as a collection of "resources" having an intrinsic value no larger than their usefulness at the moment. Thanks in part to the scientific revolution, we organize our knowledge of the natural world into smaller and smaller segments and assume that the connections between these separate compartments aren't really important. In our fascination with the parts of nature, we forget to see the whole.

The ecological perspective begins with a view of the whole, an understanding of how the various parts of nature interact in patterns that tend toward balance and persist over time. But this perspective cannot treat the earth as something separate from human civilization; we are part of the whole too, and looking at it ultimately means also looking at ourselves. And if we do not see that the human part of nature has an increasingly powerful influence over the whole of nature — that we are, in effect, a natural force just like the winds and the tides — then we will not be able to see how dangerously we are threatening to push the earth out of balance.

Our perspective is badly foreshortened in another way as well. Too often we are unwilling to look beyond ourselves to see the effect of our actions today on our children and grandchildren. I am convinced that many people have lost their faith in the future, because in virtually every facet of our civilization we are beginning to act as if our future is now so much in doubt that it makes more sense to focus exclusively on our current needs and short-term problems. This growing tendency to discount the value of investments made for the long term — whether of wealth, effort, or caution — may have begun with the realization that nuclear weaponry had introduced a new potential for an end to civilization. But whatever its genesis, our willingness to ignore the consequences of our actions has combined with our belief that we are separate from nature to produce a genuine crisis in the way we relate to the world around us. We seem to sense something of our jeopardy; we seem to share a restlessness of spirit that rises out of the lost connection to our world and our future. But we feel paralyzed, too attached to old assumptions and old ways of thinking to see a resolution to our dilemma.

I have been wrestling with these matters for a long time. My earliest lessons on environmental protection were about the prevention of soil erosion on our family farm, and I still remember clearly how important it is to stop up the smallest gully "before it gets started good." When I was a boy, there were plenty of examples elsewhere in the county of what happened when gullies got out of control and cut deep slashes through the pasture, taking the topsoil and muddying the river. Unfortunately, little has changed: even now, about eight acres' worth of prime topsoil floats past Memphis every hour. The Mississippi River carries away millions of tons of topsoil from farms in the middle of America, soil that is now gone for good. Iowa, for example, used to have an average of sixteen inches of the best topsoil in the world. Now it is down to eight inches; most of the rest of it is somewhere on the bottom of the Gulf of Mexico.

I always wondered why the families that lived on those farms never taught their boys and girls how to stop gullies before they got started. I have since learned part of the answer: people who lease the land for short-term profits often don't consider the future. From fence row to fence row, they strip-mine the topsoil and move on. And even if you own the land, it's hard to compete in the short term against somebody who doesn't care about the long term.

Our farm taught me a lot about how nature works, but lessons learned at the dinner table were equally important. I particularly remember my mother's troubled response to Rachel Carson's classic book about DDT and pesticide abuse, *Silent Spring,* first published in 1962. My mother was one of many who read Carson's warnings and shared them with others. She emphasized to my sister and me that this book was different — and important. Those conversations made an impression, in part because they made me think about threats to the environment that are much more serious than washed-out gullies — but much harder to see.

This nearly invisible poison, which had been first welcomed as a blessing, became for me a symbol of how carelessly our civilization could do harm to the world, almost without realizing its own power. But later, during the Vietnam War, I encountered an even more powerful new poison, which was also welcomed at first. I went to Vietnam with the army and vividly remember traveling through countryside that used to be jungle but now looked like the surface of the moon. A herbicide called Agent Orange had cleared the jungle, and we were glad of it at the time, because it meant that the people who wanted to shoot at us had fewer places to hide. Years later, after learning that Agent Orange was the suspected cause of chromosomal damage and birth defects in the offspring of soldiers, I came to feel differently about it. Indeed, along with many others, I started to feel wary of all chemicals that have extraordinarily powerful effects on the world around us. How can we be sure that a chemical has only those powers we desire and not others we don't? Are we really taking enough time to discover their long-term effects? Agent Orange, after all, is just one of the better-known examples of a whole new generation of powerful compounds created in the chemical revolution, which picked up speed after World War II; over the past fifty years, herbicides, pesticides, fungicides, chlorofluorocarbons (CFCs), and thousands of other compounds have come streaming out of laboratories and chemical plants faster than we can possibly keep track of them. All of them are supposed to improve our lives, and hundreds of them have done so. But too many have left a legacy of poison that we will be coming to terms with for many generations.

I carried these concerns with me to Congress, and in 1978 I received a letter from a farm family near Toone, Tennessee, about the sickness they felt was caused by pesticide waste dumped next to their land. It turned out they were right: a company from Memphis, seventy-five miles to the west, had bought up the neighboring farm and dumped several million gallons of hazardous waste into trenches that leaked into the well water for miles around. As a result, I organized the first congressional hearings on toxic waste and focused on two sites, the small rural community of Toone, Tennessee, and one other recently discovered waste dump at a little place in upstate New York, Love Canal. Subsequently, of course, Love Canal became synonymous with the problem of hazardous chemical waste. Toone didn't, but the family received one of the biggest judgments ever handed down in a lawsuit over damages from toxic waste.

But strip-mined topsoil and hazardous chemical waste — bad as they are — still represent essentially local threats to the environment. They are serious, but they are minor compared to the global threat we now face.

I was introduced to the idea of a global environmental threat as a young student when one of my college professors was the first person in the world to monitor carbon dioxide (CO_2) in the atmosphere. Roger Revelle had, through sheer persistence, convinced the world scientific community to include as part of the International Geophysical Year (1957–58) his plan for regularly sampling CO_2

Mauna Loa Observatory, Hawaii
Monthly Average Carbon Dioxide Concentration

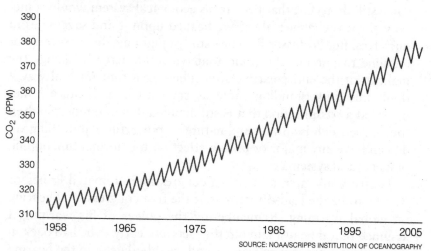

SOURCE: NOAA/SCRIPPS INSTITUTION OF OCEANOGRAPHY

The pre-industrial concentration of CO_2 was 280 parts per million. In 2005, that level, measured high above Mauna Loa, was 381 parts per million. In summer, the line goes down as vegetation in the Northern Hemisphere (with most of the earth's land area) breathes in vast quantities of CO_2. In winter, when the leaves have fallen, the line climbs again. The peak concentration has grown steadily higher because of such human activities as burning fossil fuels and the destruction of forests.

concentrations in the atmosphere. His colleague C. D. Keeling actually took the measurements from the top of the Mauna Loa volcano in Hawaii. In the middle 1960s Revelle shared with the students in his undergraduate course on population the dramatic results of the first eight years of measurements: the concentrations of CO_2 were increasing rapidly each year (see illustration). Professor Revelle explained that higher levels of CO_2 would create what he called the greenhouse effect, which would cause the earth to grow warmer. The implications of his words were startling: we were looking at only eight years of information, but if this trend continued, human civilization would be forcing a profound and disruptive change in the entire global climate.

Since that time, I have watched the Mauna Loa reports every year, and, indeed, the pattern has been unbroken — except that now the rate of increase is faster. Even twenty-five years ago, the basic premises of the greenhouse effect were never subject to serious scientific challenge, although most people assumed then, as some still do today, that the earth's ecological system would somehow absorb whatever abuse we heaped upon it and save us from ourselves. But Professor Revelle's study taught me that nature is not immune to our presence, and that we could actually change the makeup of the entire earth's atmosphere in a fundamental way. I think this understanding came as such a shock because I had inherited an assumption that is still handed down to most children today: the earth is so vast and nature so powerful that nothing we do can have any major or lasting effect on the normal functioning of its natural systems.

Twelve years later, as a young congressman, I invited Professor Revelle to be the lead-off witness at the first congressional hearing on global warming. Remembering the power of his warning, I assumed that if he just laid out the facts as clearly as he had back in that college class, my colleagues and everybody else in the hearing room would be just as shocked as I had been — and thus galvanized into action. Instead, I was the one who was shocked. Not by the evidence: it was even more troubling than I had remembered it. This time I was startled by the reaction on the part of some smart people who I thought should know better. But the unrestrained burning of cheap fossil fuels has many ferocious defenders, and this was my first encounter, though hardly the last, with the powerful and determined opposition to the dangerous truth about what we are doing to the earth.

Over the next few years, I began seriously studying global warming and several other difficult environmental issues. I held hearings, pushed for research funding and precautionary legislation, read many books and journals, and spoke to people all over the country — experts and concerned citizens alike — about how we could resolve the growing crisis. In some ways, the response was encouraging. By the late 1970s, the issue was of at least some concern to a broad segment of the population. But despite mounting evidence that the problem was truly global, few people were willing to think about the comprehensive nature of the response needed.

My own initial efforts to spread the word about global warming were an example. Most people still thought of the environment in local or regional terms, so it was impossible to get adequate funding for research on global warming. Nor was there any consensus about the need for immediate action. Even the major environmental groups resisted the issue: some told me they had other priorities. Many were cautious about what seemed like scanty evidence at the time, and a few of them were overly sensitive to an admittedly difficult political problem: if global warming were taken seriously and the world began searching for substitutes for coal and oil, nuclear power might receive a big boost. Still, an awareness of global warming as a major threat slowly began to rise, and we made real progress on several other fronts. In December 1980, for instance, in the lame duck session immediately before the Reagan inauguration, I finally succeeded, along with Congressmen Jim Florio, Tom Downey, and others, in passing the Superfund Law to clean up hazardous chemical dump sites.

Ironically, my own understanding of the global environmental crisis was greatly enhanced by my involvement in what seemed like a very different issue. Beginning in January 1981, I spent many hours each week for more than thirteen months intensively studying the nuclear arms race. In the spring of 1982, I offered a comprehensive approach to dealing with it, an approach that differed from older efforts to solve the problem in three important respects. First, it located a principal source of the nuclear impasse in the military relationship between the arsenals as it was perceived by each superpower. Second, it identified the ways in which the characteristics of particular weapons technologies affected those perceptions and influenced the ways both countries thought about the relationship between the two arsenals. Third, it prescribed a specific, simultaneous, and step-by-step evolution in both weaponry and arms control designed to eliminate the fear of a first strike on both sides. One of my chief recommendations — to ban multiple-warhead missiles and deploy a new, more stabilizing single-warhead ICBM instead — was actually adopted as a centerpiece of our nuclear strategy.

My study of the arms race led me to think about other issues, especially the global environment, in a new, more productive way. For example, I began to separate the parts of the environmental issue that were fundamentally local in nature, like hazardous waste sites, from those that represented threats to the entire globe. Going further, I began to understand the importance of looking beyond simple questions about what we are doing to the various *parts* of the environment; it became clear to me that we have to consider the complex nature of our interaction with the *whole* environment; more specifically, I zeroed in on the central importance of our way of thinking about that relationship.

I also now had a deeper appreciation for the most horrifying fact in all our lives: civilization is now capable of destroying itself. My work in Congress took on a new urgency, in part because, just as Samuel Johnson said that the prospect of hanging in a fortnight serves to concentrate the mind wonderfully, my work on nuclear arms control served to concentrate my mind on some of the larger purposes of politics. And as I began to think in broader ways about the course of our nation and our civilization, I also began to think about what role I might play in determining that course.

In March 1987, I decided to run for president. This is not the place to discuss my campaign in any detail, but a few observations may be instructive, since it taught me a great deal about the way our country perceives the environmental crisis. In fact, one of the main reasons I ran was to try to elevate the importance of the crisis as a political issue. In the speech in which I declared my candidacy, I focused on global warming, ozone depletion and the ailing global environment and declared that these issues — along with nuclear arms control — would be the principal focus of my campaign. Little did I know that even a more seasoned and experienced candidate than myself would have had a difficult time keeping his campaign focused on issues that were considered exotic at best by pollsters and political professionals. The columnist George Will, for instance, described my candidacy as being motivated by "a consuming interest in issues that are, in the eyes of the electorate, not even peripheral. These are issues such as the 'greenhouse effect' and the thinning ozone."

Worse, I started to wonder whether the issues I knew to be important really were peripheral after all. I began to doubt my own political judgment, so I began to ask the pollsters and professional politicians what they thought I ought to talk about. As a result, for much of the campaign I discussed what everybody else discussed, which too often was a familiar list of what the insiders agree are "the issues." The American people sometimes suspect that campaign agendas come straight from the pollsters and political professionals. Too often they're right.

In my own defense, I will say that throughout the campaign I did look for opportunities to return to the issue of the global environment. And even though I came to downplay it in my standard stump speech, I continued to emphasize it heavily in my meetings with editorial boards throughout the country. But the national press corps, reflecting the consensus of the political community, resolutely refused to consider the global environment as an important part of the campaign agenda. For example, the day the scientific community confirmed that the dangerous hole in the sky above Antarctica was caused by CFCs, I canceled my campaign schedule and gave a major speech outlining a comprehensive proposal to ban CFCs and take a number of other steps that would address the crisis of the global atmosphere. The entire campaign went into high gear, alerting the press, staging the speech, distributing advance copies of the text, and generally promoting the event. The result was that not a single word was written in any newspaper in America about the speech or the issue — as a campaign issue — even though the scientific finding was, as I had expected, front-page news throughout the world.

I don't want to leave the impression that the media's unwillingness to focus on the global environment was the only reason why the issue failed to ignite serious debate during the campaign. The truth is, most voters didn't consider it an overriding issue, and I didn't do a good job of convincing them otherwise. During one debate in Iowa, for example, after I had discussed the greenhouse effect in passionate detail, one of my fellow candidates derided my comments with the suggestion that I sounded as if I was running for national scientist. The harder truth is that I simply lacked the strength to keep on talking about the environmental crisis constantly whether it was being reported in the press or not.

George Will and other campaign analysts had been correct; the issue of the global environment would not help me get elected president. But when I returned to the Senate in the spring of 1988, I at least had the satisfaction of seeing what I thought were some results from the hundreds of discussions I had had with editorial boards around the country. And more people were beginning to pay attention to the issue: during that summer, temperatures reached record levels, and for the first time in what was already the hottest decade since temperatures were first written down, people began to wonder out loud whether global warming was responsible. By the fall, the issue I had sought to introduce into the campaign was, if nothing else, being discussed in public by the nominees of both major parties. For example, George Bush declared in one speech that he would, if elected, exercise leadership on global warming and "confront the greenhouse effect with the White House effect." It was, as we now know, an empty promise, but at least it could no longer be said that the global environment as an issue was "not even peripheral." It was now clearly peripheral!

My campaign gave me a fresh perspective on a lot of things, but especially useful was a new way of looking at the role I could play in Congress. I remember, for example, a long car trip one day with Tim Wirth, my colleague from Colorado, during which he and I had an exceptionally candid conversation about the politics of the global environment inside the Senate. We had worked together as close friends on other issues for a dozen years, but now we were in danger of getting in each other's way as we made similar points on the same issue. Both of us were familiar with examples of petty rivalries that interfered with the development of sound policy, and both of us felt so strongly about this issue that we wanted to figure out how to avoid destructive forms of competition. It was the kind of conversation I probably wouldn't have been comfortable having only a few years earlier, but by then it seemed entirely natural. Tim and I agreed to work together whenever it was productive; since then, we have worked closely together and with a number of others on a variety of new approaches to the issue. For example, we joined with other senators — John Chafee, Max Baucus, John Heinz, John Kerry, and Rudy Boschwitz — to create the first Interparliamentary Conference on the Global Environment, held in Washington in the spring of 1990. There, parliamentarians from forty-two nations entered into unprecedented agreements on the full range of threats to the global environment. We have worked with the majority leader, George Mitchell, and committee chairmen like

Fritz Hollings and Sam Nunn and a number of others to get started on an effective strategy.

Every education is a kind of inward journey, and my study of the global environment has required a searching reexamination of the ways in which political motives and government policies have helped to create the crisis and now frustrate the solutions we need. Ecology is the study of balance, and some of the same principles that govern the healthy balance of elements in the global environment also apply to the healthy balance of forces making up our political system. In my view, however, our system is on the verge of losing its essential equilibrium. The problem is not so much one of policy failures: much more worrisome are the failures of candor, evasions of responsibility, and timidity of vision that characterize too many of us in government. More than anything else, my study of the environment has led me to realize the extent to which our current public discourse is focused on the shortest of short-term values and encourages the American people to join us politicians in avoiding the most important issues and postponing the really difficult choices.

But the strengths of our political system ultimately rely on the strengths of its individual members, and each of us must achieve our own balance, what we hope will be a healthy integration of our hopes and fears, desires and responsibilities, needs and devotions. I am reminded of a new form of holistic photography that captures three-dimensional images of people and objects called holograms. One of the curiosities of this new science that makes it useful as a metaphor is that every small portion of the photographic plate contains all the visual information necessary to recreate a tiny, faint representation of the entire three-dimensional image. The image becomes full and vivid only when that portion is combined with the rest of the plate. Since I first heard this phenomenon described, it has often struck me that it resembles the way each individual, like a single small part of a holographic plate, reflects, however faintly, a representation of the sum total of the values, choices, and assumptions that make up the society of which he or she is a part.

But civilization is not a frozen image; it is in constant motion, and if each of us reflects the larger society, we are also carried along by it. Our ways of thinking and perceiving, our desires and behaviors, our ideologies and traditions — all are inherited in significant measure from our civilization. We may suffer the illusion from time to time that we are going to go our own way, but it is genuinely hard to break out of patterns of thought and action that are integral to our culture. Meanwhile, civilization now rushes ahead with tremendous momentum, and even the individual who believes we are on a collision course with the global environment will find it difficult to separate his or her course from that of the civilization as a whole. As always, it is easier to see the need for change in the larger pattern than to address the need for it in oneself. Nevertheless, with personal commitment, every individual can help ensure that dramatic change does take place.

I have therefore come to believe that the world's ecological balance depends on more than just our ability to restore a balance between civilization's ravenous appetite for resources and the fragile equilibrium of the earth's environment; it depends on more, even, than our ability to restore a balance between ourselves as individuals and the civilization we aspire to create and sustain. In the end, we must restore a balance within ourselves between who we are and what we are doing. Each of us must take a greater personal responsibility for this deteriorating global environment; each of us must take a hard look at the habits of mind and action that reflect — and have led to — this grave crisis.

The need for personal equilibrium can be described in an even simpler way. The more deeply I search for the roots of the global environmental crisis, the more I am convinced that it is an outer manifestation of an inner crisis that is, for lack of a better word, spiritual. As a politician, I know full well the special hazards of using "spiritual" to describe a problem like this one. For many, it is like one of those signs that warns a motorist, Steep Slope — Truckers Use Brakes. But what other word describes the collection of values and assumptions that determine our basic understanding of how we fit into the universe?

This book, and the journey it describes, is thus a search for ways to understand — and respond to — the dangerous dilemma that our civilization now faces. Looking for a map to guide me on this journey, I reluctantly concluded that I had to look inside myself and confront some difficult and painful questions about what I am really seeking in my own life, and why. I grew up in a determinedly political family, in which I learned at an early age to be very sensitive — too sensitive, perhaps — to what others were thinking, and to notice carefully — maybe too carefully — the similarities and differences between my way of thinking and that of the society around me. Now, in midlife, as I search through the layers of received knowledge and intuited truth woven into my life, I can't help but notice similar layers of artifice and authenticity running through the civilization of which I am a part. That is why this journey has taken me inside my own relationship to the environment as well as to environmental tragedies all over the globe, inside my own relationship to politics as well as to political meetings and debates about the environment in this country and around the world.

In a way, then, the search for truths about this ungodly crisis and the search for truths about myself have been the same search all along. The searching is not new — either in my personal life or where the environmental crisis is concerned. What is new in both cases is the intensity. And I know exactly when and how that started, because a single horrifying event triggered a big change in the way I thought about my relationship to life itself. One after-noon in April 1989, I walked out of a baseball stadium and saw my son, Albert, then six years old, get hit by a car, fly thirty feet through the air and scrape along the pavement another twenty feet until he came to rest in a gutter. I ran to his side and held him and called his name, but he was motionless, limp and still, without breath or pulse. His eyes were open with the nothingness stare of death, and we prayed, the two of us, there in the gutter, with only my voice. Slowly, painfully, he fought through his shock and fear and latched on to the words as a beacon to find his way back to the street, where others now gathered, including two off-duty nurses who, thank God, knew enough about the medical realities to keep him alive in spite of his massive injuries inside and out. When the ambulance finally arrived, the technicians took a long time trying to stabilize his vital signs enough to leave the scene safely; finally they raced to the emergency room and the next phase of what became an epic struggle by dozens of skilled men and women to hang on to a dear and precious life.

For the next month, my wife, Tipper, and I stayed in the hospital with our son. For many more months, our lives were completely consumed with the struggle to restore his body and spirit. And for me something changed in a fundamental way. I don't think my son's brush with death was solely responsible, although that was the catalyst. But I had also just lost a presidential campaign; moreover, I had just turned forty years old. I was, in a sense, vulnerable to the change that sought me out in the middle of my life and gave me a new sense of urgency about those things I value the most.

This life change has caused me to be increasingly impatient with the status quo, with conventional wisdom, with the lazy assumption that we can always muddle through. Such complacency has allowed many kinds of difficult problems to breed and grow, but now, facing a rapidly deteriorating global environment, it threatens absolute disaster. Now no one can afford to assume that the world will somehow solve its problems. We must all become partners in a bold effort to change the very foundation of our civilization.

But I believe deeply that true change is possible only when it begins inside the person who is advocating it. Mahatma Gandhi said it well: "We must be the change we wish to see in the world." And a story about Gandhi — recounted by Craig Schindler and Gary Lapid — provides a good illustration of how hard it is to "be the change." Gandhi, we are told, was approached one day by a woman who was deeply concerned that her son ate too much sugar. "I am worried about his health," she said. "He respects you very much. Would you be willing to tell him about its harmful effects and suggest he stop eating it?" After reflecting on the request, Gandhi told the woman that he would do as she requested, but asked that she bring her son back in two weeks, no sooner. In two weeks, when the boy and his mother returned, Gandhi spoke with him and suggested that he stop eating sugar. When the boy complied with Gandhi's suggestion, his mother thanked Gandhi extravagantly — but asked him why he had insisted on the two-week interval. "Because," he replied, "I needed the two weeks to stop eating sugar myself."

I have tried to confront in my own life the same ill habits of thought and action that I am attempting to understand and working to change in our civilization as a whole. On a personal level, this has meant reexamining my relationship to the environment in large and small ways — everything from wondering how my spiritual life can be more connected to the natural world to keeping a careful eye on our household's use of electricity, water, and, indeed, every kind of resource — and recognizing my own hypocrisy when I use CFCs in my automobile air conditioner, for example, on the way to a speech about why they should be banned. I don't claim any special skill or courage as a seeker of truth, but I'm convinced of one thing: anybody who spends serious time looking hard for the truth about anything has to become more sensitive to the many distractions and distortions that interfere with the task — whether they are obstacles in the line of sight or inside the searcher. A highly successful maverick geologist, with a reputation for finding fossil fuel reserves where others failed, once said, "In order to find oil, you've got to be honest."

On a professional level, my job happens to be politics, and I am devoting more time than ever to the effort to heal the global environment. I have discussed the issue at hundreds of town hall meetings in communities throughout Tennessee, I have put forward several legislative proposals in Congress, and I have looked for every opportunity in this country and around the world to speak up about the crisis.

But, perhaps most important, I have become very impatient with my own tendency to put a finger to the political winds and proceed cautiously. The voice of caution whispers persuasively in the ear of every politician, often with good reason. But when caution breeds timidity, a good politician listens to other voices. For me, the environmental crisis is the critical case in point: now, every time I pause to consider whether I have gone too far out on a limb, I look at the new facts that continue to pour in from around the world and conclude that I have not gone nearly far enough. The integrity of the environment is not just another issue to be used in political games for popularity, votes, or attention. And the time has long since come to take more political risks — and endure much more political criticism — by proposing tougher, more effective solutions and fighting hard for their enactment.

I guess that's actually the reason I ended up writing this book: to fully search my heart and mind about this challenge to which I feel called — and in the process to summon the courage to make a full and unreserved commitment to see it through. It didn't start that way, but as David Halberstam observed at the conclusion of *The Next Century:* "A book has a trajectory of its own." And although I didn't plan to use this book as an opportunity to offer a series of undoubtedly controversial proposals for saving the global environment, I'm glad to say that whether you agree with them or not, you will find in Part III the tough new proposals from which I have shied away — until now.

As the trajectory of this book begins, I want you to know that I've done the best I can to make it honest and true. The global environmental crisis is, as we say in Tennessee, real as rain, and I cannot stand the thought of leaving my children with a degraded earth and a diminished future. That's the basic reason why I have searched so intensively for ways to understand this crisis and help solve it; it is also why I am trying to convince you to be a part of the enormous change our civilization must now undergo. I am struggling to be a part of that same change myself, and my hope is that you will open your heart and mind to the words and ideas that follow. They represent not only an expression of my beliefs but also a deep commitment to them. I hope you, too, will make a commitment to help bring the earth back into balance, for as W. H. Murray has said: "Until one is committed there is hesitancy, the chance to draw back, always ineffectiveness. Concerning all acts of initiative . . . there is one elementary truth, the ignorance of which kills countless ideas and splendid plans: that the moment one definitely commits oneself, then providence moves too."

PART I

BALANCE AT RISK

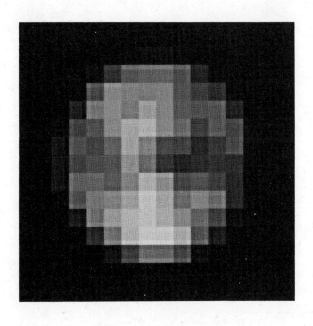

1

Ships in the Desert

I was standing in the sun on the hot steel deck of a fishing ship capable of processing a fifty-ton catch on a good day. But it wasn't a good day. We were anchored in what used to be the most productive fishing site in all of central Asia, but as I looked out over the bow, the prospects of a good catch looked bleak. Where there should have been gentle blue-green waves lapping against the side of the ship, there was nothing but hot dry sand — as far as I could see in all directions. The other ships of the fleet were also at rest in the sand, scattered in the dunes that stretched all the way to the horizon.

Oddly enough, it made me think of a fried egg I had seen back in the United States on television the week before. It was sizzling and popping the way a fried egg should in a pan, but it was in the middle of a sidewalk in downtown Phoenix. I guess it sprang to mind because, like the ship on which I was standing, there was nothing wrong with the egg itself. Instead, the world beneath it had changed in an unexpected way that made the egg seem — through no fault of its own — out of place. It was illustrating the newsworthy point that at the time Arizona wasn't having an especially good day, either, because for the second day in a row temperatures had reached a record 122 degrees.

As a camel walked by on the dead bottom of the Aral Sea, my thoughts returned to the unlikely ship of the desert on which I stood, which also seemed to be illustrating the point that its world had changed out from underneath it with sudden cruelty. Ten years ago the Aral was the fourth-largest inland sea in the world, com-

Fishing ships stranded in a desert that not long ago was part of the Aral Sea. Once the fourth-largest inland sea in the world and the most prolific source of fish in Soviet Central Asia, the Aral is drying up because of human intervention in its ecological balance. As the water retreated, fishermen dug a canal, trying desperately to reach the sea.

parable to the largest of North America's Great Lakes. Now it is disappearing because the water that used to feed it has been diverted in an ill-considered irrigation scheme to grow cotton in the desert. The new shoreline was almost forty kilometers across the sand from where the fishing fleet was now permanently docked. Meanwhile, in the nearby town of Muynak the people were still canning fish — brought not from the Aral Sea but shipped by rail through Siberia from the Pacific Ocean, more than a thousand miles away.

I had come to the Aral Sea in August 1990 to witness at first hand the destruction taking place there on an almost biblical scale. But during the trip I encountered other images that also alarmed me. For example, the day I returned to Moscow from Muynak, my friend Alexei Yablokov, possibly the leading environmentalist in the Soviet Union, was returning from an emergency expedition to the White Sea, where he had investigated the mysterious and unprecedented death of several *million* starfish, washed up into a knee-deep mass covering many miles of beach. That night, in his apartment, he talked of what it was like for the residents to wade through the starfish in hip boots, trying to explain their death.

Later investigations identified radioactive military waste as the likely culprit in the White Sea deaths. But what about all of the other mysterious mass deaths washing up on beaches around the world? French scientists recently concluded that the explanation for the growing number of dead dolphins washing up along the Riviera was accumulated environmental stress, which, over time, rendered the animals too weak to fight off a virus. This same phenomenon may also explain the sudden increase in dolphin deaths along the Gulf Coast in Texas as well as the mysterious deaths of 12,000 seals whose corpses washed up on the shores of the North Sea in the summer of 1988. Of course, the oil-covered otters and seabirds of Prince William Sound a year later presented less of a mystery to science, if no less an indictment of our civilization.

As soon as one of these troubling images fades, another takes its place, provoking new questions. What does it mean, for example, that children playing in the morning surf must now dodge not only the occasional jellyfish but the occasional hypodermic needle washing in with the waves? Needles, dead dolphins, and oil-soaked birds — are all these signs that the shores of our familiar world are fast eroding, that we are now standing on some new beach, facing dangers beyond the edge of what we are capable of imagining?

With our backs turned to the place in nature from which we came, we sense an unfamiliar tide rising and swirling around our ankles, pulling at the sand beneath our feet. Each time this strange new tide goes out, it leaves behind the flotsam and jetsam of some giant shipwreck far out at sea, startling images washed up on the sands of our time, each a fresh warning of hidden dangers that lie ahead if we continue on our present course.

My search for the underlying causes of the environmental crisis has led me to travel around the world to examine and study many of these images of destruction. At the very bottom of the earth, high in the Trans-Antarctic Mountains, with the sun glaring at midnight through a hole in the sky, I stood in the unbelievable coldness and talked with a scientist in the late fall of 1988 about the tunnel he was digging through time. Slipping his parka back to reveal a badly burned face that was cracked and peeling, he pointed to the annual layers of ice in a core sample dug from the glacier on which we were standing. He moved his finger back in time to the ice of two decades ago. "Here's where the U.S. Congress passed the Clean Air Act," he said. At the bottom of the world, two continents away from Washington, D.C., even a small reduction in one country's emissions had changed the amount of pollution found in the remotest and least accessible place on earth.

But the most significant change thus far in the earth's atmosphere is the one that began with the industrial revolution early in the last century and has picked up speed ever since. Industry meant coal, and later oil, and we began to burn lots of it — bringing rising levels of carbon dioxide (CO_2), with its ability to trap more heat in the atmosphere and slowly warm the earth. Fewer than a hundred yards from the South Pole, upwind from the ice runway where the ski plane lands and keeps its engines running to prevent the metal parts from freeze-locking together, scientists monitor the air several times every day to chart the course of that inexorable change. During my visit, I watched one scientist draw the results of that day's measurements, pushing the end of a steep line still higher on the graph. He told me how easy it is — there at the end of the earth — to see that this enormous change in the global atmosphere is still picking up speed.

Two and a half years later I slept under the midnight sun at the other end of our planet, in a small tent pitched on a twelve-foot-thick slab of ice floating in the frigid Arctic Ocean. After a hearty breakfast, my companions and I traveled by snowmobiles a few miles farther north to a rendezvous point where the ice was thinner — only three and a half feet thick — and a nuclear submarine hovered in the water below. After it crashed through the ice, took on its new passengers, and resubmerged, I talked with scientists who were trying to measure more accurately the thickness of the polar ice cap, which many believe is thinning as a result of global warming. I had just negotiated an agreement between ice scientists and the U.S. Navy to secure the release of previously top secret data from submarine sonar tracks, data that could help them learn what is happening to the north polar cap. Now, I wanted to see the pole itself, and some eight hours after we met the submarine, we were crashing through that ice, surfacing, and then I was standing in an eerily beautiful snowscape, windswept and sparkling white, with the horizon defined by little hummocks, or "pressure ridges" of ice that are pushed up like tiny mountain ranges when separate sheets collide. But here too, CO_2 levels are rising just as rapidly, and ultimately temperatures will rise with them — indeed, global warming is expected to push temperatures up much more rapidly in the polar regions than in the rest of the world. As the polar air warms, the ice here will thin; and since the polar cap plays such a crucial role in the world's weather system, the consequences of a thinning cap could be disastrous.

Considering such scenarios is not a purely speculative exercise. Six months after I returned from the North Pole, a team of scientists reported dramatic changes in the pattern of ice distribution in the Arctic, and a second team reported a still controversial claim (which a variety of data now suggest) that, overall, the north polar cap has thinned by 2 percent in just the last decade. Moreover, scientists established several years ago that in many land areas north of the Arctic Circle, the spring snowmelt now comes earlier every year, and deep in the tundra below, the temperature of the earth is steadily rising.

As it happens, some of the most disturbing images of environmental destruction can be found exactly halfway between the North and South poles — precisely at the equator in Brazil — where billowing clouds of smoke regularly blacken the sky above the immense but now threatened Amazon rain forest. Acre by acre, the rain forest is being burned to create fast pasture for fast-food beef; as I learned when I went there in early 1989, the fires are set earlier and earlier in the dry season now, with more than one Tennessee's worth of rain forest being slashed and burned each year. According to our guide, the biologist Tom Lovejoy, there are more different species of birds in each square mile of the Amazon than exist in all of North America — which means we are silencing thousands of songs we have never even heard.

But for most of us the Amazon is a distant place, and we scarcely notice the disappearance of these and other vulnerable species. We ignore these losses at our peril, however. They're like the proverbial miners' canaries, silent alarms whose message in this case is that living species of animals and plants are now vanishing around the world *one thousand times faster* than at any time in the past 65 million years (see illustration).

To be sure, the deaths of some of the larger and more spectacular animal species now under siege do occasionally capture our attention. I have also visited another place along the equator, East Africa, where I encountered the grotesquely horrible image of a dead elephant, its head mutilated by poachers who had dug out its valuable tusks with chain saws. Clearly, we need to change our purely aesthetic consideration of ivory, since its source is now so threatened. To me, its translucent whiteness seems different now, like evidence of the ghostly presence of a troubled spirit, a beautiful but chill apparition, inspiring both wonder and dread.

A similar apparition lies just beneath the ocean. While scuba

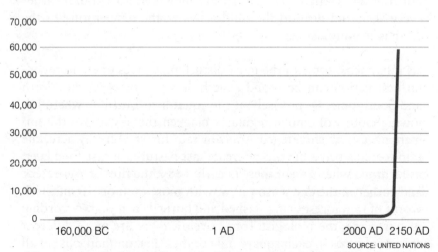

SPECIES LOST

SOURCE: UNITED NATIONS

The normal to "background" rate of extinction remained essentially unchanged for the last 65 million years—from the disappearance of the dinosaurs along with countless other species at the end of the Cretaceous era until the present century.

diving in the Caribbean, I have seen and touched the white bones of a dead coral reef. All over the earth, coral reefs have suddenly started to "bleach" as warmer ocean temperatures put unaccustomed stress on the tiny organisms that normally live in the skin of the coral and give the reef its natural coloration. As these organisms — nicknamed "zooks" — leave the membrane of the coral, the coral itself becomes transparent, allowing its white limestone skeleton to shine through — hence its bleached appearance. In the past, bleaching was almost always an occasional and temporary phenomenon, but repeated episodes can exhaust the coral. In the last few years, scientists have been shocked at the sudden occurrence of extensive worldwide bleaching episodes from which increasing numbers of coral reefs have failed to recover. Though dead, they shine more brightly than before, haunted perhaps by the same ghost that gives spectral light to an elephant's tusk.

But one doesn't have to travel around the world to witness humankind's assault on the earth. Images that signal the distress of our global environment are now commonly seen almost anywhere. A few miles from the Capitol, for example, I encountered another startling image of nature out of place. Driving in the Arlington, Virginia, neighborhood where my family and I live when the Senate is in session, I stepped on the brake to avoid hitting a large pheasant walking across the street. It darted between the parked cars, across the sidewalk, and into a neighbor's backyard. Then it was gone. But this apparition of wildness persisted in my memory as a puzzle: Why would a pheasant, let alone such a large and beautiful mature specimen, be out for a walk in my neighborhood? Was it a much wilder place than I had noticed? Were pheasants, like the trendy Vietnamese potbellied pigs, becoming the latest fashion in unusual pets? I didn't solve the mystery until weeks later, when I remembered that about three miles away, along the edge of the river, developers were bulldozing the last hundred acres of untouched forest in the entire area. As the woods fell to make way for more concrete, more buildings, parking lots, and streets, the wild things that lived there were forced to flee. Most of the deer were hit by cars; other creatures — like the pheasant that darted into my neighbor's backyard — made it a little farther.

Ironically, before I understood the mystery, I felt vaguely comforted to imagine that perhaps this urban environment, so similar to the one in which many Americans live, was not so hostile to wild things after all. I briefly supposed that, like the resourceful raccoons and possums and squirrels and pigeons, all of whom have adapted to life in the suburbs, creatures as wild as pheasants might have a fighting chance. Now I remember that pheasant when I take my children to the zoo and see an elephant or a rhinoceros. They too inspire wonder and sadness. They too remind me that we are creating a world that is hostile to wildness, that seems to prefer concrete to natural landscapes. We are encountering these creatures on a path we have paved — one that ultimately leads to their extinction.

On some nights, in high northern latitudes, the sky itself offers another ghostly image that signals the loss of ecological balance now in progress. If the sky is clear after sunset — and if you are watching from a place where pollution hasn't blotted out the night sky altogether — you can sometimes see a strange kind of cloud high in the sky. This "noctilucent cloud" occasionally appears when the earth is first cloaked in the evening darkness; shimmering above us with a translucent whiteness, these clouds seem quite unnatural. And they should: noctilucent clouds have begun to appear more often because of a huge buildup of methane gas in the atmosphere. (Also called natural gas, methane is released from landfills, from coal mines and rice paddies, from billions of termites that swarm through the freshly cut forestland, from the burning of biomass and from a variety of other human activities.) Even though noctilucent clouds were sometimes seen in the past, all this extra methane carries more water vapor into the upper atmosphere, where it condenses at much higher altitudes to form more clouds that the sun's rays still strike long after sunset has brought the beginning of night to the surface far beneath them.

What should we feel toward these ghosts in the sky? Simple wonder or the mix of emotions we feel at the zoo? Perhaps we should feel awe for our own power: just as men tear tusks from elephants' heads in such quantity as to threaten the beast with extinction, we are ripping matter from its place in the earth in such volume as to upset the balance between daylight and darkness. In the process, we are once again adding to the threat of global warming, because methane has been one of the fastest-growing greenhouse gases, and is third only to carbon dioxide and water vapor in total volume, changing the chemistry of the upper atmosphere. But, without even considering that threat, shouldn't it startle us that we have now put these clouds in the evening sky which glisten with a spectral light? Or have our eyes adjusted so completely to the bright lights of civilization that we can't see these clouds for what they are — a physical manifestation of the violent collision between human civilization and the earth?

Even though it is sometimes hard to see their meaning, we have by now all witnessed surprising experiences that signal the damage from our assault on the environment — whether it's the new frequency of days when the temperature exceeds 100 degrees, the new speed with which the sun burns our skin, or the new constancy of public debate over what to do with growing mountains of waste. But our response to these signals is puzzling. Why haven't we launched a massive effort to save our environment? To come at the question another way: Why do some images startle us into immediate action and focus our attention on ways to respond effectively? And why do other images, though sometimes equally dramatic, produce instead a kind of paralysis, focusing our attention not on ways to respond but rather on some convenient, less painful distraction?

In a roundabout way, my visit to the North Pole caused me to think about these questions from a different perspective and gave them a new urgency. On the submarine, I had several opportunities to look through the periscope at the translucent bottom of the ice pack at the North Pole. The sight was not a little claustrophobic, and at one point I suddenly thought of the three whales that had become trapped under the ice of the Beaufort Sea a couple of years earlier. Television networks from four continents came to capture their poignant struggle for air and in the process so magnified the emotions felt around the world that soon scientists and rescue workers flocked to the scene. After several elaborate schemes failed, a huge icebreaker from the Soviet Union cut a path through the ice for the two surviving whales. Along with millions of others, I had been delighted to see them go free, but there on the submarine it occurred to me that if we are causing 100 extinctions each day — and many scientists believe we are — approximately 2,000 living species had disappeared from the earth during the whales' ordeal. They disappeared forever — unnoticed.

Similarly, when a little girl named Jessica McClure fell into a well in Texas, her ordeal and subsequent rescue by a legion of heroic men and women attracted hundreds of television cameras and journalists who sent the story into the homes and minds of hundreds of millions of people. Here, too, our response seems skewed: during the three days of Jessica's ordeal, more than 100,000 boys and girls her age or younger died of preventable causes — mostly starvation and diarrhea — due to failures of both crops and politics. As they struggled for life, none of these children looked into a collection of television cameras, anxious to send word of their plight to a waiting world. They died virtually unnoticed. Why?

Perhaps one part of the answer lies in the perceived difficulty of an effective response. If the problem portrayed in the image is one whose solution appears to involve more effort or sacrifice than we can readily imagine, or if even maximum effort by any one individual would fail to prevent the tragedy, we are tempted to sever the link between stimulus and moral response. Then, once a response is deemed impossible, the image that briefly caused us to consider responding becomes not just startling but painful. At that point, we begin to react not to the image but to the pain it now produces, thus severing a more basic link in our relationship to the world: the link between our senses and our emotions. Our eyes glaze over as our hearts close. We look but we don't see. We hear but refuse to listen.

Still, there are so many distressing images of environmental destruction that sometimes it seems impossible to know how to absorb or comprehend them. Before considering the threats themselves, it may be helpful to classify them and thus begin to organize our thoughts and feelings so that we may be able to respond appropriately.

A useful system comes from the military, which frequently places a conflict in one of three different categories, according to the theater in which it takes place. There are "local" skirmishes, "regional" battles, and "strategic" conflicts. This third category is reserved for struggles that can threaten a nation's survival and must be understood in a global context.

Environmental threats can be considered in the same way. For example, most instances of water pollution, air pollution, and illegal waste dumping are essentially local in nature. Problems like acid rain, the contamination of underground aquifers, and large oil spills are fundamentally regional. In both of these categories, there may be so many similar instances of particular local and regional problems occurring simultaneously all over the world that the pattern appears to be global, but the problems themselves are still not truly strategic because the operation of the global environment is not affected and the survival of civilization is not at stake.

However, a new class of environmental problems does affect the global ecological system, and these threats are fundamentally strategic. The 600 percent increase in the amount of chlorine in the atmosphere during the last forty years has taken place not just in those countries producing the chlorofluorocarbons responsible but in the air above every country, above Antarctica, above the North Pole and the Pacific Ocean — all the way from the surface of the earth to the top of the sky. The increased levels of chlorine disrupt the global process by which the earth regulates the amount of ultraviolet radiation from the sun that is allowed through the atmosphere to the surface; and if we let chlorine levels continue to increase, the radiation levels will also increase — to the point that all animal and plant life will face a new threat to their survival.

Global warming is also a strategic threat. The concentration of carbon dioxide and other heat-absorbing molecules has increased by almost 25 percent since World War II, posing a worldwide threat to the earth's ability to regulate the amount of heat from the sun retained in the atmosphere. This increase in heat seriously threatens the global climate equilibrium that determines the pattern of winds, rainfall, surface temperatures, ocean currents, and sea level. These in turn determine the distribution of vegetative and animal life on land and sea and have a great effect on the location and pattern of human societies.

In other words, the entire relationship between humankind and the earth has been transformed because our civilization is suddenly capable of affecting the entire global environment, not just a particular area. All of us know that human civilization has usually had a large impact on the environment; to mention just one example, there is evidence that even in prehistoric times, vast areas were sometimes intentionally burned by people in their search for food. And in our own time we have reshaped a large part of the earth's surface with concrete in our cities and carefully tended rice paddies, pastures, wheatfields, and other croplands in the countryside. But these changes, while sometimes appearing to be pervasive, have, until recently, been relatively trivial factors in the global ecological system. Indeed, until our lifetime, it was always safe to assume that nothing we did or could do would have any lasting effect on the global environment. But it is precisely that assumption which must now be discarded so that we can think strategically about our new relationship to the environment.

Human civilization is now the dominant cause of change in the global environment. Yet we resist this truth and find it hard to imagine that our effect on the earth must now be measured by the same yardstick used to calculate the strength of the moon's pull on the oceans or the force of the wind against the mountains. And if we are now capable of changing something so basic as the relationship between the earth and the sun, surely we must acknowledge a new responsibility to use that power wisely and with appropriate restraint. So far, however, we seem oblivious of the fragility of the earth's natural systems.

This century has witnessed dramatic changes in two key factors that define the physical reality of our relationship to the earth: a sudden and startling surge in human population, with the addition of one China's worth of people every ten years, and a sudden acceleration of the scientific and technological revolution, which has allowed an almost unimaginable magnification of our power to affect the world around us by burning, cutting, digging, moving, and transforming the physical matter that makes up the earth.

The surge in population is both a cause of the changed relationship and one of the clearest illustrations of how startling the change has been, especially when viewed in a historical context. From the emergence of modern humans 200,000 years ago until Julius Caesar's time, fewer than 250 million people walked on the face of the earth. When Christopher Columbus set sail for the New World 1,500 years later, there were approximately 500 million people on earth. By the time Thomas Jefferson wrote the Declaration of Independence in 1776, the number had doubled again, to 1 billion. By midway through this century, at the end of World War II, the number had risen to just above 2 billion people.

In other words, from the beginning of humanity's appearance on earth to 1945, it took more than ten thousand generations to reach a world population of 2 billion people. Now, in the course of one human lifetime — mine — the world population will increase from 2 to more than 9 billion, and it is already more than halfway there (see the graph on the following pages).

Like the population explosion, the scientific and technological revolution began to pick up speed slowly during the eighteenth century. And this ongoing revolution has also suddenly accelerated exponentially. For example, it is now an axiom in many fields of science that more new and important discoveries have taken place in the last ten years than in the entire previous history of science. While no single discovery has had the kind of effect on our relationship to the earth that nuclear weapons have had on our relationship to warfare, it is nevertheless true that taken together, they have completely transformed our cumulative ability to exploit the earth for sustenance — making the consequences of unrestrained exploitation every bit as unthinkable as the consequences of unrestrained nuclear war.

Now that our relationship to the earth has changed so utterly, we have to see that change and understand its implications. Our challenge is to recognize that the startling images of environmental destruction now occurring all over the world have much more in common than their ability to shock and awaken us. They are symptoms of an underlying problem broader in scope and more serious than any we have ever faced. Global warming, ozone depletion, the loss of living species, deforestation — they all have a common cause: the new relationship between human civilization and the earth's natural balance.

There are actually two aspects to this challenge. The first is to realize that our power to harm the earth can indeed have global

POPULATION GROWTH THROUGHOUT HISTORY

World population, after remaining stable for most of history, began to grow gradually after the agricultural revolution a few thousand years ago. This slow rate of increase continued until the onset of the industrial revolution when the curve began sloping upward. In this century, the rate of increase has acclerated so rapidly that an extra billion people are being added to the population each decade.

FIRST MODERN HUMANS

160,000 BC   100,000 BC   10,000 BC   7000 BC   6000 BC   4000 BC

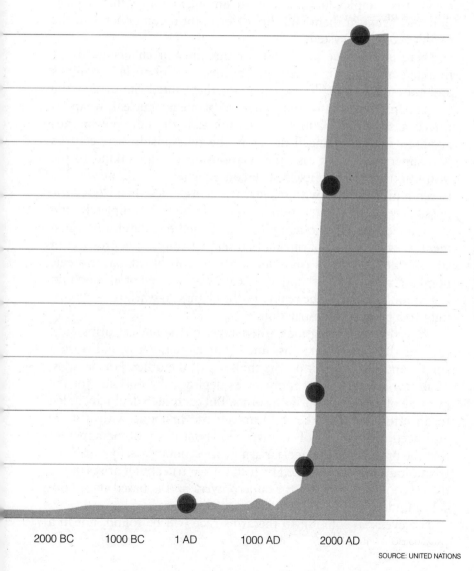

2000 BC   1000 BC   1 AD   1000 AD   2000 AD

BILLIONS OF PEOPLE

SOURCE: UNITED NATIONS

and even permanent effects. The second is to realize that the only way to understand our new role as a co-architect of nature is to see ourselves as part of a complex system that does not operate according to the same simple rules of cause and effect we are used to. The problem is not our effect *on* the environment so much as our relationship *with* the environment. As a result, any solution to the problem will require a careful assessment of that relationship as well as the complex interrelationship among factors within civilization and between them and the major natural components of the earth's ecological system.

There is only one precedent for this kind of challenge to our thinking, and again it is military. The invention of nuclear weapons and the subsequent development by the United States and the Soviet Union of many thousands of strategic nuclear weapons forced a slow and painful recognition that the new power thus acquired forever changed not only the relationship between the two superpowers but also the relationship of humankind to the institution of warfare itself. The consequences of all-out war between nations armed with nuclear weapons suddenly included the possibility of the destruction of both nations — completely and simultaneously. That sobering realization led to a careful reassessment of every aspect of our mutual relationship to the prospect of such a war. As early as 1946 one strategist concluded that strategic bombing with missiles "may well tear away the veil of illusion that has so long obscured the reality of the change in warfare — from a fight to a process of destruction."

Nevertheless, during the earlier stages of the nuclear arms race, each of the superpowers assumed that its actions would have a simple and direct effect on the thinking of the other. For decades, each new advance in weaponry was deployed by one side for the purpose of inspiring fear in the other. But each such deployment led to an effort by the other to leapfrog the first one with a more advanced deployment of its own. Slowly, it has become apparent that the problem of the nuclear arms race is not primarily caused by technology. It is complicated by technology, true; but it arises out of the relationship between the superpowers and is based on an obsolete understanding of what war is all about.

The eventual solution to the arms race will be found, not in a new deployment by one side or the other of some ultimate weapon or in a decision by either side to disarm unilaterally, but rather in new understandings and in a mutual transformation of the relationship itself. This transformation will involve changes in the technology of weaponry and the denial of nuclear technology to rogue states. But the key changes will be in the way we think about the institution of warfare and about the relationship between states.

The strategic nature of the threat now posed *by* human civilization to the global environment and the strategic nature of the threat *to* human civilization now posed by changes in the global environment present us with a similar set of challenges and false hopes. Some argue that a new ultimate technology, whether nuclear power or genetic engineering, will solve the problem. Others hold that only a drastic reduction of our reliance on technology can improve the conditions of life — a simplistic notion at best. But the real solution will be found in reinventing and finally healing the relationship between civilization and the earth. This can only be accomplished by undertaking a careful reassessment of all the factors that led to the relatively recent dramatic change in the relationship. The transformation of the way we relate to the earth will of course involve new technologies, but the key changes will involve new ways of thinking about the relationship itself.

2

The Shadow Our Future Throws

The most dangerous threat to our global environment may not be the strategic threats themselves but rather our perception of them, for most people do not yet accept the fact that this crisis is extremely grave. Of course, there is always a degree of uncertainty about complex issues, and careful study is always necessary, but it is all too easy to exaggerate the uncertainty and overstudy the problem — and some people do just that — in order to avoid an uncomfortable conclusion. Still others are genuinely troubled that for all we do know about the environmental crisis, there is a great deal about it that we do not know.

Some of the questions scientists still need to answer seem deceptively simple, such as, Where does it rain? When and how much? Questions that are easy in our own backyards are mysteries to science when posed on a global scale. Where are the clouds? How is the earth's surface changing? How wet or dry is the soil? These questions are extremely important because the answers have a direct bearing on how seriously we regard the underlying threat. Take the question about clouds, for example. A small number of scientists argue that we don't have to worry about global warming because when the greenhouse gases trap more heat from the sun in the atmosphere, the earth will automatically produce more clouds, which in turn will serve as a kind of thermostat to regulate the earth's temperature. Or take the questions about soil and rainfall. Again, a very few scientists argue that we don't have to worry about climate change causing widespread droughts in the middle of continental landmasses because the faster evaporation of moisture from the soil in a warmer atmosphere will be offset by changes in rainfall patterns.

More exotic questions with answers that are harder to obtain also demand attention. What is happening to the West Antarctic ice sheet? How much ice is melting in the Arctic Ocean? As noted in the first chapter, the navy is now helping to answer that last question by releasing data about the measurements to scientists. But there will always be more questions than answers. How, then, can we hope to act in time to address this emerging crisis if there is still so much to learn about it?

After years of debate and attempts to convince skeptics that the time for delay is over, I am resigned to the idea that even though we already know more than enough, we must also thoroughly investigate any significant scientific uncertainty that impedes our ability to come together and face this crisis. The knowledge thus gained will not only deprive the skeptics of some of their excuses for procrastination, it will also help us choose strategies for responding to the crisis, identify the most effective and least costly solutions, and solidify public support for the increasingly comprehensive changes that will be necessary.

But research in lieu of action is unconscionable. Those who argue that we should do nothing until we have completed a lot more research are trying to shift the burden of proof even as the crisis deepens. This point is crucial: *a choice to "do nothing" in response to the mounting evidence is actually a choice to continue and even accelerate the reckless environmental destruction that is creating the catastrophe at hand.*

In order to grasp why further waiting is so painfully wrong, it is important to be clear about what is still uncertain and what is already established as fact. For example, the precise effects of doubling the concentration of carbon dioxide (CO_2) in the atmosphere during the next few decades are uncertain. But it is clear that doubling CO_2 will in fact increase global temperatures and in the process subject us to the risk of catastrophic changes in global climate patterns. And the pace of these potential changes could also be of special concern — because the ecological system has difficulty adapting to rapid change.

We need to act now on the basis of what we know. Some

scientists believe that we are in danger of passing a kind of point of no return, after which we will have missed the last good opportunity to solve the problem before it spirals out of our control. If we choose not to act, will we indeed pass that point?

In Tennessee there is an old saying: When you are in a hole, stop digging. To put it another way, the truly conservative approach to the problem of global warming, for example, would be to stop thickening the blanket of greenhouse gases and try to prevent further damage while we study our options.

But our annual production of CO_2 and other greenhouse gases is already so large and is increasing so rapidly that simply stabilizing the amount already in the atmosphere would require significant changes in the technology we use and in the way we live our lives. I suspect that many of those who say that it is probably all right to run these risks — to make no change in our current pattern — are really saying that they simply do not want to think about the disruption that would accompany any serious effort to confront the problem. Our vulnerability to this form of procrastination is heightened where strategic threats to the environment are concerned because they seem so big as to defy our imagination. And since the crisis must still be described in the language of science, we are also vulnerable to the false reassurances of a tiny group within the scientific community who argue that the threats don't exist. A few scientists, for instance, believe that global warming is, in the words of Professor Richard Lindzen of MIT, "a largely political issue without scientific basis"; their views sometimes carry far too much weight.

The news media must take some responsibility for this quandary. They present scientific issues the way they present political issues: they prefer to emphasize controversy and disagreement. That approach can be welcome, for we know that truth is often best discovered through a vigorous give-and-take between people holding opposite points of view. But there is a difference between scientific uncertainty and political uncertainty. Where science thrives on the unknown, politics is often paralyzed by it. Yet the dialogue between science and politics has not yet accounted for this difference. In this case, when 98 percent of the scientists in a given field share one view and 2 percent disagree, both viewpoints are sometimes presented in a format in which each appears equally credible.

This is not to say that the 2 percent are wrong and should not be heard. But their theories should not be given equal weight with the consensus now emerging in the scientific community about the gravity of the danger we face. If, when the remaining unknowns about the environmental challenge enter the public debate, they are presented as signs that the crisis may not be real after all, it undermines the effort to build a solid base of public support for the difficult actions we must soon take.

Indeed, sometimes the remaining uncertainties are cynically used by partisans of the status quo for the express purpose of preventing the coalescence of public support for action. On the eve of Earth Day 1990, for example, the Bush White House circulated to its policy spokesmen a confidential memorandum suggesting the most effective arguments to use in trying to convince people not to support action against global warming. The memo, which was leaked to the press, advised that instead of directly arguing that there is no problem, "a better approach is to raise the many uncertainties." So much for Bush's promise to confront the greenhouse effect with the White House effect.

To counter this cynical approach, we must put in perspective all the unknowns that will continue to plague discussions of the environmental crisis. We should begin with the debate over global warming, because while it is only one of several strategic threats, it has become a powerful symbol of the larger crisis and a focus for the public debate about whether there really is a crisis at all. In fact, some people seem to hope that if the seriousness of global warming can be disproved, they will no longer have to worry about any environmental crisis.

But the theory of global warming will not be disproved, and the skeptics are vastly outnumbered by former skeptics who now accept the overwhelming weight of the accumulated evidence. In an effort to provide world leaders with a consensus on global warming, in 1989 the United Nations established the Intergovernmental Panel on Climate Change, under whose auspices a distinguished group of scientists undertook a major international review of the evidence. These scientists concluded — almost unanimously — that global warming is real and the time to act is now.

The insistence on complete certainty about the full details of global warming — the most serious threat that we have ever faced — is actually an effort to avoid facing the awful, uncomfortable truth: that we must act boldly, decisively, comprehensively, and quickly, even before we know every last detail about the crisis. Those who continue to argue that the appropriate response is merely additional research are simply seeking to camouflage timidity or protect their vested interest in the status quo.

Resistance to seeing the strategic threats often focuses on the lack of complete information and perfect understanding. We should acknowledge that we will never have complete information. Yet we have to make decisions anyway; we do this all the time. And one way we draw conclusions from incomplete information is by recognizing patterns.

It is already clear that our information about the global environmental crisis does fall into a discernible pattern. For many, this pattern has become painfully obvious. But for others, it is still invisible. Why? The answer, in my opinion, is fear: too often we don't let ourselves see a pattern because we are afraid of its implications. Indeed, sometimes the implications suggest dramatic changes in our way of life. And, of course, those who have the heaviest investment in the status quo — whether it is economic, political, intellectual, or emotional — often organize ferocious resistance to the new pattern regardless of the evidence.

Galileo was charged with a form of subversion for describing a pattern he saw in the heavens. One of the troubling implications of that pattern was that the earth was not the center of God's universe. But for his judges, the most troubling aspect of his theory was that the earth, already disconcertingly round, actually moved. During his trial, Galileo acknowledged the subversiveness of his ideas by pleading that he didn't really believe what his discoveries implied. Instead, he had merely presented a clever challenge to the existing order as a way of increasing the satisfaction and certitude with which the existing order would be accepted after it triumphed over his knowingly impertinent design. Even he had to bend to the conventions of his day.

The assumption that important things remain the same and don't move is a common source of opposition to discomfiting new ideas. I remember how one of my sixth-grade classmates pointed at a map of the world and traced his finger along the eastern coast of South America as it jutted into the South Atlantic. Moving across to Africa, he then drew his finger around its western coast, indented in a pattern that seemed to reprise the boundary between Brazil and the ocean.

"Did they ever fit together?" he asked.

"No," the teacher replied. "That's ridiculous."

Even though they did, of course, fit together, and even though continental drift has long since been accepted as scientific fact, it is worth remembering that as late as 1970, some of the most respected geologists in the world dismissed the theory in terms that echoed the confident ridicule expressed by my sixth-grade teacher in 1959. Why? Because they made an assumption about the world — continents don't move — which seemed reasonable but was actually wrong. And then they refused to challenge it. In the immortal words of Yogi Berra, "What gets us into trouble is not what we don't know. It's what we know for sure that just ain't so."

The scientists who ridiculed continental drift misunderstood·the amount of change on the earth that was possible. Similarly, in deciding how to evaluate the strategic threats to the global environment, many of the skeptics are basing their resistance to action on an assumption about the amount of change that is possible. They believe that the earth is so big and nature so powerful that we cannot possibly have any profound or lasting effect on them. To put it another way, they assume that the natural equilibrium of the global environment simply cannot move. Unfortunately, that just ain't so. It used to be, but it's not anymore.

How can this mistaken and increasingly dangerous assumption be changed? To begin with, we have to deal with the limitations imposed by our perspective, which is often sharply limited in time and space. First of all, we are accustomed to considering change in very short time spans — a week, a month, a year, or, if we're feeling especially expansive, a century. So a change that is actually quite rapid when measured in geologic time may seem to move very slowly in the context of a human lifetime. It takes a leap of the imagination to either accelerate or slow down a process of change in the environment enough to see it in a more familiar frame and thus discern its meaning.

Sometimes a television commercial will use slow-motion films of an automobile crashing into a brick wall at high speed. The suddenness of the collision in real time makes it seem like a single, immediate transformation of the car into a twisted mass of metal. But in slow motion we see a process of change in which the various parts of the car slowly crumple, one by one, colliding in seemingly logical and predictable ways with each other and with the occupants inside. The steering column, for example, may be pushed by the engine to impale one dummy while a second dummy slowly shatters the windshield with its wooden head.

What is now occurring in the global environment can be seen in similar terms. Our ecological system is crumpling as it suffers a powerful collision with the hard surfaces of a civilization speeding toward it out of control. The damage is remarkably sudden and extensive in the context of the long period of stability in the environment before the damage, but we see the destruction in slow motion. When the Aral Sea dries up and its fish all die, for example, it is as if this fragile ecosystem is gradually being crumpled by the force of civilization crashing into it. When vast areas of the rain forest are cleared and the species living there become extinct, it is as if the forest is shattered in slow motion by the impact of its collision with civilization. And when an overpopulated nation overgrazes its pastureland, causing a collapse in its ability to provide food the following year, it is as if the force of its collision with nature has pushed it abruptly backward in a crushing blow, like a dashboard striking the forehead of a child.

But most of us act as if we don't perceive a collision at all, partly because the crunching and crushing and shattering all take place over a longer time span than we associate with a violent collision. We are not unlike the laboratory frog that, when dropped into a pot of boiling water, quickly jumps out. But when placed in lukewarm water that is slowly heated, the frog will remain there until it is rescued.

The meaning of many patterns is conveyed by contrast as opposed to sameness. Sameness and gradual change often lull the senses, obscuring danger from minds that reserve their alertness for sharp contrasts. If an individual or a nation looks at the future one year at a time and sees the past in the context of a single lifetime, a great many large patterns are concealed. When one considers the relationship of the human species to the earth, not much change is visible in a single year in a single nation. Yet if one looks at the entire pattern of that relationship, from the emergence of our species until the present, a sharp and distinctive contrast beginning in the very recent past clearly signals the dramatic change to which we now must respond.

Another limiting factor is our normal spatial perspective. It is helpful to stand at some distance from any large pattern we are trying to comprehend. And that is difficult when we are standing in the middle of the pattern. As Ralph Waldo Emerson said, "The field cannot well be seen from within the field," and the cliché goes, "You can't see the forest for the trees."

In ancient Peru, artists drew very large figures on the ground that can only be recognized by us from high in the air. Since the artists did not have airplanes, how did they draw them? Leaving aside any bizarre theories, all the artists needed was enough imagination to shift their perspective and mentally gain geographic distance from the place they were standing. And we now need to do something similar in order to recognize what is happening to us and to the earth.

Hundreds of years ago, those who believed the earth was flat could point straight out to the horizon from wherever they stood and find convincing evidence from their limited perspective. Whoever challenged that prevailing notion had to somehow transcend their geographic limitations in order to imagine a global pattern much larger than the one their senses could directly perceive.

The same challenge confronts us now, as we try to comprehend what we are doing to the earth. Even though the pattern of our relationship to the environment has undergone a profound transformation, most people still do not see the new pattern — partly because it is global and we are not used to such a large, spatial perspective. The sights and sounds of this change are spread over an area too large for us to hold in our field of awareness. The only way we can hope to understand it is to imagine it from a new and distant perspective, not unlike the one from which the earth was first perceived as round instead of flat.

Specialists in the field of graphics once conducted a study on exactly how much visual information must be included in a mosaic before people looking at it can recognize the pattern it contains. They took a picture of Abraham Lincoln and, using a computer, broke up the visual information into a checkerboard pattern of squares. Each square was a different shade of gray, which represented the average intensity of the light in that area of the photograph. Starting with lots of little squares — all of which together easily preserved the clarity of the original photograph — the scientists gradually increased the size of each square until there were only a few dozen large squares of different grays, each of which reflected the average grayness of the area of the original photograph. Not surprisingly, the resulting mosaic appeared to be nothing more than an apparently random checkerboard pattern of gray squares — until viewed from a distance, when the original image of Lincoln was instantly clear.

When viewing the overall pattern of worldwide environmental degradation, we sometimes find it difficult to attain a sufficiently distant perspective from which to make sense of the confusing jumble of information. Those searching for answers in black and white see only varying shades of gray and believe no pattern exists. For example, if one looks at a map of temperature changes in the world, one sees a jumble of large squares representing the average temperature over large checkerboard areas of the surface of the earth. Its pattern is just as obscure as Lincoln's picture when viewed from a few inches away.

Those first striking pictures taken by the Apollo astronauts of the earth floating in the blackness of space were so deeply moving because they enabled us to see our planet from a new perspective — a perspective from which the preciousness and fragile beauty of the earth was suddenly clear. Archimedes, who invented the lever, is reported to have said that if only he had "a place on which to stand" at a sufficient distance from the earth, he could move the

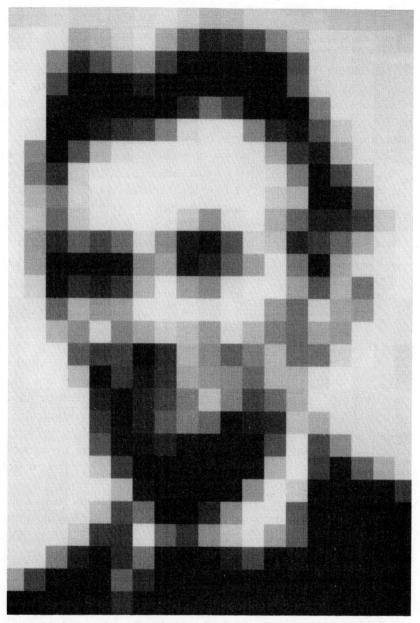

Held close to the eye, this computer-generated mosaic appears to be a meaningless jumble of light and dark squares. Held at arm's length (or at a greater distance), the face of Abraham Lincoln is clearly visible.

entire world. Our ability to see large patterns is an even more powerful tool than the lever, but, like the lever, its power is enhanced in proportion to our distance — both in time and space — from the pattern we want to understand. That is why historians often have more power to explain the meaning of a pattern in human events than do observers describing the events as they occur.

In order to recognize the pattern of destruction, we have to see it from a distance, both in time and space. Since the pattern is truly global, we have to see the entire world in our mind's eye. If we focus on just a small area of the earth, the pattern will remain invisible. (In this regard, it is worth remembering that the entire United States of America covers less than 3 percent of the surface of the earth.) Moreover, since the pattern is unfolding over time, we must find a way to see the startling contrast between the incredibly rapid changes now under way and the ordinarily glacial pace of change in the environment throughout history.

In fact, systemic changes in our way of seeing the world do sometimes occur. We are almost always surprised when we recognize profound change, perhaps because we grow so accustomed to the slow and gradual change that usually measures the pace of our lives. We find it hard to imagine, much less predict, a sudden, systemic change that moves our world from beneath us and takes us from one equilibrium to a new, profoundly different equilibrium, although it can sometimes be anticipated if we can identify a significant threshold beyond which an obviously different pattern must clearly obtain. In our personal lives, for example, the onset of puberty or the birth of a child are among the thresholds of predictable systemic change.

But it is altogether different when a civilization undergoes systemic change. It naturally seems easier to avoid even thinking about it, especially if we can argue that it still lies in the future. One reason many world leaders have difficulty responding to the environmental crisis is that the worst of the predicted effects seems decades away, and they are so unprecedented that they seem to defy common sense. After all, millions of people are suffering in poverty and dying of starvation, warfare, and preventable diseases — right now. These are urgent problems requiring urgent attention; how do we at the same time acknowledge and confront a problem that seems in the main to lie in our future? Fortunately, many people are beginning to look ahead, and there is a growing understanding that the environmental crisis must be considered in a different way. One of the philosophers of the environmental movement, Ivan Illich, explained the beginnings of activism on the global environment by saying, "What has changed is that our common sense has begun searching for a language to speak about the shadow our future throws."

Where can we find such a language? Two models from science may help us predict what will happen and tell us where we are. First, the new scientific theory of change, called Chaos Theory, is revolutionizing the way we understand many changes in the physical world. Not long after Newtonian physics led to a revolution in our understanding of cause and effect, the model of the world implied in Newton's science was lifted wholesale into politics, economics, and society at large. Many are now convinced that in a similar way, the insights of Chaos Theory will soon be absorbed into political science and social analysis.

Chaos Theory describes how many natural systems show significant changes in the way they operate even as they remain within the same overall pattern ("dynamic equilibrium"). According to this theory, certain critical boundaries define that overall pattern and cannot be exceeded without threatening the loss of its equilibrium. When large changes force it beyond these boundaries, the system suddenly shifts into an entirely new equilibrium; it adopts a new pattern with new boundaries. In some ways, the central ideas in Chaos Theory are not new at all. Aficionados of the symphony, for example, recognize a crescendo as the point of maximum instability in a piece of music, coming just at the point when the music flows to a new equilibrium with resolution and harmony. Soon we will learn to recognize crescendos in human affairs more easily — and see that they frequently signal the beginning of systemic, chaotic change from one form of equilibrium to another. Such a crescendo now seems apparent in the wave upon wave of discordant calls of distress from every corner of the globe. The relationship between human civilization and the earth is now in a state that theorists of change would describe as disequilibrium. At the birth of the nuclear age, Einstein said that "everything has changed but our way of thinking." At the birth of the environmental age, the same point holds.

Our challenge is to accelerate the needed change in thinking about our relationship to the environment in order to shift the pattern of our civilization to a new equilibrium — before the world's ecological system loses its current one. This change in thinking will also follow the pattern described in Chaos Theory, with little change evident until a threshold is passed, and then, as key assumptions are modified, a flood of dramatic changes will occur all at once.

But where is the threshold for the dramatic change in our relationship to the environment, and how can we recognize the new pattern in time to change our assumptions about how we should relate to our world? A second scientific model that may help us is Einstein's Relativity Theory. Bear with me: although complicated, Relativity Theory can easily be explained with the help of a picture showing how time and space are shaped by mass. An especially dense mass like a "black hole" is shown as a deep well, with space and time arrayed around it in a grid that slopes down to the center.

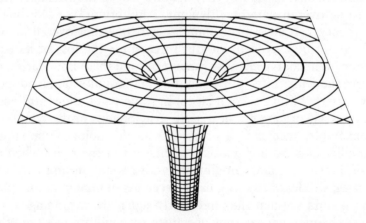

A black hole as portrayed by physicists, who explain that the continuum of space and time represented by the flat grid is bent by the dense mass of the black hole that pulls the grid down into a deep well of space and time. Large historical events shape political consciousness in much the same way.

Our political awareness often seems to be shaped exactly like this grid, within which a large historical event such as World War II is like the dense mass that exerts a powerful gravitational influence on every idea or other event close to it in time or space. In a similar way, the Holocaust shapes every idea we have about human nature. And smaller events with less historical "mass" also exert their own "pull" on our thinking, especially on our thinking about matters of comparable mass that lie in close proximity to them. Several small events clustered together in time and space may exert sufficient gravitational influence to compel us to search for a trend or common explanation for the way our experience of history has been altered by their collective mass. For example, each of the communist governments in Eastern Europe fell separately in the late summer and early autumn of 1989, but their combined impact on history was extremely powerful.

Even future events can exert a gravitational influence on our thinking. In other words, time is relative in politics just as in physics. For example, the political will that led to mass protests against escalating the nuclear arms race during the early 1980s came from a popular awareness that civilization seemed to be pulled toward the broad lip of a downslope leading to a future catastrophe — nuclear war — that would crush human history forever into a kind of black hole. There is now reason to hope that we have effectively changed our course enough to avoid that catastrophe, even though we still have to fight its gravitational pull. If we do avoid a nuclear holocaust, our success will be attributable in large part to our ability to recognize a large pattern and make adjustments in our thinking and collective behavior in time to avoid the worst.

This is not unlike the challenge we face today in the global environmental crisis. The potential for true catastrophe lies in the future, but the downslope that pulls us toward it is becoming recognizably steeper with each passing year. What lies ahead is a race against time. Sooner or later the steepness of the slope and our momentum down its curve will take us beyond a point of no return. But as the curve becomes steeper and catastrophe's pull becomes stronger, our ability to recognize the pattern of its pull is greatly enhanced. The odds that we will discern the nature of our plight improve dramatically as we get closer to the edge of history — the point from which we can gaze into the black hole's very center.

Throughout the world, we are witnessing the first stirrings of a new political will to slow down our momentum toward environmental catastrophe. The challenge now is to accelerate the widespread recognition of this pattern and organize to change our current direction — before our momentum carries us past the point beyond which an ecological collapse is inevitable.

In distinguishing between what is still uncertain and what is known about this crisis, it is important to emphasize that one thing we know quite well is that nature exhibits a recurring pattern of interdependency among the parts of the ecological system. We should assume with great confidence that if we upset the ecological balance of the entire earth in one way, we will upset it in related ways as well. Consequently, while a given behavior may at first seem harmless in that part of the environment we are able to observe, we are unlikely to know enough about the effects of what we are doing to predict the consequences for other parts of the system — precisely because all its parts exist in a delicate balance of interdependency.

This phenomenon of interdependency is probably best illustrated by what scientists call positive feedback loops, which magnify the force with which change occurs. In fact, almost everywhere you look throughout the ecological system, natural mechanisms tend to accelerate the pace of change once it is set in motion. That is one of the reasons our assault on the environment is so reckless. Since we are interfering with the operation of complex systems, the relatively simple rules of linear cause and effect cannot explain, much less predict, the consequences of our interference.

The basic principles behind positive feedback loops are easily understood. We are all familiar with so-called nonlinear systems, which can magnify the results of simple repetitive acts. For example, consider the law of compound interest and its effect on our decisions about personal finances. If I use my credit card to borrow money and then, the following month, use it again to borrow the same amount — plus an extra amount to pay the first month's interest on what I had previously borrowed — that pattern could, if continued indefinitely, feed upon itself and drive my finances toward a crisis. How long I could continue before my finances spiraled into bankruptcy would depend upon the size of the amount borrowed each month compared to my monthly income and expenses.

The law of compound interest can also magnify change in a positive direction. If I were to place the same amount of money in my savings account each month along with the steadily growing interest earned during the preceding month, the total amount saved would obviously increase at a nonlinear rate, and the rate itself would grow faster each month even if there were no change in the monthly dollar amount I added.

The same kind of positive feedback loops commonly occurs in nature as well and must be reckoned with when we try to calculate the damage that can result from a given pattern in our relationship to the global environment. Some of these feedback loops are quite complex; others are relatively simple.

When I was flying over the Amazon rain forest in a small plane, I was struck by what happened immediately after a thunderstorm moved across an area of the forest: as soon as the rain stopped, clouds of moisture began to rise from the trees to form new rain clouds that moved west, driven by the wind, where they provided the water for new rain falling out of new thunderstorms.

Any interruption of this natural process can have a magnified impact. When large areas of rain forest are burned, the amount of rainfall recycled to adjacent areas is sharply reduced, depriving those areas of the rain they need to maintain their healthy condition. If the deforested area is large enough, the amount of rainfall removed from adjacent areas will be enough to cause a reinforcing drought cycle, which slowly kills more trees, thus further reducing rainfall recycling and accelerating the death of the forest in turn. And when the overarching canopy of leaves is removed, the sudden warming of the forest floor leads to the release of huge quantities of methane and CO_2, as a kind of biochemical "burning" takes place. The massive increase in the number of dead tree trunks and branches leads to an explosion in the population of termites, which themselves produce enormous quantities of methane. Thus the destruction of forests magnifies the global warming trend in several different ways — some of them simple, some complex — few of which are taken into account when the forests are destroyed.

The overuse of pesticides presents a similar danger, again because of a feedback loop. Pesticides often leave the most resistant pests behind as the more vulnerable ones disappear. Then, when the resistant pests multiply to fill the niche left by their dead cousins, larger quantities of pesticides are used in an effort to kill the more resistant pests, and the process is repeated. Soon, enormous quantities of pesticides are sprayed on the crops to kill just as many pests as were there when the process began. Only now the pests are stronger. And all the while, the quantity of pesticides to which we ourselves are exposed continues to increase.

The examples of excessive pesticide use and poor irrigation techniques both illustrate problems that, while widespread, are usually local in their effects. But sometimes entire regions are affected. The regional catastrophe of the Aral Sea, for instance, came about mainly because of an unpredicted feedback loop which magnified the impact of a poor irrigation strategy. Similarly, while the effects of deforestation are usually local, positive feedback loops of the kind that occur in the Amazon can magnify the impact of extreme cases into regional and even global tragedies.

Other problems start out as regional questions and are then magnified by feedback loops into serious global threats. Consider, for example, the controversy over the regional impact of global warming on the vast stretches of frozen tundra in Siberia. Some argue that this will have a positive impact, perhaps even opening large areas of Siberia to the cultivation of crops. Using a simple linear model and calculating a single effect from a single cause, one might conclude that this could indeed be a benefit. One might conclude further that this supposed benefit should be balanced against any unwelcome consequences from global warming. Indeed, it is with such calculations that some of the more extreme skeptics conclude that global warming may be a good thing.

But when one looks more closely at the nonlinear effects of thawing the tundra, dangerous new risks must be added to the calculation. As frozen tundra thaws, enormous quantities of methane are expected to be produced and released into the atmosphere. In recent years, the rate of increase in methane concentrations has slowed, but since each methane molecule is twenty times more effective as a greenhouse gas than each molecule of CO_2, if these huge new quantities of methane are released from thawing tundra, there will be a significant increase in the overall concentration of greenhouse gases, and global warming will be accelerated. Then the cycle reinforces itself: more tundra thaws, releasing still more methane into the atmosphere. (It is also worth noting that for other reasons having to do with the concentrations of ice underneath the top layers of tundra, it can seldom be used as cropland, anyway, even after it thaws.)

Unfortunately, this question is not merely hypothetical. Siberia is one of the regions of the world that seems to be warming most rapidly. That is not surprising, because all of the models have predicted it on the basis of the positive feedback loop that magnifies the effect of melting snow and the consequent increase in the absorption of sunlight at the surface. But the speed of warming in some recent measuring periods has been astonishing. For example, in March 1990, the average recorded temperature throughout Siberia was an astonishing 18 degrees Fahrenheit higher than any previous March on record. For the world as a whole, the year 1990 was, of course, just the latest "warmest year on record."

Still other feedback loops are clearly strategic threats. Consider, for example, the way in which the two best-known crises, global warming and stratospheric ozone depletion, reinforce each other in a complex positive feedback loop. Global warming increases the amount of water vapor throughout the atmosphere and traps infrared heat in the lower part of the sky which would otherwise radiate back out to space, passing through the stratosphere. As a result, the stratosphere actually cools as the lower atmosphere warms. A cooler stratosphere with more water vapor means more ice crystals in the ozone layer, especially in the polar regions, where chlorofluorocarbons (CFCs) mingle with the ozone in the presence of the ice, thus depleting the ozone at a faster rate. The thinner the ozone layer, the more the ultraviolet radiation strikes the surface of the earth and all the organisms living there. The ultraviolet radiation strikes vegetation that normally absorbs vast quantities of CO_2 through photosynthesis and seems to seriously disrupt its ability to do so. As the vegetation absorbs less CO_2, more of it accumulates in the atmosphere, causing still more global warming — and still more stratospheric cooling. The cycle is reinforced and magnified. It feeds upon itself.

Some of the most dangerous and powerful feedback loops involving the oceans are still matters of intense scientific inquiry. For example, there is preliminary evidence that as the oceans get warmer they stop absorbing CO_2 at their present rate. This possibility is especially disturbing since the amount of CO_2 in the oceans is fifty times the amount currently in the atmosphere. So if only 2 percent were no longer absorbed there, the amount in the atmosphere might accordingly double — and in the process warm the oceans even more. Moreover, some argue that the warming of shallow Arctic Ocean waters will lead to almost as much new methane in the atmosphere as the warming of the tundra.

Similar phenomena are also perpetuated by human beings. When economics enters the picture, the environment can be threatened by new kinds of feedback loops that are just as complex and dangerous as some of those found in the natural world. For example, underdeveloped, impoverished nations borrow large sums of money from banks in the richer nations. In order to pay the interest in the currency of their lender, they have to sell something of value to the export market. All too often, that means converting vast areas of land from farms and gardens that were growing staples of the local diet to plantations concentrating on a single cash crop sold for export. The shift of land away from local food crops decreases the supply and increases the price, thus further impoverishing the people. If food prices are controlled, government subsidies are increased, further impoverishing the government. At the same time, the increased supply of cash crops from much of the developing world drives commodity prices down and reduces the income from export earnings these countries might otherwise receive. The money for the cash crops goes typically to a few large landowners (and corrupt government officials) who, instead of spending the money in the local economy, send it to accounts overseas in the same banks that made the original loans to their country. As the country goes further into debt, it borrows still more money to pay the interest and forces still more land into cash crops — and the cycle continues, even though everyone assumes the debt will never be paid.

In 1985, the amount of hard currency flowing out of the developing world to the nations of the developed world was larger than all of the money flowing in the opposite direction, whether in the form of loans or foreign aid or payments for exports. Moreover, because of this complicated feedback loop, that gap has been growing steadily every year since. It is, in the memorable phrase of Robert McNamara, "like a blood transfusion from the sick to the healthy."

It is the human factor in all of these feedback loops that is critical to saving the global environment. We need a positive feedback loop that feeds on itself in a good way and accelerates the pace of the positive changes now so urgently needed. That can only happen when we adopt a new, long-range global perspective and accept responsibility for confronting the problem head-on. And when we appreciate how much is already known about the problem, we will be better able to recognize the new pattern of unprecedented systemic change.

It is true, however, that large patterns are even harder to recognize if they are fundamentally new. For one thing, it is hard to gain historical perspective on an event that is completely unlike any other we have ever seen before. Indeed, some skeptics dismiss the environmental crisis precisely because of the lack of historical reference points. But they exist; they may require some extrapolation, but they can be found in the history of how human communities responded in the past when they experienced much smaller climate changes than those we are now confronting.

3

Climate and Civilization:
A Short History

Beginning in 1816, "the year without a summer," widespread crop failures led to food riots in nearly every country of Europe, producing a revolutionary fervor that swept the continent for three years. In France, for example, the existing government fell and the conservative Duc de Richelieu was asked to form a new one. Everywhere governments struggled to maintain social order as an unprecedented crime epidemic surged in the cities. The Swiss were stunned by the wave of criminal activity. Even the number of suicides increased dramatically, along with executions of women for infanticide.

Historians describe "swarms of beggars" clogging the roads and beseeching passersby. In a typical account, a traveler through Burgundy in 1817 reported that "beggars, very numerous yesterday, have increased greatly; at every stage a crowd of women and children and of old men gather round the carriage." Another observer, who was visiting Burgundy from the British Isles, added that the number, while large, was "by no means as many as besieged the traveller in Ireland." In Switzerland, eyewitnesses said the numbers of beggars thronging every highway were so huge as to resemble armies. They had desperation in their eyes and, in the words of a local chronicler, Ruprecht Zollikofer, "the paleness of death in their cheeks."

As fears of revolution mounted in several countries, military force was used to control the growing crowds demanding food. An unprecedented wave of arson began to strike in almost every country. Ominously, the first anti-Semitic riots in the history of modern Germany broke out in the Bavarian town of Wurzburg in the summer of 1819 and, after famine and revolutionary fervor had exacerbated tensions and resentments, quickly spread throughout Germany and as far north as Amsterdam and Copenhagen.

Europe was just recovering from the Napoleonic Wars and was experiencing many changes. But although no one realized it at the time, the proximate cause of this suffering and social unrest was a change in the composition of the global atmosphere following an unusually large series of eruptions of the Tambora volcano, on the island of Sumbawa, Indonesia, in the spring of 1815. Scientists estimate that 10,000 people were killed in the initial eruption and approximately 82,000 more died of starvation and disease in the following months. However, the worst effects on the rest of the world were not felt until a year later, by which time the dust ejected into the sky had spread throughout the atmosphere and had begun to dramatically reduce the amount of sunlight reaching the surface of the earth and to force temperatures down.

In New England, there was widespread snow in June of 1816 and frost throughout the summer. The *Old Farmer's Almanac* became popular when a typographical error predicted snow in July 1816 — and it happened. From Ireland across England to the Baltics, rain fell almost continuously from May to October. The disruption of reliable climate patterns had carefully documented social consequences: failed harvests, food riots, and the near-collapse of society throughout the British Isles and Europe. The historian John D. Post has called it "the last great subsistence crisis in the Western world."

The climate changes precipitating this crisis appear to have lasted fewer than three years, perhaps because much of what is blown into the atmosphere by volcanoes falls back out of the atmosphere in a relatively short period of time. That is why the effects of even the largest volcanic eruptions, while often global, typically do not persist for more than one to two years. The eruption of Mount Pinatubo in the Philippines in 1991, for example, had a significant but short-lived global impact, cooling the earth and temporarily masking the much more powerful warming caused by human civilization, and temporarily accelerating ozone depletion.

Nevertheless, the large volcanic eruptions recorded throughout history are instructive about longer-term changes in three important ways. First, they demonstrate how dependent our civilization is on stable climate conditions of the kind we have enjoyed for most of the last 10,000 years. Second, they show how tragedies striking one part of the world can be caused by climate changes originating in an entirely different part of the world. And third, they suggest the devastating consequences of a comparatively sudden and massive manmade change in the global climate pattern.

Because the ancients knew little about the world beyond their own borders, they had no way of comprehending the cause-and-effect relationships between volcanoes on the other side of the world and dramatic climate changes in their own lands. Recently, however, detailed climate records from the ice cores in Greenland and Antarctica have determined the dates of major volcanic eruptions throughout antiquity, and scientists have correlated these records with evidence from tree rings, geology and archaeology, and a meticulous analysis of documents from ancient civilizations concerning climate history. The Chinese, in particular, have preserved records dating back as far as thirty-six centuries.

Thus, records from tree rings and ice cores, along with documents left by Chinese historians, have now been combined to describe the devastating effects of one of the largest volcanic eruptions in recorded history: Santorini, seventy miles north of Crete, exploded around 1600 B.C. with a force a hundred times larger than that of the well-known eruption of Krakatoa in 1883. The climate effects of Santorini most likely contributed to the sudden disappearance not long afterward of the Minoan civilization, which had dominated the eastern Mediterranean for a thousand years during the Bronze Age. (Some historians believe that the disappearance of the Minoans was the basis of Plato's description of the loss in a single day of the fabled Atlantis.)

Five centuries later, sometime between 1150 and 1136 B.C., the Hekla 3 volcano, in Iceland, blew millions of tons of dust and particulates into the atmosphere. Contemporaneously, according to a primitive Chinese script preserved on dry bamboo strips, "it rained dust at Po." According to another Chinese writer, "For ten days the sky rained ashes. The rain was gray." And according to still a third, "it snowed in the sixth month and the snow was over a foot deep . . . frosts killed the five cereal crops . . . fiber crops did not mature . . . and there was heavy rainfall." This time, archaeologists found evidence of devastating consequences in the Western Hemisphere as well. Scottish archaeologists assert that at this same time, 90 percent of the population of Scotland and northern England disappeared. Moreover, an analysis of soil samples indicates that extremely heavy precipitation and frigid temperatures forced the temporary cessation of agriculture.

Sometime around 209 B.C. there was a huge eruption, believed to be from a volcano in Iceland, which left its evidence deep in the annual layers of snow and ice covering Greenland and the frost-damaged rings of Irish oaks. Two years later, according to the Chinese historian Szu-ma Ch'ien, "the harvest had failed" for reasons no one understood. And two years after that, the Chinese historian Pan Ku wrote in the *Han shu*, "a great famine" killed more than half the population. "People ate each other." The emperor, he wrote, lifted the legal prohibitions against the sale of children. It was during this period, according to the Chinese *Table of Dynastic Records*, in 208 B.C., that "stars were not seen for three months."

The famous eruption of Mount Etna in Sicily in 42 B.C. was chronicled by Roman poets but only recently linked with catastrophic climate changes affecting China by historians studying newly translated texts. Pan Ku described how the sun was "veiled and indistinct" and how harvests failed, sending grain prices up more than 1,000 percent. He noted an edict issued in the summer that said: "The multitudes work hard plowing and weeding, without producing results. They suffer from famine and there is no way they can be saved."

Surprisingly, small climate changes caused by volcanic eruptions may also have played a major role in one of the modern era's seminal events, the French Revolution. In his groundbreaking study of the history of climate, *Times of Feast, Times of Famine,* Emmanuel Le Roy Ladurie describes in meticulous detail the disastrous crop failures and poor harvests in France during the six years immediately preceding the Revolution of 1789, culminating in the bitter winter of 1788–89 and one of the coldest Mays in history before the storming of the Bastille. That year the wine harvest was "an utter failure."

As it happens, one of the best available reports on the weather in those years comes from Benjamin Franklin, who had been in France since December of 1776. In May 1784 he wrote,

> During several summer months of the year 1783, when the effects of the sun's rays to heat the earth in these northern regions should have been greatest, there existed a constant fog over all of Europe and parts of North America. This fog was of a permanent nature; it was dry, and the rays of the sun seemed to have little effect in dissipating it as they easily do a moist fog rising from water. They were, indeed, rendered so faint in passing through it that, when collected in the focus of a burning glass they would scarcely kindle brown paper. Of course their summer effect in heating the earth was exceedingly diminished. Hence the surface was nearly frozen. Hence the snow remained on it unmelted, and received continual additions. . . . Perhaps the winter of 1783–84 was more severe than any that happened for many years.

Franklin shrewdly speculated that "the cause of this universal fog is not yet ascertained. . . . Whether it was the vast quantity of smoke, long continuing to issue during the summer from Hekla in Iceland, and that other volcano [Skaptar Jokul] which rose out of the sea near the island, which smoke might be spread by various winds, is yet uncertain." What he could not have known was that in addition to the Icelandic eruptions, later that same year the Asama volcano in Japan registered one of the most violent eruptions in history and in all likelihood was the main source of the unusually cold years of the middle 1780s, contributing to the crop failures and social unrest preceding the French Revolution, which decisively reshaped the modern world.

The role of climate in shaping human history is, of course, extremely complex, and climate historians often debate the degree to which climate should be assigned a deterministic role. It always interacts with the social, political, and economic factors that dominate our traditional approach to history, but some climate upheavals seem from circumstantial evidence to be extremely significant, even dominant, factors in shaping public moods and attitudes immediately before political upheavals. Just as the enormous climate-induced suffering from 1816 through 1819 obviously contributed to the concurrent political unrest in Europe, it seems clear that the climate-induced suffering in France from 1783 through 1789 played a major role in worsening the political mood in which the French Revolution took place. However, it seems just as clear that climate changes were only one of many causes leading to these events. And just because climate has been largely ignored in standard histories does not mean that it should suddenly be given an exclusive explanatory role.

Nevertheless, the effects of climate change on the political and social stability of civilization are powerful, and as we consider the possibility that humankind is now changing the climate of the entire globe to a degree far greater — and faster — than anything that has occurred in human history, we would do well to examine some of the lessons provided by nature.

In addition to helping cause famine and political unrest, one of the most dramatic effects of climate change on civilization has been massive migrations from one geographic area to another. In fact, one of the greatest migrations in history — the one that introduced human beings into North America and then South America — came about as a direct result of climate change. During the last Ice Age, roughly 20,000 years ago, when vast amounts of seawater were frozen into ice, sea level was about three hundred feet lower than it is today. Large areas of those parts of the ocean bottom we call the continental shelf were exposed as dry land, and shallow ocean straits, like the Bering Strait and the Gulf of Carpentaria, were instead land bridges. These bridges served as the migratory routes for the people now known as aborigines in Australia and the Asiatic nomads now known in North America as Native Americans and in South America as Indians or indigenous people. As the glaciers retreated, the sea level rose again some 10,000 years ago, stranding the Native Americans and aborigines on their new continents. At the same time, as temperatures climbed, the global climate settled into the pattern that it has roughly maintained ever since.

In fact, the Ice Age that so profoundly affected the Americas shaped the very roots of all human civilization. The cave paintings that represent the first known graphic communication by human beings appeared 17,000 years ago, when people sought refuge and warmth during the worst and coldest millennia.

Indeed, the succession of ice ages and warming interglacial periods between 1 million and 40,000 years ago is believed by most historians to have provided the impetus for the development of rudimentary social organizations. The archaeological and anthropological records indicate that each time the ice retreated, the primitive peoples of the Eurasian landmass grew more populous and their culture more advanced.

Between 8000 and 7000 B.C., when favorable climate conditions prevailed as the glaciers were melting into their present retreats, the area we know as Mesopotamia saw the creation of agricultural surpluses. And the trading of those surpluses is believed to have been responsible for the invention of money, the first communities to use brick and stone architecture, and the development of a broad range of arts and crafts. Jericho, for example, the oldest known city, was founded in this period, while Europe was just beginning to recover from the Ice Age.

Later, smaller but still significant climate fluctuations continued to shape the beginning of more complex social forms. Some historians believe that the first appearance of highly organized societies in the fertile river valleys of the Tigris, the Euphrates, and the Nile was stimulated by a major climate transition some 3,000 years ago. A new climate pattern — characterized by drought during most of the year and annual flooding — forced communities to cluster in the river valleys. The challenge of containing and distributing the floodwaters for irrigation, storing the annual harvests, and distributing food supplies required that many basic mechanisms of human civilization be put in place. In the Bible, Joseph's warning to Pharaoh to prepare for seven lean years following seven fat years reflects the new awareness of humankind of its vulnerability to changes in the weather patterns. In turn, when Pharaoh appoints Joseph, who interpreted the ecological meaning of Pharaoh's dream, to oversee preparations for the lean years, his decision reflects the assertion by humankind of the power to anticipate and prepare for climate fluctuations.

But it's now becoming clear that climate is even more basic to the development of humankind. Anthropologists, evolutionary biologists, and climate specialists — including Elisabeth S. Vrba, Frederick E. Grine, Richard G. Klein, and David Pilbeam — have recently combined the history of climate changes with the anthropological evidence to produce a new consensus — that human evolution itself was shaped by dramatic transitions in global climate patterns during the last 6 million years. The science writer William K. Stevens describes "an outpouring of analysis" and says that "scientists are sketching out the influential roles played by climate and ecology in shaping human evolution."

The major global cooling period that gradually took place more than 5 million years ago corresponds with the appearance of the first hominids, called australopithecines. It happened because — in the view of many scientists — at least one species of tree-dwelling ape was able to adapt to the disappearance of its forest habitat by learning to forage on the ground and walk on two legs, leaving the hands — which had evolved to grasp tree limbs — free to hold and carry food and objects, some of which later became tools.

A second global cooling period about 2.5 million years ago, more extreme and abrupt, explains — in this view — the "pulse" or evolutionary stimulus that produced a new, advanced branch of robust australopithecines. They were eventually displaced by the genus *Homo,* which appeared about 100,000 years ago after four relatively short (in geologic terms) but extreme ice ages — immediately before the last Ice Age. This period of incredible ecological change put a premium on the larger brains needed to adapt to rapidly changing climate conditions. The new discoveries relating the emergence of *Homo sapiens* to global climate changes have solved one of the mysteries in the human story by providing, at least in ecological terms, the missing link in the history of evolution. Then, 40,000 years ago, the so-called cultural explosion of tools and jewelry may have coincided with an unusually warm millennium in Europe.

But within this larger glacial and interglacial pattern there have been significant fluctuations. While they are quite small compared to either an ice age or to the manmade warming period now in prospect, they have nevertheless been large enough to have dramatic effects on civilization.

For example, a climate shift known as the subatlantic deterioration, from 500 to 400 B.C., led to a change in wind and moisture distribution and lower temperatures across Europe that are generally credited with bringing about the end of the northern Bronze Age and spurring the Germanic invasions of southeastern Europe from Scandinavia. Less than a century later, in what may be more than a coincidental continuation of the southeastern thrust of migration, the Macedonians conquered Greece. It was in the very next generation, as the climate began to warm all over the world in roughly 300 B.C., that Alexander the Great conquered the "known world" and spread Greek civilization throughout the Mediterranean and beyond.

This same period of relative warming cleared the Alpine passes separating Italy from the rest of Europe and corresponded to the awakening of Rome's imperial ambition. Moreover, the simultaneous clearing of mountain passes in Asia led to the expansion of Chinese civilization and the opening of the Silk Route. Some 750 years later, the end of this warming period corresponded to the final years of the Roman Empire. To the many explanations of why Rome fell, climate historians add the sudden shift in global climate patterns between A.D. 450 and 500 that led to a prolonged freezing drought in central Europe which, they suspect, may have stimulated the corresponding onset of the massive migrations that eventually became known as the barbarian invasions.

In sixteenth-century India, the grand city of Fatepur Sikri was completely abandoned just after its completion when a sudden change in the monsoon pattern deprived it of water. The people who had planned to live there were forced to go elsewhere, merely repeating a pattern already established in the Indian subcontinent. In fact, one of the first examples of an empire collapsing largely as a result of climate changes occurred a few hundred miles west of Fatepur Sikri some twenty-four centuries earlier. For a thousand years before 1900 B.C., the great Indus civilization flourished in what is now northwestern India and Pakistan. Then, suddenly, at a time when climate historians describe a southward expansion of cold polar air into northern Canada, the climate patterns changed, and what once were large cities and settlements were buried under the sand dunes of the Rajputana Desert, forcing the people to move elsewhere. Similarly, the collapse of the Mali civilization of West Africa in the fourteenth century is among other societal declines that climate historians now suspect may have been caused by sudden shifts in climate patterns.

And then there is the mystery of Mycenae, the elaborate civilization derived from Minoan culture that was the home of King Agamemnon in Homer's epics and that — after dominating the Aegean for more than two centuries — abruptly disappeared soon after 1200 B.C. Historians and archaeologists have speculated that there was an invasion by peoples from farther north, and there is evidence that many Mycenaeans fled to the south and east, but the suddenness of the collapse has remained a puzzle. However, recent climate analyses have added a provocative piece of evidence: just before the disappearance of the Mycenaean civilization, a dramatic shift in the prevailing wind and moisture patterns throughout Europe, the Mediterranean, North Africa, and the Middle East suddenly diverted the regular rainfall on which Mycenae had always depended. The new pattern still brought moisture from the west, across the Mediterranean, but from farther south and at such low altitudes that the rain fell on the western side of the mountains at the edge of the Peloponnesian peninsula. This triggered a prolonged and unrelenting drought in Mycenae, on the eastern side of the mountains, drying up wells and streams, killing crops, and eventually forcing the people to leave.

Some climate historians also believe that this same set of changes in the Mediterranean weather patterns was largely responsible for catastrophic flooding episodes in the Hungarian plain, which led in turn to the great eruption of Bronze Age peoples across the Bosporous from the Balkans. These mass migrations by the Phrygians and other peoples from what is now Armenia caused the collapse of the Hittite civilization of Asia Minor around 1200 B.C., triggering politically and militarily disruptive mass migrations through Cyprus, Syria, Palestine, and Egypt, echoes of which are found throughout the Old Testament. In fact, the same migration from the Hungarian plain sent another group of people to the southwest, through the mountain passes into Italy, where they became known as the Etruscans and planted the seeds of what in time became the Roman civilization.

In the Western Hemisphere, a new analysis of global climate records may shed light on the mysterious rise and fall of the classic Mayan civilization, which began to flourish around A.D. 250–300 in what we call Yucatán, southern Mexico, and Central America. For reasons that are still unclear and that provoke vigorous debate among archaeologists and historians, the Mayan culture suddenly collapsed around 950. The Mayans had built fantastic cities, with elaborate underground reservoirs and giant structures as large as any in the world at that time. These included sophisticated observatories from which their astronomers calculated the precise length of the solar year and the lunar month. They knew the precise orbital path of the planet Venus and were even able to predict eclipses. Their mathematicians independently discovered the mathematical concept of zero. Yet this enormously sophisticated culture suddenly ended. Its cities were mysteriously abandoned, not destroyed. There was an abrupt end to the manufacture of fine pottery and carvings, the creation of monuments, temples, records, calendars, and writing, and the rapid depopulation of the ceremonial centers and the countryside — all within fifty to a hundred years. Scientists have produced a variety of theories — from fratricidal violence and societal breakdown to an unknown invasion to hurricanes, earthquakes, soil exhaustion, water loss, savanna grass competition, and overpopulation.

What no study has suggested is that a change in the global climate pattern may explain the Mayans' collapse. Yet the historical climate record of the Western Hemisphere suggests that around A.D. 950, temperatures increased and the climate changed; at precisely the same time as the Mayan collapse, far to the north, Leif Eriksson sailed through the Labrador Sea between the new settlements of his father, Eric the Red, in Greenland and North America and became the first European to set foot on what he called Vinland.

Thus began the global climate shift known as the medieval warm epoch. Although it is understood as a European phenomenon, it clearly seems to have been a shift in the global climate pattern, recorded in North America by the first Europeans there. Indeed, the climate shift was the reason they were able to go there at all. Up until around 900, the North Atlantic sea routes from Scandinavia and Iceland to the new communities in Greenland had been completely frozen over and impassable. And at the end of the warm epoch, around 1300, temperatures began to fall, and sea ice again blocked the routes. The sporadic trips to Vinland had already ceased; soon the ships could no longer travel from Greenland back to Iceland for supplies. A generation later the last settlers froze to death, and Leif Eriksson's voyage was eclipsed in history by that of a southern European, Columbus.

But what happened to the climate in Yucatán around 950? If a new climate pattern permitted the settlement of Greenland and — however briefly — North America, could it have made the Mayan civilization in Central America suddenly untenable as flora and fauna changed, as pests migrated north from the equator, as rainfall patterns were altered, and as the fierce tropical sunshine took its toll on a society that had grown up in a slightly cooler and more hospitable climate? That may be at least part of the solution to the mystery of the disappearing Mayans.

After the warming epoch, temperatures dropped again at the beginning of the fourteenth century, causing major problems in Europe and Asia. To begin with, the transition suddenly brought repeated waves of humidity sweeping from the North Atlantic through the British Isles and across vast areas of the continent. For almost ten years, rotted harvests and flooded rivers doomed the people of western Europe to a series of famines that reached their peak in the Great Famine of 1315–17. In 1315, Guillaume de Nangis reported from Rouen and Chartres that crowds of pitiful, emaciated men and women were coming to the churches in terrified processions to pray for relief from the unrelenting rains. "We saw a large number of both sexes, not only from nearby places but from places as much as five leagues away, barefooted, and many even, except the women, in a completely nude condition," he said, "with their priests coming together in procession at the Church of the Holy Martyrs." That year and the next, the European grain harvests were completely destroyed. Le Roy Ladurie reported that the summer of 1316 "was so damp there was not even enough good weather to shear the sheep." The repeated famines caused an unprecedented number of deaths; but worse was to come with the Black Death, thirty years later.

Just before the Black Death, four years of poor weather and crop failures caused widespread malnutrition and increased susceptibility to disease, leading some to fear a repetition of the Great Famine. These fears stimulated grain imports from Asia Minor, among other places, which brought diseased rats first to Constantinople and then to the ports of Messina and Marseilles. From there, they and the plague they carried spread in only two years to wipe out as much as a third of the population of western Europe.

The plague itself actually originated in China, where the first reported deaths occurred in 1333. One year earlier, as a result of the same global climate changes that produced constant rains in Europe, unusually heavy rainfall in China caused the repeated Yellow River floods, which had grown steadily worse since 1327. They culminated in the largest flood of the Middle Ages in 1332, when a reported 7 million Chinese people lost their lives.

"There can be little doubt that the waters had dislocated the habitats of the wildlife as well as the human settlements, including those of the plague-carrying rodents," writes the climate historian Hubert H. Lamb. He concludes, "It is probably no coincidence that the Bubonic plague epidemic, which ultimately swept the world as the Black Death, started in 1333 in China" — the year following the great flood, in areas where decomposing human corpses had been numerous.

One of the most important and well-documented climate fluctuations is known as the Little Ice Age (1550–1850), which was associated with significant social changes all across Europe. People spent more time indoors, keeping warm around suddenly popular fireplaces, and partly as a result, new patterns of social relations evolved. The exchange of ideas about subjects like science intensified. Romantic ideals took on a new significance in the arts, as did the concept of the individual in politics. Outdoors, however, the new climate realities were harsh for some in northern Europe.

Imagine the shock in Aberdeen, Scotland, in 1690, when an Eskimo in his kayak appeared in the River Don. The migration of Europeans toward Greenland had long since come to a frozen halt, but the Eskimos' favored habitat was now extending south as far as the Orkney Islands and northern Scotland.

The Scots, confronted with the failure of their cod fisheries and their crops, experienced repeated famines and began to leave their homeland. By 1691, 100,000 Scots, a tenth of the population, had settled in the part of Ireland closest to Scotland, Ulster (now known as Northern Ireland), displacing and evicting the native Irish and setting in motion the enormous problems and seemingly insoluble violence that continue to this day.

In the years following the Scottish migration, Ireland as a whole continued to grow in population. Historians generally agree that Ireland became a social and political mess; England's dominance led to a number of foolish decisions, of which the decision by King James VI to facilitate the Scottish migration was only the first. Archaic rules of land ownership helped to create a culture of poverty, which in turn encouraged early marriage and further population growth. Between 1779 and 1841 the population increased by 172 percent, making Ireland, by Disraeli's estimate, the most densely populated area of Europe. The fateful decision to rely almost exclusively on a single food crop — potatoes — for subsistence set the stage for the horrible tragedy known as the Great Potato Famine.

As the Little Ice Age drew to a close, average temperatures rose slightly, enough to create the wet and warm climate conditions conducive to potato blight. Modern laboratory studies show that the particular blight that struck Ireland, *Phytophthora infestans,* requires a period of at least twelve hours with the relative humidity at 90 percent or more with temperatures at 10 degrees Centigrade or higher and free water on the potato leaves for at least another four hours. The possibility of such conditions coming together were much lower during the Little Ice Age, when Ireland began to depend on the potato; by the mid-1840s the odds had improved with the new warming trend.

The blight seems to have originated in a new strain of potatoes from Peru; it first appeared in the northeastern United States in 1843 and in Flanders the following year. By the summer of 1845 the spores had spread to Ireland. That winter was one of the warmest the Irish could remember; the spring was also warm, and in June temperatures soared to an average three to four degrees warmer than the hundred-year average. That summer as a whole was the second warmest of the nineteenth century. On top of that, there were sixty-four days of rain in July, August, and September, twenty-four in August alone.

The blight struck with a terrible vengeance at the one crop by which Ireland lived or died. More than a million people died in Ireland during the next few years of starvation and diseases related to malnutrition. The horrific reports of the survivors give us some sense of what the famine meant in human terms. In December 1846, the father of two very young children in County Cork died of starvation (as their mother had previously). According to the inquest, "His death became known only when the two children toddled into the village of Schull. They were crying of hunger and complaining that their father would not speak to them for four days; they told how he was 'as cold as a flag.' The other bodies on which an inquest was held were those of a mother and child who had both died of starvation. The remains had been gnawed by rats."

A contemporary newspaper report recorded this typical eyewitness account: "In a cabbage garden I saw the bodies of Kate Barry and her two children very lightly covered with earth, the hands and legs of her large body entirely exposed, the flesh completely eaten off by the dogs, the skin and hair of the head lying within a couple of yards of the skull, which, when I first threw my eyes on it, I thought to be part of a horse's tail. . . . I need make no comment on this, but ask, *are we living in a portion of the United Kingdom?*"

The practice of growing a single crop over vast areas instead of a variety of plants is known as monoculture. The problem is the risk of vulnerability to a plant disease or a resistant pest that suddenly wipes out the whole crop. This vulnerability is even greater when a single strain of a single crop is used. The Irish had come to rely on a single strain of potatoes as virtually their sole source of food, a strain that maximized yields in the climate conditions that had prevailed for the previous 300 years. The story of the potato famine is a lesson in how artificial modifications to our relationship with nature, like monoculture, that fail to take into account the vagaries of natural climate change can increase the vulnerability of a society attempting to feed its people. It also shows how a rapid warming can cause catastrophe.

Historically, climate tragedies like the one that caused the potato famine have led to massive migrations toward wealthier countries, especially the United States. Three decades earlier, the great subsistence crisis of 1816–17 had also stimulated a flood of migration, not only from Europe to the United States but — because the effects of the climate change were felt well beyond Europe — also within the United States. For example, historical accounts of the westward migration from Maine indicate that after "the uncommonly cold and unpropitious" springs of 1816 and 1817, a terrible fear of famine lent "a fresh impulse to the enchanting spirit of emigration. Hundreds who had homes, sold them for small considerations, and lost no time in hastening away into a far country." The connection between the migration from Maine and the unusual climate patterns caused in 1816–17 by the Tambora volcano tends to be bolstered by the statistics: they show that once the unusual climate patterns ended (when the dust of the volcano fell back out of the atmosphere), in 1818, Maine resumed its steady population growth. An identical pattern was documented in New Hampshire, Vermont, Connecticut, and the Carolinas. One eyewitness wrote that "a sort of stampede took place . . . during the summer of 1817."

Perhaps the largest forced migration in American history was the mass departure from Kansas, Oklahoma, Texas, parts of New Mexico, Colorado, Nebraska and other Plains states during the period of the early 1930s referred to as the Dust Bowl years. Like the Great Potato Famine, the Dust Bowl resulted from unwise land use, which heightened the vulnerability of the land and its people to unexpected climate changes. During the 1920s, there was a revolution in agriculture throughout the High Plains states. Mechanization led to the development of the tractor, the combine, the one-way plow, and the truck. These, in turn, led to the "great plow-up" of the late 1920s. Agricultural experts mistakenly believed that the repeated plowing of land until it was smooth and pulverized made it better able to absorb and hold rainwater. Agronomic research, focusing on different ways to increase water absorption, completely overlooked the problem of wind erosion, which became a far more serious threat because of these very changes in agricultural methods.

For a few years there were record crops, and the early warning signs of wind erosion were ignored. Even when acreage was left fallow, the farmers continued to plow it as a way of discouraging weeds and, again, encouraging the absorption of moisture, to assure a good start when the wheat was planted.

The fall of 1930 and the spring and summer of 1931 brought heavy rains and hardship but a record harvest nonetheless. After a dry winter, in March 1932 strong winds began to blow through and take some of the topsoil with them. The spring rains were scattered and deficient, and then, in early summer, flooding from hard rains eroded the soil, punctuating a drought that had made the summer unusually dry overall. The fall was quite dry, and by the onset of winter many fields had been abandoned.

The big dust storms began in January 1933 and continued off and on for more than four years, devastating the crops, dispiriting the people, and creating nightmarish conditions, leading many to pick up stakes for California or back east. In 1934, Secretary of the Interior Harold Ickes advised the people of the Oklahoma panhandle to simply leave their homes. Only 15 percent of the acreage between Texas and Oklahoma would be harvested that year.

Those who stayed, actually the majority, also suffered. In Colorado, the editor of the *Morton County Farmer* wrote in the spring of 1935,

> We can see nothing out our windows but dirt, every time our teeth (or the dentist's, or maybe you have your store teeth paid for) come together, you feel dirt and taste it; haven't heard a thing for hours, my ears are full, can't smell, my nose is full, can't walk, my shoes are full but not of feet . . . we are and have been having a dirt storm. It hasn't been real life for two days. Everything is covered with a little of Old Mexico or Texas or Colorado or what have you . . . The earth looks hard and barren — everybody has a dirty face, even your creditors hardly know you. But there is no way out — not even out of our front door. We live in a dugout and slide down the steps now. Diving out the window is fun after you get used to it.

Emergency hospitals were set up to treat the many cases of "dust pneumonia," a collection of bronchial and other respiratory diseases caused and made worse by the constant inhalation of dust. The dust and dirt from the continuing storms blew all the way to the Atlantic Ocean. Not until 1937 did conditions finally stabilize.

Of course, the history of climate change is also the history of human adaptation to climate change. During the subsistence crisis of 1816–17, for example, the bureaucratic, administrative tendencies of the modern state were given great impetus. In virtually every European country, central governments organized and distributed the scarce supplies of food and imported new stocks from Odessa, Constantinople, Alexandria, and America. For the first time, large-scale public works projects were organized chiefly to provide employment in the hope of staving off the popular disturbances and food riots that accompanied the subsistence crisis. In the 1930s, the Dust Bowl was among the many disruptive social and economic problems that led to an even more complex version of the administrative state, Franklin Roosevelt's New Deal.

All of these changes in climate patterns took place during temperature variations of only 1 to 2 degrees Centigrade. Yet today, at the close of the twentieth century, we are in the process of altering global temperatures by up to three to four times that amount and causing changes in climate patterns that are likely to have enormous impacts on global civilization. Among the most dramatic effects, if the historical record is any guide, will be massive migrations of people from areas where civilization is disrupted to other areas where they hope to find the means for survival and a better way of life — but with unpredictable consequences for those areas.

About 10 million residents of Bangladesh will lose their homes and means of sustenance because of the rising sea level, due to global warming, in the next few decades. Where will they go? Whom will they displace? What political conflicts will result? That is only one example. According to some predictions, not long after Bangladesh feels the impact, up to 60 percent of the present population of Florida may have to be relocated. Where will they go?

Florida has already borne the brunt of one of the largest ecologically induced migrations of this century: some 1 million people emigrated from Haiti to the United States in the last decade — not only because of political oppression but also because the worst deforestation and soil erosion in the world made subsistence farming impossible for them. Although some of the Haitians have been absorbed, most have not, and all have suffered greatly, enduring perilous journeys and uncertain futures.

Sir Crispin Tickell, a leading British diplomat and environmentalist, noted in a speech to the Royal Society in London in 1989 that "a heavy concentration of people is at present in low-lying coastal areas along the world's great river systems. Nearly one third of humanity lives within sixty kilometers of a coastline. A rise in mean sea level of only twenty-five centimeters would have substantial effects . . . a problem of an order of magnitude which no one has ever had to face . . . in virtually all countries the growing numbers of refugees would cast a dark and lengthening shadow."

In the developed world, we now have the ability to insulate most people from the kind of suffering, disease, famine, and forced migration that, in the ancient world, often accompanied fluctuations in the global climate equilibrium and the attendant disruptions of the weather patterns upon which those fragile civilizations depended. But we insulate ourselves by burning even more fossil fuels and creating still more CO_2. And as we continue to expand into every conceivable environmental niche, the fragility of our own civilization becomes more apparent. Moreover, as the world population surges, our resilience in the face of climate variability is diminishing. In any event, the climate changes that we are now bringing about by modifying the global atmosphere are likely to dwarf completely the ones that caused the great subsistence crisis of 1816–19, for example, or those that set the stage for the Black Death.

In the course of a single generation, we are in danger of changing the makeup of the global atmosphere far more dramatically than did any volcano in history, and the effects may persist for centuries to come. The global temperature changes for which we are responsible are likely to be five times larger than the fluctuations that produced the Little Ice Age, for example, or the global climate change that led to the Great Famine of 1315–17.

As increased ultraviolet radiation weakens the human immune system, especially in the tropics, and as explosive population growth and urbanization continue to disrupt traditional cultural patterns, hundreds of millions of people may well become even more susceptible to the spread of diseases when populations of pests, germs, and viruses migrate with the changing climate patterns.

Our increasingly aggressive encroachment into the natural world and the resulting damage to the ecological systems of the earth have weakened the resilience of the global environment itself and threatened its very ability to maintain its equilibrium.

And how will the world respond? During the Irish potato famine, a combination of blind devotion to laissez-faire economics, cold indifference to suffering, anti-Irish racism, and anti-Catholic bigotry contributed to the United Kingdom's tragic failure to respond humanely. Given the advances of civilization since then, it is difficult to imagine that such a horror would be tolerated today. However, the number of children dying of starvation on an average day in our modern world is more than forty times greater than the number who starved on an average day at the height of the famine. The scenes going on right before our eyes today are just as terrible as the accounts from 1846. A combination of blind devotion to laissez-faire responses, political ineptitude in the countries affected, the paralysis that even a little bit of racism can promote, and the willful blindness of "denial" are promoting the continuation of our own great famine right now. It would not be at all surprising to hear an eyewitness in Ethiopia or the Sudan echo the observer of the famine victims and cry out, "Are we living in a portion of the same planet that has the United States and Europe and Japan?"

Actually, new climate analyses now show conclusively that the dramatic increase of famine in those areas of Africa that include Ethiopia, the Sudan, and Somalia coincides with a dramatic shift in rainfall patterns. "There was little trend in precipitation until the 1950s when, after a relatively wet episode, precipitation [in Northern Africa and the Middle East] declined drastically," a decline that has continued and accelerated for the last forty years, accompanied by concurrent "significant increases in [European] precipitation." So reported a team of researchers in *Science* in 1987, after an extensive array of climate measurements covering a century and a half identified large shifts in rainfall patterns over recent decades. Their study found that while rainfall steadily *decreased* in the Sahel portion of Africa and the Middle East, it steadily *increased* proportionally in Europe.

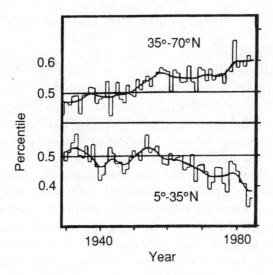

The bottom line here shows the steady decline of rainfall in the latitudes of Africa where repeated droughts and famine have killed tens of millions of people in recent years. The top line shows a mirroring increase in precipitation over the same period in the latitudes that include Europe.

These researchers are worried that this forty-year trend, which has been one of many factors in producing repeated and persistent famines, is an early consequence of global warming; if so, it may indicate even more disruptive changes in climate patterns as the warming continues. Another climate expert, Hubert Lamb, writing about the recent forty-year trend in the Sahel and the mass famines and immigrations that have accompanied it, said, "Some complete national territories may, in the long run, become more or less uninhabitable if the development continues and goes further." However, in spite of the circumstantial evidence, climate researchers are reluctant to link global warming definitively to these catastrophic changes because the phenomena involved are so complex.

It is possible, however, to draw some inescapable conclusions from what they have observed. This much is certain: fragile societies in the midst of a modern, affluent global civilization are undergoing enormous suffering partly as a consequence of a shift in climate patterns, whatever the cause of the shift may be. Meanwhile, the rest of the world has been unable to provide more than stopgap solutions for their suffering.

Moreover, even after highly publicized warnings from virtually the entire global scientific community that the current pattern of our civilization is creating dramatic changes in global climate patterns, likely to be many times larger than any experienced in the last 10,000 years, we are doing virtually nothing to address the principal causes of this catastrophe in the making. We know from the history of climate changes that they can cause unprecedented social and political upheavals, especially in fragile, densely populated societies. Ironically, we are ignoring the lessons of the Irish famine and shifting global agricultural patterns toward an unprecedented and increasing dependence on monoculture.

The lessons of the Dust Bowl are also being ignored. Sweeping changes in patterns of land use that turn out to be disastrously inappropriate are far more common today than in the decade before the Dust Bowl. The massive clearing of tropical rain forests is, of course, an ecological catastrophe of the first magnitude, beside which the Dust Bowl pales in comparison — not least because the earth could at least recover from the latter in a few generations, whereas the damage from the former could last for tens of millions of years. The sudden irrigation of the vast desert areas surrounding the Aral Sea in Soviet Central Asia represents another tragic mistake from which recovery may be difficult, if indeed it is possible at all.

Sometimes the damage from inappropriate land use is more subtle. In California, for example, using massive quantities of water from the northern part of the state to irrigate rice fields in southern areas reclaimed from the desert seemed like a good idea — until a new drought cycle began to hit the West in the late 1980s. During the last drought that approached the severity of the recent one (in the 1930s), California had 18 million people and proved resilient enough to endure the climate extreme. In 1991, with 32 million people, California may have been just as resilient, but fewer than 80,000 farmers use 85 percent of the state's water. As a result, the effects of the drought have been extremely disruptive.

In this period of extraordinary population growth, we have grown accustomed to the idea that the pressure of population on the environment is something new. But it is actually a recurrent theme in the history of environmental change. For example, climate historians have speculated that a similar pattern of expansion beyond the environment's carrying capacity may provide an explanation for the mysterious disappearance around the year 1280 of southwestern Colorado's Anasazi civilization, which occupied the spectacular cliff dwellings of Mesa Verde. Fairly reliable evidence indicates that its disappearance corresponded with a drought which, while severe, was not dissimilar from earlier droughts that the cliff dwellers successfully endured. According to the archaeological record, however, there was one crucial difference this time around: the Anasazi population had grown significantly larger just before its disappearance.

The lesson from this experience is almost unbearably obvious. Our global civilization, which after the many thousands of generations up to the end of World War II had reached a population of fewer than 2.5 billion people, may, by quadrupling in the space of a single lifetime, dramatically increase our vulnerability to the extreme climate changes that we ourselves are now setting in motion.

The signs of increased vulnerability are already evident, not only in the Sahel and the Amazon and the Aral Sea but in California and Florida and the High Plains states, which are now using up their underwater aquifers just as surely as Kansans once pulverized their topsoil until it blew away. The pressure of the population at the foothills of the Himalayas has led in the last few decades to such extensive deforestation that the rains now rush wildly down the slopes, across Bangladesh and eastern India, carrying an enormous tonnage of topsoil to silt up the Ganges River system and worsen the flooding that results. The Bay of Bengal is almost perpetually brown with the soil that ought to be growing crops. In my own state of Tennessee, the same phenomenon is occurring in a different form: new subdivisions strip the hillsides of the vegetation that used to soak up the rain; the creeks and rivers silt up and in some counties what used to be called a hundred-year flood now comes every few years.

It is now clear that the relationship between humankind and climate change has been reversed: where civilization once feared Nature's whim, the earth must now suffer ours — though we may yet relearn a healthy fear of disturbing nature's balance.

It is worth noting as well that the relationship between humankind and evolution has also begun to reverse. The current "era" in which we are living is described by geologists as the cenozoic era. Beginning 65 million years ago with the disappearance of the dinosaurs, the cenozoic era has been characterized by the flourishing of a larger number of more varied life forms than during any previous era in the earth's 4.6-billion-year history. Now, as the theologian Thomas Berry notes, human civilization, because it is destroying as many as half of all the living species on earth during the lifetimes of persons now living, is in effect bringing about the end of the cenozoic era — in our lifetimes.

What comes next? The "year without a summer" in 1816 produced massive famines and helped stimulate the emergence of the administrative state. What will global warming produce — a new worldwide bureaucracy to manage the unimaginable problems caused by massive social and political upheavals, mass migrations, and the continuing damage to the global environment by civilization itself? Is that what we want? Wouldn't it be better to prevent the chaos instead of scrambling to cope with it after it occurs?

The story of humankind and our relationship to the earth may be seen as a continuing adventure or a tragedy shrouded in mystery. The choice is ours. The "year without a summer" teaches us how vulnerable civilization is to small global climate changes. In the lifetimes of people now living, we may experience a "year without a winter." But, unlike the transient climate changes associated with volcanic emissions, we are carelessly initiating climate changes that could well last for hundreds or even thousands of years. The ancient civilizations that disappeared during significant natural climate changes in the past could tell us a great deal that we seem not to want to hear. What if our children, because of our actions, face not just a year without a winter but a decade without a winter? Will that be our most significant legacy? The answer may well depend on whether we can learn from the ancient cultures that disappeared.

If we do not, if we instead persist in our willful ignorance of the powerful changes we are setting in motion, we may ultimately leave little more than a mystery to puzzle some new human community in the distant future, trying to understand what happened to the ancient lost civilization that made such grand structures of concrete and steel and plastic so long ago.

4

Buddha's Breath

The magnitude of the changes we are imposing on the world's climate pattern is made obvious from the perspective of history, but in any given year our attention is likely to focus on the swirl of contemporary events — and specific problems with pollution, particularly of the air. No sooner had the political dust of Eastern Europe's revolution against communism settled in 1989 than the world gasped in horror at the unbelievable levels of pollution — especially air pollution — throughout the communist world. We learned, for example, that in some areas of Poland, children are regularly taken underground into deep mines to gain some respite from the buildup of gases and pollution of all sorts in the air. One can almost imagine their teachers emerging tentatively from the mine, carrying canaries to warn the children when it's no longer safe for them to stay above the ground.

One visitor to the Romanian "black town" of Copsa Mica noted that "the trees and grass are so stained by soot that they look as if they had been soaked in ink." A local doctor reported that even horses can stay only for two years in the town; "then they have to be taken away, or else they will die."

In the northern reaches of Czechoslovakia, the air is so badly polluted that the government actually pays a bonus to anyone who will live there for more than ten years; those who take it call it burial money. To the east, in the Ukraine, that one republic puts eight times as many particulates into the air each year as does the entire United States of America.

Throughout the developing world, similar nightmares are found on every continent. In Ulan Bator in Outer Mongolia, the local beverage, curdled mare's milk, has to be protected from the black flecks in the air that settle on every surface. Mexico City suffers the worst air pollution, day in and day out, of any city on earth. There are occasional tragedies, such as the accidental release of poisonous gas into the air above Bhopal, India, which capture the whole world's attention. But the constant deadly levels of air pollution in cities throughout the developing world do not, even though on a "normal" day they are responsible for the deaths of more people than died at Bhopal.

The developed world, including the United States and Japan, has its own problems with air pollution, of course, in cities like Los Angeles and Tokyo. But there have also been some dramatic successes. Pittsburgh, once legendary for its thick, soupy air, is now one of the most livable cities in the world. Most residents of Nashville don't even know that their city was once called Smokey Joe. London still has serious problems, but nothing that compares with the "killer smogs" of the 1950s. And since the Atmospheric Nuclear Test Ban Treaty stopped most aboveground nuclear explosions in the 1960s, the levels of deadly strontium 90 in the air have fallen dramatically.

Some of the successes in dealing with air quality have created new problems. For example, the use of tall smokestacks to reduce local air pollution has helped to worsen regional problems like acid rain. The higher the air pollution, the farther it travels from its source. Some of what used to be Pittsburgh's smoke is now Labrador's acidic snow. Some of what Londoners used to curse as smog now burns the leaves of Scandinavian trees.

And while many of the measures that control local and regional air pollution also help reduce the global threat, many others actually increase that threat. For example, energy-consuming "scrubbers," used to control acid precipitation, now cause the release of even more carbon dioxide (CO_2) into the atmosphere. A power plant fitted with scrubbers will produce approximately 6 percent more global air pollution in the form of CO_2 for each BTU of energy generated. Moreover, the sulfur emissions from coal plants partly offset, and temporarily conceal, the regional effects of the global warming these plants help to produce worldwide.

It is this problem — global air pollution — that presents the true strategic threat to which we must now respond. The political battles against local air pollution are the easiest to organize because the direct effect on human health can be seen vividly in the hazy, smog-choked skies and heard loudly in the hacking and coughing of the affected citizenry. The battles to control regional air pollution are more complex because the people who are most affected often live in a different, downwind region from the people most responsible for causing it. Still, this problem is finally being addressed, even as heated arguments continue over cause and effect.

However, the political struggle to control atmospheric pollution at the global level has barely begun. Every person on earth is part of the cause, which makes it difficult to organize an effective response. But every person on earth also stands to suffer the consequences, which makes an effective response essential and ought to make it possible to find one — once the global pattern is widely recognized.

One threshold that must be crossed before we can recognize the global pattern is the prevailing notion that the sky is limitless. Some of the pictures brought back from space by the astronauts and cosmonauts show that in fact the atmosphere is a very thin blue translucent blanket covering the planet itself. The earth's diameter is one thousand times greater than the width of the translucent atmosphere surrounding it; to put it in perspective, the distance from the ground to the top of the sky is no farther than an hour's cross-country run. The total volume of all the air in the world is actually quite small compared to the enormity of the earth, and we are filling it up, profoundly changing its makeup, every hour of every day, everywhere on earth.

We would prefer not to believe this, but consider the North Pole, far from any factory or freeway, where pollution known as Arctic Haze now reaches levels during the winter and spring that are comparable to the levels of pollution in many large industrial cities. Scientific analysis indicates that most of the Arctic Haze originates in northern Europe, making it, in effect, a particularly extensive example of regional pollution. Nevertheless, it illustrates the point that air pollution now reaches every place on earth. Air samples in Antarctica make the same point.

But the most troubling strategic air pollution threats are those that are truly ubiquitous and uniform throughout the world. Ironically, these threats are least likely to cause anyone immediate and direct personal harm, and consequently they are often perceived as benign. However, they are the changes most likely to do serious and lasting damage to the ecological balance of the earth itself.

The molecules of the air exist in a state of equilibrium; similarly, the atmosphere exists in a state of dynamic equilibrium with itself and with life on the planet. Dramatic changes in that equilibrium in just a few decades can threaten the balancing role played by the atmosphere within the larger global ecological system.

Most things on earth have adjusted through the eons to an amazingly persistent and stable balance in the makeup of the global atmosphere. The relatively small number of air molecules in the atmosphere have been continuously recycled through animals and plants since oxygen was first produced in large volume by photosynthetic microorganisms almost 3 billion years ago. Those animals and plants have adapted, over long periods, to the precise combination of molecules that have been present in the air throughout most of evolution, and they have, in turn, affected the composition of the atmosphere.

In every breath we take, we bathe our lungs in a homogeneous sample of that same air — many trillions of molecules of it — with at least a few in each breath that were also breathed by Buddha at some point during his life, and a like number that were breathed by Jesus, Moses, Mohammed — as well as Hitler, Stalin, and Genghis Khan. But the air we breathe is profoundly different than was theirs. For one thing, mixed in with the air molecules are a variety of pollutants, which vary according to where we live. More important, however, the concentration of some natural compounds has been artificially changed everywhere on earth. For example, every single person alive now inhales with each breath 600 percent more chlorine atoms than did Moses or Mohammed. The chemicals responsible for all this extra chlorine — now ubiquitous in the world's air — were first used in world commerce less than sixty years ago. That extra chlorine doesn't directly affect human health as far as we know, but it has a dangerous and debilitating strategic effect on the healthy functioning of the atmosphere. Like an acid, it burns a hole in the earth's protective ozone shield above Antarctica and depletes the ozone layer worldwide.

Ozone depletion is, in fact, the first of three strategic, as opposed to local or regional, air pollution threats; the other two are diminished oxidation of the atmosphere (a little-known but potentially serious threat) and global warming. All three have the power to change the makeup of the entire global atmosphere and, in the process, disrupt the atmosphere's crucial balancing role in the global ecological system. Ozone depletion changes the atmosphere's ability to protect the earth's surface from harmful quantities of short-wave (ultraviolet) radiation. Decreased oxidation potential damages the atmosphere's ability to constantly cleanse itself of pollutants like methane. Global warming increases the amount of long-wave (infrared) radiation retained in the lower atmosphere and thereby inhibits the atmosphere's ability to maintain global temperatures within the relatively constant range that provides stability for the current global climate system. In all three cases, the changes are ubiquitous and persistent. Let's consider each one in turn.

A thinner ozone layer allows more ultraviolet radiation to strike the earth's surface and all living things on or near the surface. Many life forms are vulnerable to large increases in this radiation, including many plants, which normally remove huge quantities of CO_2 from the atmosphere through photosynthesis. But the scientific evidence now indicates that these plants, when exposed to increased ultraviolet radiation, can no longer photosynthesize at the same rate, thus raising the levels of CO_2 in the atmosphere.

We too are affected by extra ultraviolet radiation. The best-known consequences include skin cancer and cataracts, both of which are increasingly common, especially in areas of the Southern Hemisphere such as Australia, New Zealand, South Africa, and Patagonia. In Queensland, in northeastern Australia, for example, more than 75 percent of all its citizens who have reached the age of sixty-five now have some form of skin cancer, and children are required by law to wear large hats and neck scarves to and from school to protect against ultraviolet radiation. In Patagonia, hunters now report finding blind rabbits; fishermen catch blind salmon.

Less well known are the effects of extra ultraviolet radiation on the normal functioning of the human immune system. Although the specific effects are still being investigated and debated, it's becoming clear that increased levels of radiation can indeed suppress the immune system and so may actually help to increase our vulnerability and to hasten the emergence of several new diseases of the immune system.

Each September and October, an enormous gap in the stratospheric ozone layer appears above Antarctica and the Southern Ocean, and at least one city on earth — Ushuaia, Argentina, in Patagonia — lies inside the edges of the famous ozone hole. The chemicals that cause ozone depletion — such as chlorofluorocarbons (CFCs) — have a much greater impact on the ozone layer above Antarctica for three reasons. First, since the air above Antarctica is much colder than anywhere else on earth, clouds form at much higher altitudes, putting tiny ice particles made of nitric acid and water in the stratosphere, where the ozone layer is found. The chlorine from CFCs can destroy ozone molecules much more effectively in the presence of these ice crystals than in the free-floating air.

Second, above Antarctica strong winds form a circular pattern that resembles a vortex, like the whirlpool formed by water draining from a bathtub. This vortex holds the frigid chemical brew — chlorine, bromine, ozone, and ice crystals — tightly in place, as if in a bowl, until the sun comes up.

Third, when the sun finally does come up, it ends the six-month darkness of the Antarctic night, which produces the coldest temperatures, the highest clouds, and the strongest circular wind pattern by September, just before the six-month sunshine of the Antarctic day. When the first rays of that long-awaited dawn strike the icy bowl of ozone and chlorine, they trigger a chain reaction of ozone destruction until virtually all of the ozone inside the bowl has been eaten by the chlorine and bromine. That is when the "hole" in the ozone appears. Gradually, as the constant sunlight heats up the air, the winds slow down and the bowl loses its integrity, with air from the rest of the world spilling over the edges and filling the hole. In the process, the concentration of ozone in the rest of the world's air is diluted by ozone-poor air flowing out of the bowl and mixing with the ozone-rich air outside.

Almost every year since it was discovered, the ozone hole has grown deeper and now covers an area three times as big as the forty-eight contiguous United States. Ominously, scientists have discovered the beginnings of a similar ozone problem above the Arctic, much closer to populated areas, although the whirlpool of wind above the North Pole is looser and the air is warmed by storms from the south even before the sun's rays break across the horizon. While the Antarctic is land surrounded by ocean, the Arctic is ocean surrounded by land, which conveys warmer air northward before the dawn. However, some scientists point out that on an average of every fifth year, the vortex of wind remains colder for a much longer time. If this is true, and if the concentrations of chlorine and bromine continue to increase, scientists believe it is only a matter of time before significant ozone depletion occurs in the Northern Hemisphere.

When the Antarctic ozone hole breaks up each year around the middle of November, sometimes large pieces of the hole break off like bubbles and float northward, creating dangerous risks in populous areas of the Southern Hemisphere. If an ozone hole began to appear in the Arctic, such bubbles would threaten much larger numbers of people. Already, even without a northern ozone hole, the stratospheric ozone layer has been depleted or thinned by almost 10 percent in just four decades, at least in the winter and early spring. For every 1 percent decrease in ozone, there is a corresponding 2 percent increase in the amount of ultraviolet radiation bathing our skin and a 4 percent increase in skin cancer. In the fall of 1991, scientists disclosed alarming new evidence that the ozone shield above the United States is now thinning in summer, when the sun's rays are far more dangerous, not just in winter, when they are weak. As a result, significant changes in behavior are warranted. Children, especially, should now be urged to minimize exposure to the sun.

Ironically, as the amount of ozone in the stratosphere declines, the extra ultraviolet radiation streaming through also interacts with the local air pollution above cities and increases the amount of smog — including the amount of low-level ozone. While ozone in the stratosphere protects us by absorbing ultraviolet radiation before it can reach the surface, ozone at ground level is a harmful pollutant that irritates our lungs.

Although other chemicals have contributed to the ozone depletion crisis, the principal damage has been done by CFCs. The fact
that CFCs have been produced for fewer than sixty years and yet
have already had such a sweeping impact on the atmosphere should
make us consider how many of the other twenty thousand chemical
compounds introduced each year may, when mass-produced, cause
other significant changes in the environment. Very few of them are
extensively tested for environmental effects before they are used —
although, ironically, CFCs were. It was their benign chemical stability in the lower atmosphere that enabled them to float slowly,
unimpeded, to the top of the sky, where the ultraviolet rays finally
sliced them into corrosive pieces.

What does it mean to redefine one's relationship to the sky?
What will it do to our children's outlook on life if we have to teach
them to be afraid to look up? Residents of the city already inside
the ozone hole, Ushuaia, have been advised by the Argentine
Health Ministry to stay indoors as much as possible during September and October. Sherwood Rowland notes that, ironically, the
second-largest employer in town is a company that manufactures
CFCs!

Our tendency to ignore the effect of any chemical changes in the
atmosphere has also led to the second strategic threat. Normally,
the atmosphere cleanses itself of gases and particles that interfere
with its healthy functioning. Through a process called oxidation,
substances like methane and carbon monoxide react chemically
with a natural "detergent" known as hydroxyl. But we are putting
so much more carbon monoxide into the upper atmosphere —
mostly by burning fossil fuels and forests — that it has begun to
overwhelm the small amount of hydroxyl available. And since the
atmosphere uses its supply of hydroxyl first to cleanse itself of
carbon monoxide and only then to cleanse itself of methane, the
hydroxyl is now being used up before it can get to the methane.
Partly as a result, the concentration of methane in the atmosphere
has been increasing rapidly and is now third (after CO_2 and water
vapor) as a greenhouse gas.

Many scientists now believe that the atmosphere's loss of its
self-cleansing ability is a strategic threat that may ultimately be as

serious as ozone depletion, because it attacks what is, in a sense, the autoimmune system of the atmosphere itself. But the third strategic threat, global warming, is the most dangerous of all.

By now, of course, the basic mechanism called the greenhouse effect, which causes global warming, is well understood. Long before civilization intervened, the thin blanket of gases that surround the earth was efficiently trapping a tiny portion of the sun's heat and keeping it near the surface to warm up the air just enough to prevent temperatures from plunging to frigid extremes every night — which, of course, is exactly what happens on the moon and on planets like Mars that have very thin atmospheres. On earth, the sun radiates energy in the form of light waves that slice easily right through the atmosphere to the surface, where they are absorbed by the land, water, and life forms. (As noted earlier, the upper part of the atmosphere screens out a great deal of the ultraviolet portion of the light spectrum, and as discussed later, clouds in the lower atmosphere reflect and scatter some of the incoming sunlight before it reaches the surface, although the atmosphere is still heated slightly in the process.) Much of the heat absorbed during the daytime is radiated back — out toward space — in the form of longer infrared waves, which are less energetic and thus do not slice through the atmosphere quite as easily as sunshine. As a result, some of them don't make it through the blanket, and this heat is trapped in the atmosphere.

The problem is that civilization is adding many more greenhouse gases to the atmosphere and making the "thin blanket" significantly thicker. As a result, it traps more of the heat that would otherwise escape.

There really is no remaining dispute about these basic mechanisms. The argument — to the extent that there is one anymore among reputable scientists — is instead about three unproven assertions by those who are trying to justify a decision to do nothing.

First, the skeptics argue that some feature of the global climate system may serve as a kind of thermostat to regulate temperatures and keep them within the narrow range we are used to — in spite of our apparent willingness to allow the blanket of greenhouse gases to thicken. Second, they argue that even if temperatures do go up, they probably won't go up more than a few degrees, and that won't make much difference. Indeed, it may even be beneficial — especially in those parts of the world that are presently too cold for our liking. Third, they argue that even if the changes we are causing are likely to be significant, we should wait until they occur and then adapt to them rather than move now to prevent the worst consequences by ceasing or modifying the activities causing them.

None of these arguments justifies the complacency of their proponents. As for the first, I think this vain hope for a magic thermostat goes back to an unwillingness to recognize the new relationship between humankind and the earth, in which we really can now affect the entire global environment. And thus far, the search for a powerful enough thermostat has proved to be fruitless. For example, speculation that the cloud system might somehow cancel the effects of all the extra greenhouse gases has not withstood analysis. It is true that the water vapor in clouds both contributes to the greenhouse effect by absorbing radiant energy and plays a cooling role by scattering light, in part back to space; as a result, any change in the number and distribution of clouds would have a big impact. But the evidence to date strongly points to the conclusion that water vapor seems, unfortunately, to amplify the warming trend as it traps even more infrared heat that might otherwise escape from the atmosphere. Though there is more uncertainty where clouds themselves are concerned, most water vapor is outside clouds, and clouds too may magnify the warming rather than lessen it. Indeed, the leading proponent of the idea that water vapor serves as a cooling thermostat, Richard Lindzen, publicly withdrew his hypothesis on how that might happen in 1991.

Other candidates for the role of magic thermostat appear to have been urged upon the public for political reasons. For example, three scientists with the Marshall Institute conjectured that the sun will in the near future suddenly cool down exactly enough to make global warming just the ticket. Unfortunately, neither measurements of the sun's radiation nor the accepted understanding of solar physics lends any credence whatsoever to their speculation.

An increasingly imaginative search for excuses to do nothing continues, but thus far the accumulated evidence suggests that the only thermostat capable of counteracting these reckless environmental changes is the one in our heads and hearts — and it is under our control.

As for the assertion that global warming may turn out to be a good thing, the evidence suggests first of all that even small changes in global average temperatures can have enormous effects on climate patterns. And any disruption in climate patterns can dramatically affect the distribution of rainfall, the intensity of storms and droughts, the directions of prevailing winds and ocean currents, and the appearance of erratic weather patterns, with extremes of both heat and cold.

Those of us who live in temperate latitudes are used to the annual temperature shifts that give us hot summers and cold winters. As a result, it is hard to get excited about a change in global temperatures that, even in the most extreme predictions, is less than the seasonal change we adapt to every year. But a change in the global average temperature is very different. The first time I began to think differently about global warming was when I learned from Roger Revelle that the massive transformation of the earth's climate system that we call the ice ages took place after average global temperatures dropped by only a few degrees. What is now New York City was covered by one kilometer of ice, even though world temperatures were only 6 degrees Celsius colder than today. If such a small change on the cold side caused the ice ages, what can we expect from a change of that size on the warm side? Moreover, while those changes took place over thousands of years, the ones now predicted may occur in a single lifetime. And as noted in the last chapter, even a single degree's difference in the global average temperature — if it occurs rapidly — can be incredibly disruptive to civilization. Once again, the burden of proof ought to be with those who claim that the most likely outcome is something that will be good for us.

Finally, the argument that it makes more sense to adapt to these changes rather than try to prevent them ignores the harsh truth that if we continue to tempt fate, these shifts in the climate pattern may occur so swiftly that effective adaptation may well be impossible. Moreover, the longer we wait, the more unpleasant our choices become. We are used to adapting, but never in human history have we had to adapt to anything remotely like what may be in store if we continue destroying the global environment.

We are in fact conducting a massive, unprecedented — some say unethical — experiment. As we contemplate a choice between adapting to the changes we are causing and preventing those changes, we should bear in mind that our choice will bind not only ourselves but our grandchildren and their grandchildren as well. And of course many of the changes — such as the predicted extinction of half the living creatures on earth — would be irreversible.

The chemical and thermal dynamics of global warming are extremely complex, but scientists are looking especially carefully at the role played by one molecule: carbon dioxide (CO_2). Since the beginning of the industrial revolution, we have been producing increasing quantities of CO_2, and we are now dumping vast amounts of it into the global atmosphere. Like the CFC compounds, CO_2 has been thoroughly studied and has an effect that is well understood. Unlike CFCs, however, it was already part of the atmosphere. But as a percentage of the total atmosphere, CO_2 represents only about .03 percent of the molecules that make up the air, or 355 parts per million. Even so, it has always played a critical role as the greenhouse gas that triggers enough warming to increase the amount of water vapor that evaporates from the oceans into the atmosphere. This extra water vapor, in turn, traps nearly 90 percent of the infrared rays radiated from the surface of the earth back toward space, retaining them long enough to maintain the earth's temperature in rough equilibrium.

The correlation between CO_2 levels and temperature levels over time is well established. The greenhouse effect is, after all, a natural phenomenon that has been understood for more than a century. Venus, which has much more CO_2 in its atmosphere, traps much more solar heat close to its surface and is, predictably, much hotter than the earth.

The amount of CO_2 in the earth's atmosphere has fluctuated significantly over time in cycles lasting tens of thousands of years. The ice ages, for example, correspond to periods when CO_2 concentrations were relatively lower than they have been for the last 15,000 years. A few years ago, scientists from the Soviet Union and France conducted an extensive analysis of the tiny bubbles of atmosphere trapped in the ice in a deep hole they drilled in Antarctica two miles down through 160,000 years' worth of ice. After learning to read the ice the way foresters read tree rings, they found a striking correlation between the ups and downs of CO_2 and of temperature during all that time. As can be seen on the illustration on the next page, CO_2 levels fluctuated between 200 parts per million (ppm) during the last two ice ages and 300 ppm during the period of great warming between the two ice ages. The global average temperature rose and fell along a line that seems to match the line measuring CO_2.

Surprisingly, however, the range of this natural fluctuation is quite small compared to the changes caused by humankind. We are driving CO_2 from its warm level of 300 to over 600 ppm — with most of that change occurring since World War II. In fewer than fifty years, we will have doubled the amount of CO_2 in the atmosphere when this century began. For not only are we putting huge amounts of CO_2 into the atmosphere, we are also interfering with the normal way CO_2 is usually removed from the atmosphere.

The human lung inhales oxygen and exhales carbon dioxide, and the engines of civilization have, in effect, automated the process of breathing. The wood that fuels our fires, the coal and oil and natural gas that feed our furnaces, the gasoline that runs our cars — all convert oxygen into CO_2, enormous quantities of it. It is as if CO_2 has become the exhalation of our entire industrial civilization. Trees and other plants pull CO_2 out of the atmosphere and replace it with oxygen, transforming the carbon into wood, among other things. By rapidly destroying the forests of the earth, we are damaging its ability to remove excess CO_2.

One theory suggests optimistically that the oceans of the world may serve as a thermostat by absorbing more CO_2 as the amount in the atmosphere rises. But there is no supporting evidence. It is true that the slowness with which the oceans usually respond to atmospheric changes builds a time lag into the climate system, but the evidence indicates, unfortunately, that as temperatures increase, the oceans may actually absorb less CO_2. Similarly, some skeptics have

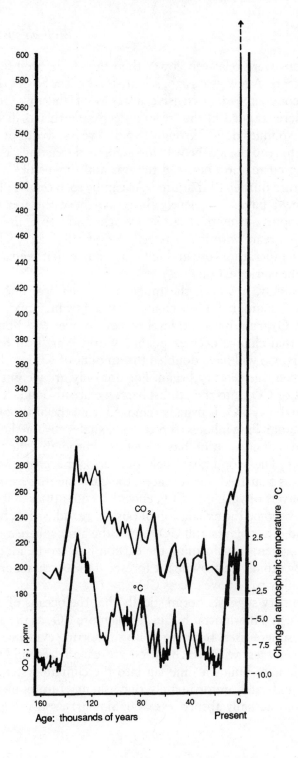

suggested that plants and trees may serve as the magic thermostat by gobbling up extra CO_2. But plants cannot grow faster without more nutrients and sunlight, regardless of how much CO_2 is in the air. Moreover, the evidence suggests that because increased temperatures speed the decay of organic matter and accelerate the respiration of plants, they actually result in a net increase in atmospheric CO_2 which might otherwise be absorbed in plant growth.

It has been said that the earth has two lungs, the forests and the ocean. Both are now being seriously impaired, and thus, so is the ability of the earth to "breathe." As it happens, the annual fluctuation in CO_2 levels makes it seem as if the entire earth breathes in and out once a year (see illustration on page 5). Since three quarters of the earth's land is north of the equator, approximately three quarters of the earth's vegetation is in the Northern Hemisphere. When the Northern Hemisphere tilts toward the sun during its spring and summer, the amount of CO_2 in the atmosphere drops

◀ This graph has been compiled from the information contained in ice cores drilled in Antarctica. By drilling two miles deep, scientists have been able to analyze the tiny bubbles of air trapped in each year's ice for the last 160,000 years. The bottom line shows the fluctuation of global atmospheric temperature from the next to last Ice Age (in the bottom left-hand corner), though the period of intense warming in between the last two ice ages, roughly 130,000 years ago, through the last Ice Age, which was at its coldest approximately 17,000 years ago (bottom right-hand corner). Global temperatures then shot upward to relatively constant levels, which have persisted more or less for the last several thousand years. The top line shows concentrations of CO_2 in the global atmosphere, which moved from less than 200 ppm (at the far left side of the graph) during the next to last Ice Age to 300 ppm during the period of warming in between the two ice ages. CO_2 levels then began to decline again throughout the last Ice Age until its end around 15,000 years ago, when the levels of CO_2 began to increase once more. In this century, human activities are adding so much CO_2 to the atmosphere that the level of concentration is expected to reach 600 ppm in less than 40 years. At the beginning of 1992, the level was already 355 ppm. The facts portrayed here are not in dispute; their implications are. If temperature and CO_2 have moved in tandem for as far back as we can measure, does that mean that the dramatic changes in CO_2 now under way (represented by the line moving up the right-hand side of the graph) will lead to rapid changes in temperature on the warm side of a magnitude that on the cold side produced the ice ages?

significantly. When the same hemisphere tilts away from the sun during its fall and winter, the deciduous plants lose their leaves and stop absorbing CO_2, thus driving the global concentration back up. But each winter, the peak level of CO_2 in the atmosphere gets higher and higher. And the rate of increase is also growing.

Given the apparent close relationship between CO_2 and temperatures in the past, it hardly seems reasonable — or even ethical — to assume that it is probably all right to keep driving up CO_2 levels. In fact, it is almost certainly *not* all right. Isn't it reasonable to assume that this unnatural and rapid change in the makeup of a key factor in the environmental equilibrium could have sudden and disastrous effects? Indeed, rising CO_2 levels may well lead to the kind of unwelcome surprise we received with the sudden opening of the ozone hole after a rapid and unnatural increase in the concentration of chlorine in the atmosphere.

The fact that the global atmosphere operates as a complex system makes it difficult to predict the exact nature of the changes we are likely to cause. Indeed, we can't even find a major part of the carbon cycle. But that is not to say that change is unlikely; rather,

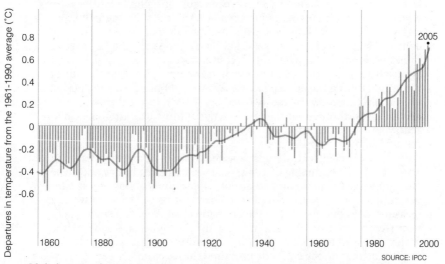

GLOBAL TEMPERATURE SINCE 1860: COMBINED ANNUAL LAND, AIR, AND SEA SURFACE TEMPERATURES FROM 1860 TO 2005

SOURCE: IPCC

Global atmospheric temperature has increased in the last hundred years by almost 1 degree Centigrade. The pattern of short-term fluctuations may be tied to variations in solar intensity, but the overall upward trend appears to be accelerating as CO_2 concentrations increase.

the change could well be sudden and systemic. And because our civilization is so carefully configured to the contours of the global environment as we know it — a relatively stable system throughout the history of civilization — any sudden change in global patterns would have disruptive and potentially catastrophic effects on human civilization.

In fact, the real danger from global warming is not that the temperature will go up a few degrees, it is that the whole global climate system is likely to be thrown out of whack. We are so used to taking the weather for granted that we forget that our climate operates in a state of dynamic equilibrium. One day may be cold, the next day hot; one season may be rainy, the next one dry. But the earth's climate follows a relatively predictable pattern in the sense that even though there are constant changes, they always fall within the boundaries of the same overall pattern. This equilibrium reflects a relatively constant relationship between several large elements of the climate system. For example, the amount of light and heat coming from the sun varies slightly over time — but not much. The orbit of the earth around the sun, the speed of the earth's rotation, and the tilt of the earth's axis all change over time — but not much.

Inside the atmosphere, the earth's weather system operates like an engine. Through wind and ocean currents, including the jet stream and the Gulf Stream, and through evaporation and precipitation, our climate transfers heat from the equator toward the poles and cold from the poles toward the equator. Just as the tilt of the earth's axis toward or away from the sun determines the arrival of summer and winter, the extent of the difference in temperature between the poles and the equator determines how much energy is needed to transfer heat in one direction and cold in the other. That means that the ratio of temperatures at the poles to those at the equator is one of the main pillars of the current climate equilibrium. If we remove that pillar — and past a certain point, higher levels of CO_2 may do exactly that — we will cross an important threshold of change, and the entire pattern of our climate system may shift from one equilibrium to another.

As global temperatures increase, the warming is not uniform throughout the earth. Different parts absorb more or less heat from the sun depending on the angle at which the rays of the sun strike the surface. The tropics on both sides of the equator get more heat because the sun's rays strike them directly from the middle of the sky. The polar regions get less heat because the sun's rays strike glancingly at the surface and are spread thinly over a much larger swath of the earth. But another important factor also determines the amount of heat absorbed by different parts of the earth: the extent to which the surface reflects the sun's rays back into space. Ice and snow glare back at the sun almost like mirrors, reflecting more than 95 percent of the heat and light that strike them. By contrast, the partly transparent blue-green water of the ocean absorbs more than 85 percent of the heat and light it receives from the sun.

This critical difference between reflective and absorptive surfaces has the most impact on the climate at the two poles. The freezing point is a threshold of change marking the boundary between two different states of equilibrium for H_2O: water above and ice below. At the edge of the polar region, at the boundary of the ice-covered surface, there is another threshold of change. Wherever the temperature pushes above the freezing point and the edge of the ice begins to melt, that tiny change transforms the relationship between that part of the earth's surface and sunlight, which the earth now absorbs instead of reflects back into space. As it absorbs more heat, the retreating edge of the ice is pressed by the accumulating warmth to melt at a faster rate. Though clouds can mitigate the effect, the process tends to feed upon itself, leading to a faster increase in temperatures at the poles than at the equator, where the sunlight-absorbing quality of the surface is mostly unaffected by increasing temperatures.

As the poles warm up at a faster rate than the equator, the difference in temperatures between them lessens, as does the amount of heat that must be transferred. As a result, the artificial global warming we are causing threatens far more than a few degrees added to average temperatures: it threatens to destroy the climate equilibrium we have known for the entire history of human civilization. As the climate pattern begins to change, so too do the movements of the wind and rain, the floods and droughts, the grasslands and deserts, the insects and weeds, the feasts and famines, the seasons of peace and war.

5

If the Well Goes Dry

Taken as a whole, our civilization has adapted over the last 9,000 years to the distinctive — and relatively constant — pattern by which the earth continuously recycles water between the oceans and the land through evaporation and runoff, distributes it in the form of precipitation, river flow, and the movement of creeks and springs, then gathers and stores it in lakes, swamps, wetlands, underground aquifers, glaciers, clouds, forests — indeed, in all forms of life. Fresh water in particular, and lots of it, has always been essential to the viability and success of any civilization. From the first irrigation networks along the Nile more than 5,000 years ago, to the Roman aqueducts and Masada cisterns, to the monumental system of tunnels that bring fresh water daily to New York City, human civilization has displayed remarkable ingenuity in assuring an adequate supply of water.

Human beings are made up mostly of water, in roughly the same percentage as water is to the surface of the earth. Our tissues and membranes, our brains and hearts, our sweat and tears — all reflect the same recipe for life, in which efficient use is made of those ingredients available on the surface of the earth. We are 23 percent carbon, 2.6 percent nitrogen, 1.4 percent calcium, 1.1 percent phosphorus, with tiny amounts of roughly three dozen other elements. But above all we are oxygen (61 percent) and hydrogen (10 percent), fused together in the unique molecular combination known as water, which makes up 71 percent of the human body.

So when environmentalists assert that we are, after all, part of the earth, it is no mere rhetorical flourish. Our blood even contains roughly the same percentage of salt as the ocean, where the first life forms evolved. They eventually brought onto the land a self-contained store of the sea water to which we are still connected chemically and biologically. Little wonder, then, that water carries such great spiritual significance in most religions, from the water of Christian baptism to Hinduism's sacred water of life.

We depend especially on fresh water, which is only 2.5 percent of the total amount of water on earth. Most of that is locked away as ice in Antarctica and to a lesser extent in Greenland, the north polar ice cap, and mountain glaciers. Groundwater makes up most of what remains, leaving less than .01 percent for all the lakes, creeks, streams, rivers, and rainfalls. This still leaves more than enough water to meet all our needs, both now and in the foreseeable future, but it is distributed unevenly throughout the world. As a result, human civilization has been restricted to more or less the same geographic pattern that conforms to the distribution of fresh water around the planet. Any lasting alteration of that pattern would therefore pose a strategic threat to global civilization as we have known it.

Unfortunately, the dramatic change in our relationship to the earth since the industrial revolution, especially in this century, is now causing profound damage to the global water system. The health of our planet depends on our maintaining a complex balance of interrelated systems, so it's not surprising that our alteration of the global atmosphere is changing the way water is transferred from the oceans to the land and back again. Warmer temperatures speed both evaporation and precipitation, accelerating the entire cycle. In addition, the increased warmth also increases the amount of water vapor in the atmosphere, which magnifies the greenhouse effect and speeds the process still further.

Moreover, because global warming heats up the polar regions faster than the tropics, it may also be changing the way the earth achieves a balance between hot and cold. The ocean helps to maintain the global equilibrium by constantly pushing toward a more even distribution of temperatures; in a distinctive and relatively stable pattern, the ocean transfers heat from the equator to the poles in huge currents near the surface such as the Gulf Stream. As the warm ocean water from the tropics moves northward, some of it evaporates along the way. When it hits the cold polar winds between Greenland and Iceland, the evaporation accelerates, leaving behind much saltier seawater, which becomes denser and heavier. This rapidly cooling water sinks to the bottom at the rate of 5 billion gallons a second, forming a deep current just as powerful as the Gulf Stream, if not as well known, which flows south underneath the Gulf Stream, near the ocean floor. In the process, it transfers cold from the poles back toward the equator.

Many scientists are worried that as the polar regions warm up faster than the tropics and the temperature differences between the two get smaller, these ocean currents, which are driven in large part by those differences, may slow down or seek a new equilibrium. If the circulatory pattern changes, the climate pattern will also change: some regions will get more rain, others less; some areas will get warmer, others colder.

In 1991 Peter Schlosser, a scientist at Columbia's Lamont-Doherty Geological Observatory, and his coworkers announced that during the 1980s, a key component of the "ocean heat pump" that drives the Gulf Stream and its colder, deeper counterpart abruptly and inexplicably slowed by almost 80 percent, to a rate "not significantly different from . . . a stagnant water body." Schlosser suspects that somehow the waters northeast of Iceland became less salty and consequently sank less rapidly. Although it is impossible to say that global warming has caused this dramatic change or that it will be more than a temporary phenomenon, it is consistent with the effects predicted with warmer world temperatures. Schlosser did say that "the cause is unknown, but whatever did this, it just shows how delicately balanced the system is."

Scientists are especially interested in the possible effects of climate changes on this ocean heat pump between Greenland and Iceland, because about 10,800 years ago, a sudden slowing in the rate of this pump is known to have caused one of the most dramatic and abrupt changes in climate history.

Wallace Broecker, a geochemist at Lamont-Doherty, did the work that led to this view. When I visited him, he explained that around 8750 B.C., as the world was emerging from the last ice age, a huge quantity of fresh water that had melted from the retreating glaciers formed an enormous inland sea in central Canada; it was

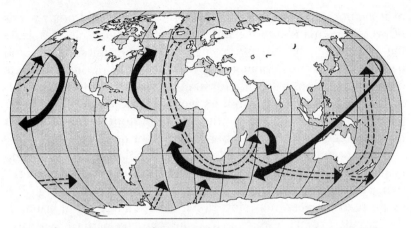

According to scientists, the primary ocean currents make up a "conveyor belt" that begins with the warm Atlantic currents, of which the Gulf Stream is a part (shown here as a solid arrow, as are the other warm currents), colliding with the cold air currents from the North Pole near Greenland and Iceland. The resulting evaporation leaves a much higher concentration of salt in the dense cold water which then sinks rapidly to form a deep cold current headed back south (shown here as an arrow of dotted lines, as are the other cold currents).

kept there in part by a huge ice dam across what we now call Lake Superior. As temperatures continued to increase, however, the dam suddenly broke, flooding the North Atlantic with vast amounts of fresh water through what is now the St. Lawrence River. Since the waters between Greenland and Iceland were suddenly not salty enough to sink, the ocean pump abruptly switched off.

What happened next should remind us that major climate changes do not always occur over thousands of years. In this case, in only a few decades, the global climate pattern dramatically changed. No longer warmed by the Gulf Stream, the North Atlantic froze over and the European continent, which had been emerging from the ice age, experienced a new and extended cold wave — and actually returned to ice age conditions, even as the rest of the world continued to warm. Then, abruptly, the ocean pump started working again and the warming trend in Europe resumed with a 10-degree Fahrenheit jump. By that time the first signs of organized civilization, cities, had appeared — not in Europe, so recently frozen, but far to the south, in Mesopotamia and the Levant, where for several centuries climate conditions had been ideal for the discovery and development of agriculture. Is it possible that early bands of humans fleeing the sudden reversal of favorable climate trends in Europe migrated south and combined their ideas about living with those of others already there and arrived at a synthesis that lead to the first organized human communities? The dates are right, but little is known about the people who fled when the ice age did an unexpected encore before leaving the European stage.

We sometimes underestimate the vulnerability of our civilization to even small changes in climate patterns — of the kind that have in the past accompanied tiny shifts in global average temperatures — much less our vulnerability to the enormous changes we are now setting in motion. For example, California depends for its water supply on a heavy snowfall in the mountains during the winter. If a small shift in the climate pattern pushes the snow line higher up the mountains and moisture that used to come in the form of snow falls as rain, the entire water distribution system will be altered. In the last few years, as global temperatures have reached record high levels, California has in fact begun to experience a sharp drop in the amount of snowfall. In 1990–91, the snow pack was less than 15 percent of its normal volume. Not surprisingly, California is in the midst of a severe drought.

These recent changes may, of course, not be related to global warming, but their impact on California is an indication of the disruptions that could accompany longer-term, more significant warming trends. In fact, studies by Charles Stockton and William Boggess of a 2-degree Celsius increase and a 10 percent decrease in precipitation show that the effects could include — because of less snow in the mountains — drops in water supplies of 40 to 76 percent throughout the river basins in the western United States. Such regional predictions are generally regarded as speculative, but warmer temperatures have, in recent years, been associated with extreme water shortages in the West and with collateral effects such as increased forest fires due to the drier conditions. Destructive fires have also become more common and more extensive in Florida's Everglades due to hotter and drier conditions that now bring the fire season earlier each year.

* * *

If the first strategic threat to the global water system is a redistribution of fresh water supplies, the second, and perhaps the most widely recognized, is the rise of sea levels and the loss of low-lying coastal areas around the world. Since one third of humankind lives within sixty kilometers of the coastline, the number of refugees likely to be created will be unprecedented.

Although the sea level has risen and fallen through different geological periods, never has the change been anywhere near as rapid as that now expected as a consequence of global warming. Nations like Bangladesh, India, Egypt, Gambia, Indonesia, Mozambique, Pakistan, Senegal, Surinam, Thailand, and China, not to mention island nations like the Maldives and Vanuatu (formerly New Hebrides), will be devastated if the projections now being made by scientists turn out to be accurate. Moreover, experts note that every coastal country will suffer adverse effects. Just as the Netherlands has been able to hold back the North Sea with an elaborate, expensive network of dikes, some wealthy nations will be better able than others to cope with the effects of the rising sea level and rapidly changing patterns of water availability. But the poor nations most at risk may watch helplessly as millions of their citizens become refugees, crossing borders into wealthier nations.

Global warming raises the sea level in several ways. Higher average temperatures result in the melting of glaciers, in ice being discharged into the oceans from the ice caps of Antarctica and Greenland, and in the thermal expansion of the volume of the sea as its water warms.

The melting of sea ice like that covering the Arctic Ocean or the icebergs in the North Atlantic will not affect the sea level, because their mass is already displacing an equivalent amount of seawater (just as a single ice cube in a glass of water will not change the water level as it melts). But an ice cube stacked on top of others, with its mass resting on them rather than floating in the water, will, when melted, raise the water level, sometimes overflowing the glass. Similarly, when land-based ice melts, the sea level goes up. The large majority of the ice on earth is in Antarctica, resting on top of a landmass or — in the case of the huge West Antarctic ice sheet — perched on the tops of several islands. This massive ice sheet is believed to have broken up and slid into the ocean during an interglacial warming period 125,000 years ago, raising sea levels twenty-three feet. Scientists have tended to discount the possibility of that catastrophe's recurring for another 200 to 300 years, but in 1991 one of the leading experts on West Antarctica, Dr. Robert Bindschadler of NASA, testified before my subcommittee that he had just been surprised to find that new samplings from the bottom of the ice sheet now show dynamic and dangerous changes. As a result, he has shortened his earlier estimate of how soon it will break up if global temperatures continue to climb.

Most of the remaining ice in the world is also on top of land: in Greenland, where the world's second-largest ice sheet plays an important role in the climate balance of the Northern Hemisphere, and in mountain glaciers throughout the world. Two of the leading experts on glaciers, Lonnie and Ellen Thompson of the Byrd Polar Research Center at Ohio State University, reported early in 1992 that all mid- and low-latitude mountain glaciers are now melting and retreating — some of them quite rapidly — and that the ice record contained in these glaciers shows that the last 50 years have been much warmer than any other 50-year period in 12,000 years. One sign that this is true appeared in 1991, when the "four-thousand-year-old man" was discovered in the Alps; he was suddenly revealed when the ice retreated for the first time since he died.

The net effect of all the warming and melting is a steadily rising sea level, almost one inch per decade now, with collateral effects such as the invasion by salt water of freshwater aquifers in coastal areas and the loss of coastal wetland areas. I studied one such area in Louisiana in 1989, Bayou Jean Lafitte, where a strip of land no more than two feet high and five feet wide at places separates the salt water of the rising ocean from the fresh water in one of the most productive breeding grounds in the United States. The next storm surge may breach the barrier and destroy the freshwater ecosystem of the bayou. The combination of storms and higher sea levels has caused steadily worsening erosion in virtually all coastal areas.

In some coastal cities like Miami, the freshwater aquifer on which it relies for its drinking water actually floats on salt water, so that rising seas would push the water table up — in some cases, to the surface. As noted in a recent WorldWatch Institute study of rising sea levels, Bangkok, New Orleans, Taipei, and Venice are among the other major cities facing similar problems. Other large cities like Shanghai, Calcutta, Dacca, Hanoi, and Karachi, which are located on low-lying riverbanks, will be among the first heavily populated areas to be flooded out.

Warming oceans are also likely to cause the average hurricane to be more powerful, scientists say, because the depth and warmth of the ocean's top layer is the single most important factor in determining the speed of a hurricane's winds. More powerful and more frequent storms coming into the land from the ocean would in turn greatly exacerbate the damage from rising sea levels, for it is during storm surges that the sea advances farthest inland from the coast.

A third strategic threat to the earth's water system involves massive changes in land use patterns, especially widespread deforestation. The destruction of a forest can affect the hydrological cycle (the natural water distribution system) in a given area just as surely as the disappearance of a large inland sea. More water is stored in the forests of the earth — especially the tropical rain forests — than in its lakes. Forests themselves produce rain clouds, partly because of evapotranspiration (transpiration is the plant equivalent of sweat; add to it the evaporation from surfaces like broad leaves). Indeed, almost immediately after rain falls on a rain forest, a fine mist begins to float back into the sky, increasing both the humidity in the air and the odds of more rain just downwind. Forests may also attract rain by producing gases called terpenes and small amounts of a compound called dimethylsulfide, which float into the atmosphere as a gas, undergo oxidation, and are transformed into an aerosol of sulfate particles which then serve as the tiny "grains" around which droplets of rainwater form — the same way a pearl forms around a tiny grain of sand or shell in an oyster.

Although we still have a great deal to learn about the symbiosis between forests and rain clouds, we do know that when the forests are destroyed, the rains eventually taper off and bring less moisture. Ironically, the heavy rains continue to fall for a while where the forest used to be, washing away the topsoil that is no longer protected by the canopy of the trees or held in place by the root system. Nearby areas are sometimes flooded by the runoff, which the forest used to soak up, and nearby rivers are often silted with the topsoil, gradually becoming clogged. Thus the rivers get shallower, their capacity to drain the floodwaters is impaired, and the flooding along the banks becomes even worse.

One tragic example of the loss of forests and then water is found in Ethiopia. The amount of its forested land has decreased from 40 to 1 percent in the last four decades. Concurrently, the amount of rainfall has declined to the point where the country is rapidly becoming a wasteland. The effects of the prolonged drought that has resulted have combined with the incompetence of its government to produce an epic tragedy: famine, civil war, and economic turmoil have wreaked havoc on an ancient and once-proud nation.

In South America, some people now fear that continued massive burning of the Amazon rain forest will interrupt the hydrological cycle that carries rainfall westward across the Amazon Basin toward Peru, Ecuador, Colombia, and Bolivia, creating future droughts in the deforested regions.

The fourth strategic threat to the global water system is the worldwide contamination of water resources with the chemical pollutants produced by industrial civilization. Unlike the global atmosphere, which is a single giant reservoir of air that is constantly and thoroughly "stirred" into an homogeneous mixture, the global water system contains a number of large reservoirs and stores that are not always thoroughly mixed with all the other water on earth. Because molecules circulate freely throughout the global atmosphere, contaminants like CFCs, which break down into chlorine atoms, can become ubiquitous in the atmosphere everywhere on earth. That is not true with the global water supply.

Nonetheless, a number of dangerous pollutants have become extremely widespread in much of the world's water resources. For example, radioactive particles left over from the explosions of nuclear weapons in the Atmospheric Testing Program are found broadly distributed in most of the world's water, albeit usually in minute amounts. These particles do not yet pose a strategic threat, but a few contaminants that have been widely dispersed in some areas — like PCBs and DDT — can be ecologically dangerous even in tiny amounts. The great Russian environmentalist Alexei Yablokov has pointed out that some powerful pesticides can cause behavioral changes in animals in extremely low concentrations. He notes, for instance, "a pesticide called *sevin,* [which] even in an infinitesimal concentration of one billionth, can change the behavior of large schools of fish: their movement becomes uncoordinated. This toxic concentration creates a chemical background in our biosphere."

One of the most visible contaminants in the oceans and in some inland river systems is oil spills. Particularly large spills, like the one intentionally released by Saddam Hussein into the Persian Gulf or the one negligently released by the *Exxon Valdez* into Prince William Sound, capture our attention, but the far more numerous, smaller spills that take place, unnoticed, every year probably do more cumulative damage to the oceans. Jacques Cousteau, the ocean explorer, claims that pollution in the oceans has now damaged the ultra-thin membrane on the ocean's surface — called the neuston — which plays a crucial role in capturing and stabilizing the food supply for the tiniest sea organisms, phytoplankton, which actually constitute the neuston and which begin the food chain. The consequences of this damage are not yet known, but phytoplankton play a significant role in ocean ecology and in coupling the ocean to the atmosphere. Pollution also poses a serious threat to another living system in the ocean, coral reefs; they too help maintain ecological stability.

The global water system, like the global atmosphere, has natural mechanisms by which it regularly cleanses itself of pollutants. But different bodies of water cleanse themselves at widely varying rates. The waters of turbulent and open bodies like the heavily polluted North Sea turn over quickly and in the process can be partially cleansed. But the waters of slow-moving, mostly enclosed seas, lakes, and aquifers turn over very slowly — the Baltic Sea, for instance, is replenished only once every eighty years. As a result, pollution dumped into the Baltic by the tsars still mingles with the enormous quantities of toxic contaminants dumped there after the Bolsheviks took over and will remain long after *their* recent overthrow. Similarly, while fast-flowing rivers wash most pollutants fairly quickly downstream, some large underground aquifers flow only a few dozen feet per year. It is thus nearly impossible to remove contaminants once they pollute these reservoirs.

In the industrial world, significant progress has been made in recent years toward cleaning up the water. In the United States, for example, the Clean Water Act of 1972 has reduced pollution levels markedly. Twenty-five years ago the Cuyahoga River in Cleveland became so polluted, it actually caught fire. Today, while it is still polluted, it is no longer flammable. In the Soviet Union, where virtually no progress has been made, rivers still catch fire. In July 1989, as a farmworker in the Ukraine named Vasili Primka was walking by the side of the River Noren picking mushrooms, he tossed his cigarette butt into the water. The river exploded and burned for five hours, largely because of a recent oil spill. Water pollution in Eastern Europe is just as bad; the Vistula River in Poland has so many poisonous and corrosive pollutants gurgling toward Gdansk that much of its water cannot even be used to cool factory machinery.

In Western Europe, as in the United States, some progress has been made in the wake of vigorous public protests, especially over outrageous incidents such as the massive spill in 1986 of toxic metals, dyes, and fertilizers into the Rhine by one company and the deadly herbicide spill by another company. Similarly, London no longer has worms in its drinking water as it did thirty years ago. And in Japan, the horrible effects of the mercury dumped in the water at Minimata, movingly shown in the photographs of W. Eugene Smith, helped catalyze tougher water pollution standards.

Overall, however, the pollution of global water resources has been growing steadily and dramatically worse. Despite the progress made in the industrial world, many problems remain, from high concentrations of lead in the drinking water of some cities to the common practice in most older American cities of mixing waste water with drainage runoff whenever it rains heavily, forcing a bypass of sewage treatment facilities that cannot handle the combined volume; the rainwater and sewage are then dumped, untreated, into creeks, rivers, and the ocean. According to an Environmental Protection Agency survey, almost half of all American rivers, lakes, and creeks are still damaged or threatened by water pollution.

It is in the Third World, however, where the effects of water pollution are most keenly and tragically felt in the form of high rates of death from cholera, typhoid, dysentery, and diarrhea from both viral and bacteriological sources. More than 1.7 billion people do not have an adequate supply of safe drinking water. More than 3 billion people do not have proper sanitation and are thus at risk of having their water contaminated. In India, for example, one hundred and fourteen towns and cities dump their human waste and other untreated sewage directly into the Ganges.

In Peru, the cholera epidemic in 1991 was an example of a similar phenomenon that is becoming increasingly common throughout the Third World. According to a study by the United Nations Environment Programme, "Four out of every five common diseases in developing countries are caused either by dirty water or lack of sanitation, and water-borne diseases cause an average of twenty-five thousand deaths a day in the Third World." Moreover, industrial waste, which is often regulated and monitored in the developed world, is generally handled more recklessly in underdeveloped countries eager to reach a Faustian bargain with polluters who are themselves sometimes eager to relocate from countries that have tighter controls. The New River, for example, which flows from northern Mexico into southern California before reaching the Pacific, is generally regarded as the most polluted river in North America, due to lax enforcement of environmental standards by Mexico.

The pressure of rapid population growth, especially in the Third World, represents the fifth major strategic threat to the global water system. In many parts of the world, groundwater is being extracted from aquifers at rates that far exceed the ability of nature to refill, or recharge, them. Yet because these underground reservoirs are out of sight, they remain out of mind — until they begin to dry up or until the ground above them begins to sink or "subside." California's Sacramento River delta, which supplies the canal system known as the California Aqueduct with half its water, is sinking about three inches each year, perhaps because it is getting less sediment. As a result, this area — which already had to be protected by a network of levees from being flooded by the ocean — is becoming much more vulnerable to the consequences of the kind of earthquake common in the adjacent earthquake zone.

The Ogallala Aquifer in the High Plains states is being drained so rapidly that many thousands of agricultural jobs will soon be at risk. And in nearby Iowa, nitrate-laden agricultural runoff has poisoned so many wells that rural areas of the state are less resilient during dry spells. In 1989, the Iowa National Guard was called out to deliver water supplies during the drought.

In Mexico City, the water level of the main aquifer is now dropping by as much as eleven feet each year, and in Beijing, the water table is dropping annually by as much as six and a half feet. Israel's Gaza Strip, the home of 750,000 Palestinians, is facing a water "catastrophe," according to the country's water commissioner, Zemah Ishai. Egypt, whose 55 million people rely almost exclusively on the Nile for drinking water, will have, by conservative estimates, a population of at least 100 million people within thirty-five years. Yet the Nile will still have no more water than it did when Moses was found in the bulrushes — in fact, it will have less, because Ethiopia and Sudan are upstream and have even faster rates of population growth.

In most of the world, the pressure of population growth on the water system is also exacerbated by increasing per capita use. One of the main reasons for this is the growing reliance on irrigation for agriculture to feed the growing populations. Of all the fresh water used by human beings throughout the world, almost three quarters (73 percent) is used for irrigation. And tragically, three fifths of all irrigation water is wasted due to inefficient and environmentally harmful techniques. In spite of the great hopes behind their construction, many large dams, such as the Aswan Dam in Egypt, have had perverse effects on the hydrological system around them, destroying valuable ecological niches, interrupting the flow of aquifers, and seriously damaging nutrient and sediment balances.

But of all the works of civilization that interfere with the natural water distribution systems, irrigation has been by far the most pervasive and powerful. In this century alone, the amount of irrigated farmland in the world has increased by 500 percent. When used correctly, irrigation is extremely effective in increasing agricultural productivity. For instance, even though only 15 percent of the world's farmland is irrigated, 33 percent of the world's crops come from that land. Unfortunately, however, much of the world relies on a method called open ditch irrigation, which not only loses 70 to

80 percent of the water to evaporation and seepage through the typically unlined ditches, but also can lead to the accumulation of large quantities of salt in the irrigated areas. This process, salinization, occurs because the salts become concentrated after the liquid containing them is reduced by evaporation. In countries that use this method, vast, once-productive areas have been abandoned due to the buildup of salt. The Aral Sea region of the Soviet Union is one example: flying over it in a small plane, I noticed first the glistening whiteness of the fields, which looked as if they had been sprinkled with a giant salt shaker.

Open ditch irrigation also typically leads to the waterlogging of the "root zone," just below the surface, which paradoxically deprives the plants of oxygen and stunts their growth. Sandra Postel, an authority on irrigation with the WorldWatch Institute, says that in addition to the Aral Sea region, other areas severely affected by salinization include Afghanistan, Turkey, the Tigris and Euphrates River basins in Syria and Iraq, 20 million hectares in India (in addition to the 7 million that have had to be abandoned to the salt), 7 million hectares in China, and 3.2 million in Pakistan. In Egypt, an estimated 50 percent of the cropland has a diminished yield due to salinization; the problem is also severe in Mexico.

Irrigation patterns sometimes lead to political disputes when water consumers upstream make profligate use of more than their share, thus depriving downstream consumers of adequate access. The impulse to make productive use of available water is, of course, as old as irrigation itself. In the twelfth century, Parakrama Bahu I, the king of Sri Lanka, said, "Let not a single drop of water that falls on the land go into the sea without serving the people." Unfortunately, as populations continue to grow, the need for water is liable to breed conflict, as many different people and communities place unreasonable demands on the sources.

In California, the residents of Los Angeles live at the tail end of a massive system for distributing water from the wetter north to the drier south. During the continuing drought of 1991, they began to question the fairness of a relatively small group of farmers using the vast majority of the state's water in a state of 32 million people. This growing dispute is not unlike the conflict between Colorado and its downstream neighbors, who feel deprived of the water that would otherwise drain out of the watersheds in Colorado. The plight of so-called tail-enders — those communities far downstream from the headwaters of a water distribution system — is becoming more severe, especially where the population is growing the fastest. These arguments and others like them in the United States will be resolved through political dialogue and legal battles, although our long-held assumption that fresh water is free and available in unlimited quantities has been shaken; our sudden realization of the need to account for the economic value of water and to measure its use is being interpreted as a sign of shortages to come.

In some volatile areas of the world, however, these conflicts over water may not be peacefully resolved and have the potential for leading to war. In 1989, I cosponsored with the water specialist Joyce Starr a series of international meetings to explore possible remedies for such conflicts. During the Persian Gulf crisis of 1990–91, there was open speculation about whether Turkey might cut off the flow of the Tigris to Iraq as a weapon of war. And Iraq sought to foul the pipes carrying drinking water to desalinization plants in Saudi Arabia with the massive oil spill it released into the Gulf. On a more hopeful note, Israel and Jordan are trying, in spite of nearly insoluble political problems, to find some means of avoiding conflict over the water in the Jordan River as the population continues to grow rapidly in both countries. Meanwhile, the same conflict is building between India and Bangladesh.

These geopolitical conflicts over water will be greatly exacerbated if global climate change alters the pattern of water distribution to which nations have carefully and painfully adjusted. The financial cost of modifying irrigation systems to new climate patterns could be staggering, especially for countries already so burdened with debt that they have difficulty affording the education and training expenses to ensure the proper operation of the systems already in place. Debt obligations drive many of these same countries to cut down their remaining forests for the hard currency they can get from selling the wood and raising cash crops on the land; in the process, their water shortages become worse.

Some people hope that desalinization plants may someday become cheap enough to provide water to the poor countries that need it the most, but this technology, like the schemes to lasso icebergs and pull them from the polar regions to the populous tropics, is unlikely to solve the underlying problem because of the enormous energy — and CO_2 — costs involved.

We need instead to lasso our common sense. The rains bring us trees and flowers; the droughts bring gaping cracks in the world. The lakes and rivers sustain us; they flow through the veins of the earth and into our own. But we must take care to let them flow back out as pure as they came, not poison and waste them without thought for the future.

6

Skin Deep

The surface of the earth is, in a sense, its skin — a thin but crucial layer protecting the rest of the planet contained within it. Far more than a simple boundary, it interacts in complex ways with the volatile atmosphere above and the raw earth below. It may seem hard to imagine it as a critical component of the ecological balance, but in fact, the health of the earth's surface is vital to the health of the global environment as a whole.

To use our own skin as an analogy, we may be surprised when anatomists describe it as the largest organ of our bodies; our skin seems to be first of all merely the boundary of our physical being and far too thin and attenuated to qualify as anything quite so complex as an organ. Yet it constantly renews itself and plays a complex role in shielding us from the harm we would otherwise suffer from the world around us; without it, even the air would corrode our raw innards.

Similarly, the surface of the earth — though it seems to be an insignificant layer of soil and rock, forest and desert, snow and ice, water and living things — serves as a vital protective skin. Just below the surface, the roots draw their nutrients from the soil and, in the process, hold it firmly in place, allowing it to absorb moisture and preventing the wind and the rain from carrying it toward the sea. Aboveground, the characteristics of the surface determine how much light is absorbed or reflected and thus help to define the planet's relationship to the sun.

Those portions of the earth that are covered with forests play a critical role in maintaining its ability to absorb carbon dioxide

(CO_2) from the atmosphere and are thus essential in stabilizing the global climate balance. As we saw in the previous chapter, the forests play a vital role in regulating the hydrological cycle. They also stabilize and conserve the soil, recycle nutrients through the shedding of their leaves and seeds (and eventually their trunks when they die), and provide the most prolific habitats for living species of any part of the earth's land surface. As a result, when we scrape the forests away, we destroy these critical habitats along with the living species that depend on them. The controversy over the destruction and loss of wetlands, which also serve as irreplaceable habitats for a disproportionate number of living species, is fueled by the same concern: that many vulnerable species will quickly become extinct when the wetlands are gone.

The most dangerous form of deforestation is the destruction of the rain forests, especially the tropical rain forests clustered around the equator. These are the most important sources of biological diversity on earth and the most vulnerable ecosystems now suffering the effects of our determined onslaught. Indeed, as many as half of all the living species on earth — some experts actually claim more than 90 percent of all living species — find their homes in tropical rain forests and cannot survive anywhere else. For that reason, most biologists believe that the rapid destruction of the tropical rain forests, and the irretrievable loss of the living species dying along with them, represent the single most serious damage to nature now occurring. While some of the other injuries we are inflicting on the global ecological system may heal over the course of hundreds or thousands of years, the wholesale annihilation of so many living species in such a breathless moment of geological time represents a deadly wound to the integrity of the earth's painstakingly intricate web of life, a wound so nearly permanent that scientists estimate that recuperation would take *100 million years.*

The ecosystems of tropical rain forests and temperate deciduous forests are entirely different. The temperate forests are all in areas that have endured several ice ages, long stretches of time during which vast, mile-high sheets of ice swept across the northern latitudes and spread out around mountain ranges from the northern and southern Andes to the Alps and Pyrenees, the Himalayas and

Pamirs, while smaller ice sheets fanned out from mountains in central East Africa, southern Australia, and New Zealand. These huge glaciers intermittently wiped out the high-latitude forests, but as they scraped across the land, they also ground up great quantities of rock and deposited rich minerals in the soils. As a result, temperate forests typically retain 95 percent of their nutrients in the soil and only some 5 percent in the forest itself, which allows them to regenerate fairly quickly.

This pattern is completely reversed in the tropical rain forests. These forests were almost untouched by the ice sheets, and their fantastic diversity of plant and animal life seems to have resulted from the uninterrupted coevolution of millions of species over tens of millions of years. But rain forests typically take root in thin, nutrient-poor soils: without the churning, fertilizing glaciers, only 5 percent of the nutrients are found in the soils, with roughly 95 percent in the forest itself. (The Amazon is a special case: scientists discovered in 1990 that it is regularly showered with fertilizing minerals carried across the Atlantic in high westerly wind currents laden with sand lifted from the dunes of the Sahara. Unusual "wind funnels" high above the Amazon seem to pull these sands down from the jet stream to the forest floor at the rate of about 100 pounds per acre per year.) Little wonder, then, that while temperate forests support thriving communities of flora and fauna, rain forests hold an absolute riot of life, with myriad species seeming to burst forth from every niche and pore.

There are three great stretches of rain forest left in the world: the Amazon rain forest, which is by far the largest; the central African rain forest in Zaire and surrounding countries; and the Southeast Asian rain forests, which are now largely concentrated in Papua New Guinea, Malaysia, and Indonesia. Other important remnants of rain forests are found in Central America, along Brazil's Atlantic coast, along the southern edge of the sub-Saharan portion of Africa's bulge, on the eastern coast of Madagascar, in parts of the Indian subcontinent and the Indochina peninsula, in the Philippines, and on the northeastern edge of Australia. Still smaller remnants can be found on islands from Puerto Rico to Hawaii to Sri Lanka.

Wherever rain forests are found, they are under siege. They are

When the rain forest is clear-cut or burned, the thin soils of the forest floor soon become vulnerable to erosion. This area of Brazil near Manaus illustrates the aftereffects.

being burned to clear land for pasture; they are being clear-cut with chain saws for lumber; they are being flooded by hydroelectric dams to generate power. They are disappearing from the face of the earth at the rate of one and a half acres a second, night and day, every day, all year round. And, for a number of reasons, the destruction of tropical rain forests is still picking up speed: the rapid population growth in tropical countries is leading to relentless pressure for expansion into marginal areas; shortages of fuel confronting an estimated 1 billion people in large areas of the Third World lead many to ravage the surrounding forests; mounting debts owed by developing countries to the industrial world encourage the exploitation of all available natural resources in a short-term effort to earn hard currency; massive, often misguided development projects that are inappropriate for tropical countries are opening formerly inaccessible vast areas to the civilized world; and livestock farming, with its insatiable demand for cleared pastureland every year, continues to spread. The list of reasons is long and complex, but the essential point is simple: in the daily battle between a growing, always ravenous civilization and an ancient ecosystem, the ecosystem is losing badly. So are the indigenous cultures who depend on the forest. Disappearing along with the trees and the living species are the last remaining ancient societies — an estimated 50 million tribal people who still live in the tropical rain forests — whose cultures have in some cases remained essentially unchanged since the Stone Age.

At the current rate of deforestation, virtually all of the tropical rain forests will be gone partway through the next century. If we allow this destruction to take place, the world will lose the richest storehouse of genetic information on the planet, and along with it possible cures for many of the diseases that afflict us. Indeed, hundreds of important medicines now in common use are derived from plants and animals of the tropical forests. When President Reagan was struggling to survive his would-be assassin's bullet, one of the critical drugs used to stabilize him was a blood pressure medication from an Amazonian bush viper.

Most of the species unique to the rain forests are in imminent danger, partly because there is no one to speak up for them. In contrast, consider the recent controversy over the yew tree, a temperate forest species, one variety of which now grows only in the Pacific Northwest. The Pacific yew can be cut down and processed to produce a potent chemical, taxol, which offers some promise of curing certain forms of lung, breast, and ovarian cancer in patients who would otherwise quickly die. It seems an easy choice — sacrifice the tree for a human life — until one learns that three trees must be destroyed for each patient treated, that only specimens more than a hundred years old contain the potent chemical in their bark, and that there are very few of these yews remaining on earth. Suddenly we must confront some tough questions. How important are the medical needs of future generations? Are those of us alive today entitled to cut down all of these trees to extend the lives of a few of us, even if it means that this unique form of life will disappear forever, thus making it impossible to save human lives in the future? News reports about the yew and its special properties have provoked healthy debate, but who will debate the loss of species unique to the rain forest? Scientists are a long way from even identifying all the species of plants and animals in the rain forests, much less discovering their possible uses in medicine, agriculture, and the like. So as we destroy huge tracts of rain forests each year, we are also destroying thousands of species that may have just as much value as the endangered yew.

There is no way of estimating the value to future generations of a resource as rich and complex as the rain forest. But José Lutzenberger, Brazil's minister of the environment, puts it this way when speaking of cutting down the rain forest and selling it for the present value of its wood — which is often used for disposable chopsticks and cheap furniture. It is, he says, "like auctioning the Mona Lisa to a roomful of shoeshine boys: many would-be bidders, like those in future generations, are not able to bid."

After the tropical rain forests are gone, the thin soils from which they rose like giant living cathedrals are suddenly bare and astonishingly vulnerable to the rain and wind. According to a study by the Wadebridge Ecological Center in the United Kingdom, scientists working in the sub-Saharan African nation of the Ivory Coast have carefully recorded incredible differences between rates of erosion before and after deforestation. Even on steep slopes, the rate of soil erosion in forested land was found to be as low as .03 tons per hectare per year. But once the land was cleared, the rate rose to an extraordinary 90 tons per hectare. India, for example, now loses an estimated 6 billion tons of topsoil each year, much of it as a direct result of deforestation. Deforestation also wreaks havoc on the hydrological cycle, eventually causing a sharp decline in the amount of rainfall in the areas where the forest once grew, and in adjacent areas downwind. It typically results first in flooding and soil erosion, then a sharp drop in rainfall.

In some countries, deforestation is also followed by the emigration of its people, first to any adjacent remaining area, where the cycle of destruction is repeated, and then sometimes across national borders. This forced migration may help send an urgent message to the northern industrial nations. In the Western Hemisphere, for example, it was the deforestation of Haiti, perhaps as much as the repression of the Duvalier regime, that led to the sudden arrival of 1 million Haitians in the southeastern United States.

The developed nations, however, have massive deforestation problems of their own. Airborne pollution has devastated European forests such as Germany's beloved Black Forest. *Waldsterben* is the name Germans have coined for the widespread phenomenon, which is even worse in heavily polluted Eastern Europe. And in the United States, particularly in heavily logged regions like the Pacific Northwest and Alaska, there is a renewed assault on the great stretches of temperate forest that are so important to us. The statistics about forests can be deceptive too: although the United States, like several other developed nations, actually has more forested land now than it did a hundred years ago, many of the huge tracts that have been "harvested" and replanted have been converted from diverse hardwoods to a monoculture of softwood conifer forests that no longer support the species that once thrived in the woods. In national forests throughout the country, logging roads are being built in order to facilitate the more rapid logging, even clear-cutting, of public lands under contracts that require the sale of the trees at rates far below market prices. This enormous taxpayer subsidy for the deforestation of public land contributes to both the budget deficit and an ecological tragedy.

It was partly for this reason that many people drew the line at protecting an endangered species — the spotted owl — in Oregon and Washington. I helped lead the successful fight to prevent the overturning of protections for the spotted owl. In the spirited Senate debate, it became clear that the issue was not just the spotted owl but the "old growth" forest itself. The spotted owl is a so-called keystone species, whose disappearance would mark the loss of an entire ecosystem and the many other species dependent upon it. Ironically, if those wishing to continue the logging had won, their jobs would have been lost anyway as soon as the remaining 10 percent of the forest was cut. The only issue was whether they would shift to new employment before or after the last remnant of forest was gone.

Whether in the tropics or the temperate zones, forests represent the single most important stabilizing feature of the earth's land surface, and they cushion us from the worst effects — particularly those associated with global warming — of the environmental crisis. But local and regional problems contribute to the strategic threats brought on by our destruction of the environment. For instance, many forests now absorb huge quantities of CO_2, but they will not do so once they are gone. Also, the current widespread burning of tropical forests adds a significant amount of the CO_2 to the atmosphere each year, and the denuded forest floor then becomes a significant new source of methane, another important greenhouse gas. In fact, the dying forests are like a giant "keystone species": much depends on their health, and if they are all cut and burned to the ground, the future of our own species is thereby endangered.

Disappearing forests are not our only concern, however. The problems of the expanding desert, soil erosion, the degradation and contamination of arable land, the destruction of wetlands and drylands alike, and the loss of habitats that results — all are different aspects of the systematic process by which we are threatening the face of the earth.

Visitors to the coast of Maine are sometimes impressed by how powerfully the glaciers scraped across that rocky land, but their force cannot compare to the effect of the worldwide scraping of our planet's surface by the cumulative power of industrial civilization. Indeed, some researchers claim that we now exploit the surface of the earth so completely, we actually consume — directly or indirectly — 40 percent of the net photosynthetic energy that results from the sunlight striking the planet. It's good to be efficient, but if true, that is too efficient; our demands are now out of balance with the needs of the rest of the earth's surface. And the results are growing to catastrophic dimensions in many places.

Next to deforestation, the most visible problem for the earth's land surface is the abuse of drylands, especially those near the edge of deserts in a pattern that often hastens the deserts' expansion — a process some call desertification. Although deserts typically undulate — two steps forward and one step back — recent decades have been marked by significant overall increases in the amount of land covered by deserts. And in some areas, deserts are advancing almost as rapidly as the glaciers once moved across the land. At their edges, impoverished and growing populations of nomadic peoples gather firewood and graze their scrawny herds of goats and sheep and cows, thereby denuding the land and inviting the desert's continued advance, especially in years when rainfall is scarce.

In Mauritania, for instance, the advance became so rapid during the 1980s that homes and businesses were literally buried by sand dunes rolling south at the rate of several kilometers a year in some years. Although the Sahara regularly expands and contracts, the expansions in the last half century have far outpaced the contractions, and the desert has grown significantly larger. Now — because of persistent hot, dry years — the great Sahara, the largest expanse of sand on earth, is advancing into Europe, especially into Spain and Italy. (Europeans do not look on it as the northern edge of the Sahara, but satellite photos make it appear so.) Regardless, in 1990 the European Community allocated $8.8 billion to combat the spread of the desert. In addition, the first desert ever seen in Eastern Europe has now appeared in the Caucasus region of the Soviet Union, due in part to unprecedented overgrazing by huge herds of sheep, kept hidden from Moscow until satellite photos revealed them to surprised central planners in the Kremlin.

A long-term shift in climate patterns that causes persistent long-term drought over an entire geographic area can have devastating effects. It is worth noting that 6,000 years ago, under a climate equilibrium that persistently brought more moisture to the northern half of Africa, cattle grazed throughout what we call the Sahara.

The drylands, which cover 18 percent of the land area in developing countries (25 percent in Africa), are most in danger of becoming desert. Though population density is normally somewhat lower in drylands, they nonetheless house more than 300 million people — and that number is growing quickly. As this population grows, so does the pressure on the drylands, for cultivation, grazing, and the gathering of wood for fuel is relentlessly degrading huge areas. According to a joint study by the World Resources Institute, the International Institute for Environment and Development, and the United Nations Environment Programme, the dryland regions of the Third World are approaching a state of acute crisis: an estimated 60 percent of the dry croplands and 80 percent of the dry rangelands are now spiraling downward in productivity as a result of overexploitation.

A study by Amadou Mamadou, an agri-economist in Niger, voices this concern. Mamadou describes the Sahel, the area running east to west across Africa from the Red Sea to the Atlantic Ocean, as "the interface between the great African desert, the Sahara, and the tropical humid zones . . . a fragile and unstable ecosystem where only an adequate vegetation cover can maintain the fertility of the soils by recycling the nutrients." And he notes that the Sahel is a network of "markedly arid ecosystems where periods of drought, once sporadic, are now appearing at shorter and shorter intervals." A similar process is now under way in Central America, where drylands make up 28 percent of the land, and in areas of South America and Central Asia, where populations are also growing rapidly.

Another kind of land especially vulnerable to degradation is found in mountainous areas of the developing world. There too, burgeoning populations are putting pressure on the delicate but vital vegetative cover that for millennia has protected the thin soils from erosion. The absorption of rainwater by vegetation is especially important in these areas because the runoff can pick up speed and power quickly if it drains unimpeded down the long, steep slopes, cutting deep gullies and stripping away the fragile topsoil. As in the drylands, population densities in these marginal areas have tended to be somewhat lower than in other regions. However, exploding birth rates throughout the developing world have pushed more and more people onto these less productive lands, which then, in turn, become highly vulnerable to erosion. Some of the worst damage is occurring in the Himalayan nations of Nepal, Bhutan, Tibet, and areas of India, including Sikkim and Kashmir. These mountain lands, which boast some of the most spectacular natural beauty on earth, are now being devastated to briefly quench the needs of a single generation. This degradation has far-reaching effects. The giant rivers that drain the Himalayas of their snowmelt and rainfall are now filling up with silt and losing their capacity to carry within their channels the same volumes of water they once transported easily to the Bay of Bengal and the South China Sea. No longer efficiently drained, these areas are now routinely vulnerable to horrible flooding of the sort that only recently claimed hundreds of thousands of lives in Bangladesh.

But the destruction of the earth's surface is hardly limited to the Third World. Indeed, here in the United States, the productivity of even some of our best land is being steadily damaged by those who have no qualms about maximizing short-term gains at the expense of long-term sustainable use. The resulting degradation of cropland takes many forms. Improper irrigation, for instance, coupled with poor drainage, leads to at least three problems. The first is water-logging of the root zone, which, in effect, drowns the plants and destroys the roots' ability to "breathe." Often, though not always, appearing as part of waterlogging is salinization, in which the evaporation of irrigation water leaves behind the potentially deadly salts on the surface and around the roots. (Excessive concentrations of salt have built up in more than 30 percent of the world's potentially arable land.) A third problem, alkalinization, seals the "pores" of the soil in a chemical reaction caused by a buildup of particular sodium salts common in some irrigation water, stunting or totally preventing the growth of crops. Other problems — a few of which I will touch on in the next chapter — lead to further exhaustion of vital nutrients and contribute to the steady decline in productive capacity.

Happily, there is some good news. Those lands that have been degraded frequently offer some of the best opportunities for restoration of the environment in a way that not only halts the destruction but reverses it and begins the process of recovery. Specifically, reforestation programs offer one of the most accessible and effective strategies for removing carbon dioxide from the environment, halting soil erosion, and restoring habitats for living species. Similarly, problems like salinization can also be reversed with the proper techniques (like drip irrigation) and careful attention over time.

But the key to reversing the current pattern of destruction and beginning the process of restoration and recovery is to dramatically change attitudes and to remove the constant pressures exerted by population growth, greed, short-term thinking, and misguided development.

7

Seeds of Privation

Nothing links us more powerfully to the earth — to its rivers and soils and its seasons of plenty — than food. It is a daily reminder of our connection to the miracle of life. Little wonder, then, that most of the world's religions require the consecration of food before it is transformed into the stuff of our lives.

But how many people still feel that sense of connection with food? Most of us no longer produce our own food, but rely instead on a huge and complex apparatus that places an amazing variety of foods from every corner of the world in our supermarkets.

The struggle to wrest an adequate supply of food from the earth has always been a central concern of the human race. Indeed, many historians believe that the first rudimentary civilizations were organized around the new strategy for acquiring food that we now call agriculture. Even before the invention of agriculture, some of the first known forms of human communication, such as the cave paintings at Lascaux, appear to have dealt with the subject of food — specifically, how to acquire it through cooperative hunting.

No one knows exactly how and why the shift from hunting and gathering to settled agriculture took place. One theory gaining currency notes that the first appearance of domesticated seeds about 12,000 years ago — near Jericho, in the area surrounding the Dead Sea — coincided with a period of climate change that rendered the Jordan River valley much drier and hotter than it had been, which in turn may have stimulated the planting of crops as an alternative to hunting and gathering. But whether its invention was due to climate change, overhunting and overgathering, population growth, or simply the slow evolution of knowledge about seeds and the accumulation of experience through trial and error in cultivating wild plants, agriculture inexorably became the preferred method of obtaining food from the environment. And from the very beginning, as we shall see, the secret of success lay in keeping track of seeds.

The history of agriculture is intertwined with the history of humankind. Each increase in the size of human settlements was accompanied by further sophistication in the cooperative effort to produce, store, and distribute ever-larger quantities of food. New technologies, like the plow and the irrigation ditch, led to new abundance but also new problems, like soil erosion and the buildup of salt in the soil. Progress was slow but steady. Through the centuries, the ratio of population to food supply remained relatively stable, with both growing at a roughly equal rate. But with the scientific revolution in the seventeenth and eighteenth centuries, the human population began surging, and for the first time it seemed possible that the population might soon outstrip the ability of the environment to yield enough food. This fear was articulated at the beginning of the nineteenth century by the English political economist Thomas Malthus; that he was famously wrong has been due to a series of remarkable innovations in the science of agricultural production. Malthus was right in predicting that the population would grow geometrically, but he didn't foresee our ability to make geometric improvements in agricultural technology. Even today, with several countries in the world suffering massive famines, there is little doubt that a commitment to use more land and newer agricultural methods could vastly increase the amount of food produced on earth. The problem we now face is therefore more complicated than the one Malthus identified. In theory, the food supply can keep up with the population for a long while yet, but in practice, we have chosen to escape the Malthusian dilemma by making a set of dangerous bargains with the future worthy of the theatrical legend that haunted the birth of the scientific revolution: Doctor Faustus.

Some of these bargains have already been exposed, and we are beginning to understand that many of the most widespread modern techniques used to squeeze more food from each season's harvest have done so at the expense of future productivity. For example, the high-yield methods frequently used in the American Midwest loosen and — over time — pulverize the soil to the point that large amounts of topsoil wash away with each rain, a process that leads inevitably to a sharp reduction in the ability of future generations to grow similar quantities of food from the same land. The extensive use of inappropriate irrigation techniques often leads to such large accumulations of salt in the soil that it becomes useless and barren. And the huge amounts of fertilizer and pesticides now routinely used in agriculture frequently drain off into the groundwater beneath the fields, contaminating them for many centuries to come.

But these problems are local and regional and can be solved by changing our agricultural methods. Now, however, the global system that gathers the enormous harvests now required faces a true strategic threat. Malthus worried about our food supply; today, we should be worrying even more about our seed supply. Every seed (and seedling) carries what is known as the germplasm; it contains not only genes but all of the special qualities that control inheritance, define the way genes work, and fix the patterns by which they combine and express their characteristics — in the words of expert Steve Witt, "the stuff of life." But the future health of the food supply depends on a wide variety of this irreplaceable stuff, and we now risk destroying the germplasm that is essential to the continued viability of our crops. Crucial to any food supply is the genetic resistance of crops to massive destruction from blights, pests, and changing climate. Maintaining genetic resistance requires the constant introduction of new strains of germplasm, many of which are found only in a few wild refuges around the world. These fragile places serve as the nurseries and reservoirs of genetic robustness, vitality, and resilience, but all of them are now in serious jeopardy. Indeed, the primary sources for all of our principal food crops are being systematically destroyed. This danger is only just now being understood by agronomists; one who does is Te-Tzu Chang, the head of the International Storage Center for Rice Genes in the Philippines, where 86,000 varieties of rice are kept. As he told *National Geographic,* "What people call progress — hydroelectric dams, roads, logging, colonization, modern agriculture — is putting us on a food-security tightrope. We are losing wild stands of rice and old domesticated crops everywhere."

Biotechnology is, to be sure, creating new crop varieties with impressive characteristics, such as uniformity, high yield, and even natural resistance to crop diseases and pests. But we have been blind to the harsh truth that the new crop varieties we create in our laboratories quickly become vulnerable to their rapidly evolving natural enemies, sometimes after only a few growing seasons. Although their genetic resistance is reinforced with new genes that are spliced into the commercial varieties every few years, many of the genes available for replenishing the vitality of food crops exist only in the wild.

Crops growing in the wild naturally proliferate into countless varieties, each with a slightly different size, shape, color, and yield and each with a different natural genetic resistance to the incredible range of predators — from insects to fungus — that constantly test them. An elaborate dance between predator and prey is played out everywhere in the natural world, a struggle in which the delicate balance of power depends upon the ability of a species to constantly sift through a vast genetic reservoir and find new characteristics that some distant cousin has used successfully to fight off the threat. When we intervene in the process of evolution by directing the selection of those genetic characteristics that will be passed on from one generation to the next, the choices are usually based on the maximum yield and current market value of the varieties in question rather than their overall genetic resilience. The vitality of the germplasm is therefore diminished while the rate of evolution among pests and blights continues unabated. Moreover, because the pests and blights are no longer aiming at a rapidly moving target, they can systematically search their own genetic arsenals for an offensive strategy that works. And when they find one, it works not only against the individual plant that is first attacked, but, because so many of our new plants are genetically identical, against billions of other suddenly vulnerable plants as well.

Of course, this is not to say that plant selection is inherently dangerous; on the contrary, it is one of history's greatest innovations, and without some interference in the natural evolution of plants, Malthus's prediction of disaster would almost certainly have come true. Indeed, plant breeding is almost as old as civilization itself. Humankind began collecting and planting valuable seeds more than 10,000 years ago, and throughout recorded history, people have carried plants from one location to another. In 1500 B.C., for example, Pharaoh Hatshephut, the world's first known female head of state, sent an expedition to the area now known as Somalia to bring back "incense trees," cedars, to plant in Egypt. More recently, Christopher Columbus took the first corn to Europe on his first return voyage from the New World; the following year, he brought European wheat and sugarcane back across the Atlantic. A few decades later, the conquistadors took potatoes from Peru to Europe. American leaders have long understood the importance of plant breeding. President Thomas Jefferson instructed all U.S. diplomats to send home seeds of potentially valuable plants from everywhere they visited; Benjamin Franklin, as an emissary to London, introduced soybeans into America. A hundred years later, the U.S. Department of Agriculture was created mainly to distribute seeds. Although it has since taken on other activities, finding and storing new varieties of seeds is still one of its most important tasks.

But we have now taken the ancient process of seed and plant selection to a technological extreme, splicing genes and consciously choosing exactly those characteristics that we believe are ideal for this year's crop. Each year's harvest of corn, for example, now comes not from thousands of genetic varieties, but from only a handful. Each variety carries a set of genes that have been carefully chosen to produce maximum yields, and billions of these seeds are cloned to produce a nearly uniform crop. If we were clever enough to anticipate nature's twists and turns, we might be able to store all the genes we need. But we have overestimated our own omniscience and underestimated the complexity and subtlety of the natural system with which we are interfering.

As we have seen, the ability of a food crop to survive depends on the richness and variety of its genetic resources. From the earliest times, domesticated crops have been threatened by disease. The ancient Romans, for instance, held a feast in late April which featured the sacrifice of a red dog to the god Robigus in a plea for protection from wheat rust. Despite their superstitions, the Romans had one great advantage over us: enough time to rely on the natural ability of plants to evolve and survive. Now, when most of our crops are grown from engineered, monocultured varieties, it may be only a matter of time before predators discover a weakness in the genetic defenses of these crops for which our artificial storehouse of genes contains no remedy.

Twenty years ago, a study by the National Academy of Sciences called *Genetic Vulnerability of Major Crops* saw the inherent danger of modern agricultural methods. It described America's principal crops as "impressively uniform and impressively vulnerable. . . . The market demands a uniform product — the farmer must produce it, and the plant breeder must produce the variety uniform in size, shape, maturity date, and the like. Uniformity in produce means uniformity in the genetics of the crop. This, in turn, means that a genetically uniform crop is highly likely to pick up any mutant strain of organism that chances to have the capacity to attack it." Since that study came out, some precautions have been taken, but during the same period, the world's population has grown by 1.5 billion people and the challenge of feeding all these mouths has created unrelenting pressures to get higher yields from ever larger and more uniform harvests. Still other pressures for uniformity come from the need for plants that can be productively frozen, tolerate heavy doses of farm chemicals, fit within the limits of special packaging, and meet the needs of specialized machinery used in the mass processing of food. As a result, the underlying problem of genetic erosion appears to be worse now than ever before. In fact, as one specialist recently put it, "The average life of a new crop variety is now roughly equivalent to the life of a new pop record."

Modern crops are genetically paralyzed, and because their natural predators are so efficient at finding their weaknesses, even the most productive new varieties must one day be discarded. In order to keep up with the rapidly evolving pests and blights, scientists are forced to constantly search through their greenhouses and seedbanks for new genetic traits that will allow the next "miracle crop" to fend off the current "miracle predator" — and simultaneously produce higher yields for an even larger number of people. But every so often a new blight or pest emerges which cannot be countered by any of the genes in their captive reservoirs of plant characteristics. At that point, their only recourse is to look to nature itself for a new and sufficiently robust "wild relative" of the domesticated plant. By virtue of its fierce struggle in a natural setting — surrounded by multiple predators and unaided by pesticides, herbicides, fungicides, and the like — this wild plant has acquired the genetic resistance that its tamed and coddled city cousin has failed to produce.

Finding these wild strains is often not a simple matter. Plant geneticists must literally return to the specific place on earth where the endangered crop makes its genetic "home" and search through the countryside — sometimes on hands and knees — for a wild relative. These genetic homelands are also known as centers of genetic diversity, or Vavilovian centers, in honor of Nikolai Ivanovich Vavilov, the Russian geneticist who discovered and described them. As it happens, there are only twelve such centers in the world, each the ancestral home of a dozen or so of the most important plants to modern agriculture (see the map on the following pages). The total number of important crops is remarkably small: virtually all of the world's food crops and feed grains come from only about 130 plant species, the vast majority of which were first cultivated in the Stone Age.

Most centers of diversity are found, as Vavilov said, "in the strip between twenty degrees and forty-five degrees north latitude, near the higher mountain ranges, the Himalayas, the Hindu Kush, those of the Near East, the Balkans, and the Apennines. In the Old World, this strip follows the latitudes, while in the New World, it runs longitudinally, in both cases conforming to the general direction of the great mountain ranges." The ancestral home of wheat, for example, is the mountainous terrain of northern Iraq, southern Turkey, and eastern Syria, squarely within the strip described by Vavilov. Many strains of wheat grow naturally here, but this variety is not reflected in domesticated wheat. Indeed, less than 10 percent of the genetic variety in wheat is found in the plants currently grown as crops. According to the biologist Norman Myers, another 30 percent of the genetic diversity in wheat can be found in the various seedbanks around the world. But almost two thirds of all wheat strains are found only in the wild, most of them still in the original Vavilovian center.

The center of diversity for coffee is in the Ethiopian highlands. But coffee is now grown in many areas of the world — the Andes region of Colombia and Brazil is one — and every once in a while, when a new pest or blight cannot be met with genetic resistance from readily available seeds, coffee growers must return to the Ethiopian highlands in search of wild relatives that can combat the new threat. A few years ago, this reliance on coffee's genetic home-land took an ironic twist. As Brazil was coming under international criticism for its tolerance of widespread destruction of the Amazon rain forest, a small group of Brazilians went to Addis Ababa to express their concern about the progressive deforestation of areas in Ethiopia vitally important to the future viability of the coffee crop.

In the case of corn, elevated portions of Mexico and Central America are home, while the potato is native to specific areas of the Andes in Peru and Chile. For centuries, even millennia, these re-mote centers of diversity were safe. Vavilov speculated that the Stone Age crops on which we so utterly depend today were able to survive in these mountainous regions because of their great diver-sity of soil types, topography, and climate. Furthermore, the inac-cessibility of the mountains and the isolation of the valleys between them provided relatively good protection from the disruptions of civilization and commerce.

Unfortunately, our global civilization has now acquired such enormous power and reach, and the demands of a burgeoning population for land, firewood, and resources of every kind and description are now so ravenous, that communities are encroach-ing rapidly on every single one of the twelve Vavilovian centers of diversity — even the most remote. For example, in Mesopotamia, the homeland of wheat, virtually the only areas where wild rela-tives of wheat can now be found are in graveyards and castle ruins. They survive because civilization, which often shows little regard for nature, at least sets aside tiny plots to commemorate its own past. But that is protection by accident, and too often these days we are relying on luck, not careful planning.

In an episode reported by Norman Myers, virtually the entire

The Great Genetic Treasure Map

This map shows the twelve areas around the world—called centers of diversity—that hold the greatest concentration of germplasm important to modern agriculture and world food production. While evidence indicates that some of the crops listed originated in their respective centers, no one knows for sure exactly where most crops first got started.

New World Centers

1 Mexico-Guatemala
Amaranth
Beans (various)
Corn
Cacao
Cashew
Cotton
Guava
Papaya
Red pepper
Squash
Sweet potato
Tobacco
Tomato

2 Peru-Ecuador-Bolivia
Beans
Cacao
Corn
Cotton
Guava
Papaya
Red pepper
Potato
Quinine
Quinoa
Squash
Tobacco
Tomato

3 Southern Chile
Potato
Chilean strawberry

4 Brazil-Paraguay
Brazil nut
Cacao
Cashew
Cassava
Para rubber
Peanut
Pineapple

5 United States
Blueberry
Cranberry
Jerusalem artichoke
Pecan
Sunflower

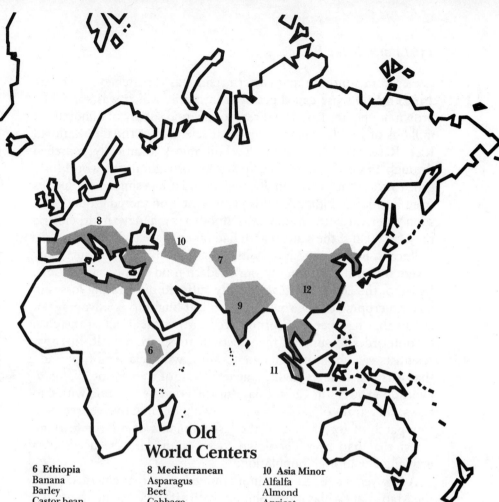

Old World Centers

6 Ethiopia
Banana
Barley
Castor bean
Coffee
Flax
Okra
Onion
Sesame
Sorghum
Wheat

7 Central Asiatic
Almond
Apple
Apricot
Broad bean
Cantaloupe
Carrot
Chick pea
Cotton
Flax
Grape
Hemp
Lentil
Mustard
Onion
Pea
Pear
Sesame
Spinach
Turnip
Wheat

8 Mediterranean
Asparagus
Beet
Cabbage
Carob
Chicory
Hops
Lettuce
Oat
Olive
Parsnip
Rhubarb
Wheat

9 Indo-Burma
Amaranth
Betel nut
Betel pepper
Chick pea
Cotton
Cowpea
Cucumber
Eggplant
Hemp
Jute
Lemon
Mango
Millet
Orange
Black pepper
Rice
Sugar cane
Taro
Yam

10 Asia Minor
Alfalfa
Almond
Apricot
Barley
Beet
Cabbage
Cherry
Date palm
Carrot
Fig
Flax
Grape
Lentil
Oat
Onion
Opium poppy
Pea
Pear
Pistachio
Pomegranate
Rye
Wheat

11 Siam, Malaya, Java
Banana
Betel palm
Breadfruit
Coconut
Ginger
Grapefruit
Sugar cane
Yam

12 China
Adzuki bean
Apricot
Buckwheat
Chinese cabbage
Cowpea
Sorghum
Millet
Oat
Orange
Peach
Radish
Rhubarb
Soybean
Sugar cane
Tea

rice crop in southern and eastern Asia was threatened in the late 1970s by a disease called grassy stunt virus, which was spread by brown hopper insects. The threat to the food supply of hundreds of millions of people was so potent that scientists at the International Rice Research Institute in the Philippines frantically searched through 47,000 varieties in gene banks throughout the world for a gene to resist the virus. Finally, they found it in a single wild species from a valley in India. But this plant wasn't on sacred ground, and soon afterward, the valley was flooded by a new hydroelectric project. What if the same search had taken place today?

Recent history abounds with situations that show just how severe the strategic threat to our modern food supply has become. In 1970, the United States suddenly suffered devastating losses in its corn crop when the southern corn leaf blight took advantage of a trait that had been uniformly bred into virtually all of the corn crop in order to simplify the genetic manipulation itself. In 1977, scientists searching in Ecuador found a wild relative of avocado that was resistant to blight, a genetic trait of tremendous value to avocado growers in California. But the good news came with bad news: this strain of avocado was growing on only twelve trees in a tiny patch of forest, one of the last remnants of a large lowland forest that had been cut down to accommodate the needs of a growing Ecuadorian population.

A few years ago, an even more immediate threat emerged when the Maoist guerrillas of the Shining Path attacked the International Potato Center in the Peruvian Andes. They dynamited buildings, took workers hostage, and murdered a guard, thus threatening the continuing viability of the 13,000 specimens in the World Potato Collection. Though the collection survived, the attack dramatically illustrated the vulnerability of these repositories — and the system that depends on them. In yet another example, part of the world's collection of wheat germplasm had to be evacuated from Syria in 1991 just before the Iraqi war began, and a diverse seed collection in Ethiopia was threatened by the civil war there the same year.

The short-term threat is, of course, not the extinction of important food crops, at least not as extinction is commonly understood. (Extinction is more a process than an event.) The way a plant or animal avoids extinction is by retaining enough genetic variety to adapt successfully to changes in its environment. If its range of genetic diversity is narrowed, then its vulnerability is correspondingly increased, sometimes to the point that a threshold is crossed and the complete disappearance of the species becomes inevitable. In every case, long before the last representative of an endangered species succumbs to its fate, the species itself becomes functionally extinct. The steady loss of genetic diversity in a species is called genetic erosion, and an astounding number of important food crops now suffer from it at a high rate. Among those listed by the United Nations International Board for Plant Genetic Resources as most at risk are the apple, avocado, barley, cabbage, cassava, chickpea, cocoa, coconut, coffee, eggplant, lentil, maize, mango, cantaloupe, okra, onion, pear, pepper, radish, rice, sorghum, soybean, spinach, squash, sugarbeet, sugarcane, sweet potato, tomato, wheat, and yam.

During most of the history of agriculture, genetic diversity has been found not only among wild relatives of food crops but also among so-called landraces (also called primitive cultivars). These are plants genetically related to the food crops used in the global agricultural system that have been developed in more primitive agricultural systems. Neither as wild as their uncultivated relatives in the mountain valleys nor quite as refined as their modern hybrid cousins, they nevertheless contain a much wider range of genetic diversity than do advanced breeding lines. Unfortunately, many landraces are now also endangered because of the spread of modern, higher-yielding varieties. An international conference in 1990 in Madras, India, sponsored by the Keystone Center, concluded that "it is an unfortunate reality that many nations have knowingly or unknowingly lost their traditional landraces due to the spread of high-yielding varieties, thereby increasing genetic homogeneity." In the United States, for example, of all the varieties of vegetables listed by the Department of Agriculture in 1900, no more than 3 percent now remain, according to one estimate.

The United States, however, has only one center of diversity (the upper Midwest, where the blueberry, cranberry, Jerusalem artichoke, pecan, and sunflower originated). Virtually all of the other centers lie in Third World countries, surrounded by exploding populations searching for firewood, food, and land — even heretofore remote land — on which to live. In order to earn hard currency from the sale of exports and thereby finance their enormous debts to the industrial nations, these impoverished countries are taking land formerly used for subsistence agriculture — many of them with genetically rich landraces — and using them instead for growing monocultured hybrid crop varieties for sale overseas. (This pattern has precedents. During the Great Potato Famine, for example, Ireland was growing a lot of wheat, but almost all of it had to be exported to pay its debts to England.) To be sure, these same new "miracle crops" also provide higher yields for domestic markets and have temporarily conquered hunger in a few of the Third World nations. But the much-heralded Green Revolution has, in most countries, failed to address fundamental economic problems, such as those caused by inequitable land ownership patterns, which often allow a wealthy elite to control a huge percentage of the productive land. And some of the ballyhooed development programs organized and funded by international financial institutions become part of the problem as well: in too many cases they turn out to be wildly inappropriate for the culture or ecology of the region in which they are placed. Moreover, the higher yields made possible by genetically altered crop strains often cannot be sustained over time, as the pests and blights catch up to them and as overirrigation and overfertilizing take their toll on soil productivity.

Meanwhile, the perceived inequity of the current arrangements in the global food system has led to the Third World's distrust of efforts on the part of multinational corporations to continue retrieving wild crop relatives from their centers of genetic diversity. There have been, after all, a number of historical examples of advanced nations taking genetic treasures from developing countries without proper compensation. The first steamship ever to sail up the Amazon River to Manaus, Brazil, left in the middle of the night with a cargo of rubber tree plants — at that time the principal source of Brazil's income. Because the voyage back to England was much faster by steam power than under sail, the plants survived — with the help of a new invention, the portable terrarium. After being nurtured in greenhouses, they were transplanted to the British colony of Ceylon the following year. Its monopoly in the rubber market broken, Brazil saw its economic fortunes plummet. Manaus, which had been the richest city in the New World, with dazzling electric lights and even a famous opera house, literally turned out its lights less than two years later.

Though much of the current suspicion of plant breeders by the Third World is unjustified, it is also not hard to understand. Developments such as the new U.S. laws providing patent protection and private ownership of new crop varieties, along with protectionism by the European common market, Japan, and others, have fueled cynicism in the developing world and led to new efforts to move toward more equitable economic relationships.

It is virtually impossible to calculate the value of maintaining the rich diversity of genetic resources on earth. And indeed, their value cannot be measured by money alone. But where food crops are concerned, we at least have some yardsticks with which to approximate the value of genes that are now endangered. The California Agricultural Lands Project (CALP) recently reported that the Department of Agriculture searched through all 6,500 known varieties of barley and finally located a single Ethiopian barley plant that now protects the entire $160 million California barley crop from yellow dwarf virus. Similar wild genes have contributed to increases in crop yields — more than 300 percent in many crops — in the last few decades. Among the many examples of the value of wild genes found by the CALP are "a seemingly useless wild wheat plant from Turkey [that] passed disease resistant genes to commercial wheat varieties worth $50 million annually to the United States alone; and, a wild hop plant [that] gave 'better bitterness' to English beer and in 1981 brought $15 million to the British brewing industry."

The value of genetic diversity has been noted, of course, by those who invest in global agriculture as well as by plant geneticists. For that reason, there is now another source of diversity besides the wild relatives and landraces: gene banks, an amazing variety of them. Some are managed by governments, some by private seed companies and multinational corporations, some by universities, and a surprising number by individuals, many of whom are merely dedicated hobbyists. The current system is in scandalous condition, with insufficient government attention and money, little coordination between different repositories, grossly inadequate protection and maintenance of national collections, and a missing sense of urgency where such a precious resource is concerned — especially with respect to the many vegetables and grains that presently play a smaller role in world agricultural trade and are thus at even greater risk.

Moreover, the entire landscape of the seed industry is changing. Multinational chemical companies have been purchasing seed companies and other sources of genetic diversity and are either marketing or preparing to market new plant varieties that are compatible with larger quantities of pesticides and fertilizers — which makes them money but further damages the global environment.

Based on the 1991 rankings of the world's largest seed companies, two of the five largest are agrichemical companies. Many of the others, including Pioneer Hi-Bred, the world's largest, have entered into agreements with chemical and biotechnology companies to help develop herbicide-resistant plant varieties.

In a few instances this could be beneficial. For example, Monsanto has cloned a gene for resistance to its environmentally less hazardous herbicide Roundup. But more often the results are more ominous. Calgene, a California biotechnology company, is working directly for the chemical company Rhone-Poulenc to develop cotton varieties resistant to bronoxynil, a reproductive toxin thought to pose hazards to farmworkers. And a German chemical company has developed plants resistant to 2,4-D, which has been shown to cause cancer in farmers. Both were scheduled for field testing in the summer of 1991. Heightening the problem, two U.S. government agencies are promoting this trend. The Department of Agriculture has declared herbicide-resistant plants a research priority and is actively supporting the field test of bronoxynil- and 2,4-D-resistant potatoes, and the U.S. Forest Service is encouraging the use of herbicide resistance in forestry, greatly expanding the market for these toxic chemicals.

What is troubling about such developments is not the involvement of multinational chemical companies per se. They have management skills, resources and global capabilities that could be useful in addressing some of the strategic problems that affect the world food system. However, the strategies that some companies are now pursuing reflect an assumption that we are clever enough to direct the evolutionary path of important plants and achieve large short-term benefits without paying a steep long-term price.

But we are not that clever, and we never have been. Indeed, agriculture is still haunted by Faustian bargains made when older technologies were introduced, many of them much less sophisticated than modern genetic engineering. Take pesticides, for example: they not only kill harmful pests, they kill many helpful ones as well, often disrupting the natural pattern of an ecosystem and so doing more harm than good. The environmentalist Amory Lovins tells the particularly troubling story of how a powerful pesticide was used in Indonesia to kill mosquitoes that were spreading malaria; the spraying also killed the tiny wasps that controlled the insect population in the thatched roofs of the houses. Before long, the roofs all fell in. Meanwhile, thousands of cats were also poisoned by the pesticide, and after they died the rat population burgeoned, which in turn brought on an epidemic of bubonic plague.

Even without catastrophic side effects, harmful pests often quickly develop immunities and encourage farmers to use larger, more deadly doses of pesticide. And agricultural runoff carries the residue into groundwater reservoirs and surface streams and into birds and fish. These hazards are not news: Rachel Carson's epochal book, *Silent Spring*, eloquently warned America and the world in 1962 of the dangers posed to migratory birds and other elements of the natural environment by pesticide runoff. But according to the National Coalition Against the Misuse of Pesticides, we produce pesticides today at a rate thirteen thousand times faster than we did when *Silent Spring* was published.

Do we really need all these poisons? One of the most extensive studies of pesticide use ever conducted, by Cornell University, concluded in 1991 that farmers who used natural alternatives to chemical control of pests (such as integrated pest management and crop rotation) could abandon many pesticides and herbicides without reducing yields at all and without significant increases in the price of food. And according to the study, in the case of those pesticides for which no substitute has yet been identified, the volume of chemicals used could, in most cases, safely be cut in half.

In addition to pesticides, some livestock farmers routinely use hormones and antibiotics. In congressional hearings I chaired in order to look at this problem in 1984, we learned the astonishing fact that 45 percent of all the antibiotics used in the United States are fed in small doses to livestock — not because farmers are that concerned about bacteria hurting their livestock, but because sub-therapeutic doses of antibiotics added to livestock feed promote faster rates of growth (for reasons that are not yet fully under-stood). Once again, however, there is a price to pay: germs that are routinely and constantly bombarded with low doses of antibiotics develop very strong defenses. And the antibiotics used to make a fast dollar when fattening livestock are exactly the same antibiotics doctors rely upon to kill germs when they attack people. The germs are almost never consumed with the meat from livestock because normal cooking kills them. But there are pathways — biologists call them vectors — by which some bacteria with superimmunities to common antibiotics cross over to attack human populations. (Salmonella, for example, is a bacteria that survives in both live-stock and humans.) Moreover, even bacteria that do not travel between livestock and human populations can in some cases trans-fer the specific genes that confer the resistance to antibiotics through "plasmids" to other kinds of bacteria that then become immune themselves. And some of these are believed to pose a growing threat to human beings.

Fertilizers, too, demand that we make a difficult bargain. Recent studies have shown that the widespread use of nitrogen fertilizer can stimulate oxygen deprivation and cause the soil to produce excessive methane and nitrous oxide. As it happens, concentrations of both methane and nitrous oxide in the atmosphere are increasing and together now account for more than 20 percent of the cause of global warming; although there are other sources of both gases, the use of nitrogen fertilizer is now considered one of the major causes of increased emissions. Fertilizers also affect genetic diversity: by compensating for differences in local environments and soil types, today's powerful fertilizers discourage diversity among crop va-rieties. So while today's higher yields are certainly desirable, even seemingly benign interventions carry price tags we have scarcely paused to consider.

Modern growing methods are hardly the only source of abuse in the global food system. Overgrazing is a major cause of desertification, as is the collection of firewood to cook the food for growing populations. The genetic engineering of animals, which has not yet been developed as much as in plants, is nevertheless starting to provoke similar concerns, as is the use of engineered hormones in livestock.

Especially alarming is the increasing evidence that we are now depleting many of the world's most important fisheries: since 1950 the total annual catch worldwide has increased by 500 percent and is now assumed to be higher than the replenishment rate in most areas. And a growing number of valuable food species are disappearing entirely. The use of thirty-five-mile-long, fine-mesh driftnets to strip-mine the oceans has recently — and rightly — provoked a great deal of public protest; even without driftnets, fishing fleets throughout the world are making an all-out assault on the productivity of the oceans. According to a California fishing authority, Duane Garrett, the new technology means that fish no longer have a chance: "Virtually every species has its *Thermopylae* — a narrow stretch of ocean through which it migrates, or an ancient spawning ground — and with advanced sonar and spotter planes, they've all been discovered and are being exploited worldwide with neither mercy nor foresight." I am particularly haunted by satellite photographs of the ocean east of New Zealand, taken at night, which show a necklace of light draped across the powerful current that races through the Cook Strait, separating the North Island and South Island. The spiraling current carries an astonishingly rich load of fish and squid, and its swirls and whorls are visible from space at night because the boats of Asian fishing fleets track the fish so accurately that the lights themselves precisely replicate the spirals of the current.

The global food supply may also be damaged by other strategic threats to the rest of the world's ecological system. For instance, dramatic increases in ultraviolet radiation due to the destruction of the ozone layer also present a serious if not fully understood threat to all crops as well as to critical links in the food chain, particularly in the oceans. Changing climate patterns due to global warming — especially the shifting distribution of rainfall — will also pose problems for food production, as will the attendant rise in sea level and northward migration of blights and pests. Moreover, the effect of several of these changes occurring simultaneously also presents unpredictable global risks.

For example, in late 1991, 325 scientists from 44 countries met in Rhode Island to investigate what most of them suspect are multiple causes of a new threat to food from the oceans in the form of a sudden worldwide explosion of algae blooms, including a high incidence of toxic "red tides." Citing the danger to fisheries and to aquaculture, Lars Edler, an algae specialist from Lund University in Sweden, told the *Boston Globe,* "I think we can safely compare the explosion of blooms we are seeing to gasps from the proverbial canary in the coal mine. There is no doubt something very significant is happening." At another conference one year earlier, experts on amphibians gathered to compare notes on the simultaneous and mysterious sharp declines in frogs, toads, and salamanders on every continent — believed to be due to multiple causes.

But the single most serious strategic threat to the global food system is the threat of genetic erosion: the loss of germplasm and the increased vulnerability of food crops to their natural enemies. Ironically, this loss of genetic resilience and flexibility is occurring at precisely the moment when those who believe that we can adapt to global warming are also arguing that we can genetically engineer new plants that will thrive in the unpredictable conditions we are creating. But scientists have never created a new gene. They simply recombine the genes they find in nature, and it is this supply of genes that is now so endangered.

Our inability to provide adequate protection for the world's food supply is, in my opinion, simply another manifestation of the same philosophical error that has led to the global environmental crisis as a whole: we have assumed that our lives need have no real connection to the natural world, that our minds are separate from our bodies, and that as disembodied intellects we can manipulate the world in any way we choose. Precisely because we feel no connection to the physical world, we trivialize the consequences of our actions. And because this linkage seems abstract, we are slow to understand what it means to destroy those parts of the environment that are crucial to our survival. We are, in effect, bulldozing the Gardens of Eden.

8

The Wasteland

One of the clearest signs that our relationship to the global environment is in severe crisis is the floodtide of garbage spilling out of our cities and factories. What some have called the "throwaway society" has been based on the assumptions that endless resources will allow us to produce an endless supply of goods and that bottomless receptacles (i.e., landfills and ocean dumping sites) will allow us to dispose of an endless stream of waste. But now we are beginning to drown in that stream. Having relied for too long on the old strategy of "out of sight, out of mind," we are now running out of ways to dispose of our waste in a manner that keeps it out of either sight or mind.

In an earlier era, when the human population and the quantities of waste generated were much smaller and when highly toxic forms of waste were uncommon, it was possible to believe that the world's absorption of our waste meant that we need not think about it again. Now, however, all that has changed. Suddenly, we are disconcerted — even offended — when the huge quantities of waste we thought we had thrown away suddenly demand our attention as landfills overflow, incinerators foul the air, and neighboring communities and states attempt to dump their overflow problems on us.

The American people have, in recent years, become embroiled in debates about the relative merits of various waste disposal schemes, from dumping it in the ocean to burying it in a landfill to burning it or taking it elsewhere, anywhere, as long as it is somewhere else. Now, however, we must confront a strategic threat to our capacity to dispose of — or even recycle — the enormous quantities of waste now being produced. Simply put, the way we think about waste is leading to the production of so much of it that no method for handling it can escape being completely overwhelmed. There is only one way out: we have to change our production processes and dramatically reduce the amount of waste we create in the first place and ensure that we consider thoroughly, ahead of time, just how we intend to recycle or isolate that which unavoidably remains. But first we have to think clearly about the complexities of the predicament.

Waste is a multifaceted problem. We think of waste as whatever is useless, or unprofitable according to our transitory methods of calculating value, or sufficiently degraded so that the cost of reclamation seems higher than the cost of disposal. But anything produced in excess — nuclear weapons, for example, or junk mail — also represents waste. And in modern civilization, we have come to think of almost any natural resource as "going to waste" if we have failed to develop it, which usually means exploiting it for commercial use. Ironically, however, when we do transform natural resources into something useful, we create waste twice — once when we generate waste as part of the production process and a second time when we tire of the thing itself and throw it away.

Perhaps the most visible evidence of the waste crisis is the problem of how to dispose of our mountains of municipal solid waste, which is being generated at the rate of more than five pounds a day for every citizen of this country, or approximately one ton per person per year. But two other kinds of waste pose equally difficult challenges. The first is the physically dangerous and politically volatile material known as hazardous waste, which accompanied the chemical revolution of the 1930s and which the United States now produces in roughly the same quantities as municipal solid waste. (This is a conservative estimate, one that would double if we counted all the hazardous waste that is currently exempted from regulation for a variety of administrative and political reasons.) Second, one ton of industrial solid waste is created each week for every man, woman, and child — and this does not even count the gaseous waste steadily being vented into the atmosphere. (For example, each person in the United States also produces an average of twenty tons of CO_2 each year.) Incredibly, taking into account all three of these conservatively defined categories of waste, every person in the United States produces *more than twice his or her weight in waste every day.*

It's easy to discount the importance of such a statistic, but we can no longer consider ourselves completely separate from the waste we help to produce at work or the waste that is generated in the process of supplying us with the things we buy and use.

Our cavalier attitude toward this problem is an indication of how hard it will be to solve. Even the words we use to describe our behavior reveal the pattern of self-deception. Take, for example, the word *consumption,* which implies an almost mechanical efficiency, suggesting that all traces of whatever we consume magically vanish after we use it. In fact, when we consume something, it doesn't go away at all. Rather, it is transformed into two very different kinds of things: something "useful" and the stuff left over, which we call "waste." Moreover, anything we think of as useful becomes waste as soon as we are finished with it, so our perception of the things we consume must be considered when deciding what is and isn't waste. Until recently, none of these issues has seemed terribly important; indeed, a high rate of consumption has often been cited as a distinguishing characteristic of an advanced society. Now, however, this attitude can no longer be considered in any way healthy, desirable, or acceptable.

The waste crisis is integrally related to the crisis of industrial civilization as a whole. Just as our internal combustion engines have automated the process by which our lungs transform oxygen into carbon dioxide (CO_2), our industrial apparatus has vastly magnified the process by which our digestive system transforms raw material (food) into human energy and growth — and waste. Viewed as an extension of our own consumption process, our civilization now ingests enormous quantities of trees, coal, oil, minerals, and thousands of substances taken from their places of discovery, then transforms them into "products" of every shape, kind, and description — and into vast mountain ranges of waste.

The chemical revolution has burst upon the world with awesome speed. Our annual production of organic chemicals soared from 1 million tons in 1930 to 7 million tons in 1950, 63 million in 1970, and half a billion in 1990. At the current rate, world chemical production is now doubling in volume every seven to eight years. The amount of chemical waste dumped into landfills, lakes, rivers, and oceans is staggering. In the United States alone, there are an estimated 650,000 commercial and industrial sources of hazardous waste; the Environmental Protection Agency (EPA) believes that 99 percent of this waste comes from only 2 percent of the sources, and an estimated 64 percent of all hazardous waste is managed at only ten regulated facilities. Two thirds of all hazardous waste comes from chemical manufacturing and almost one quarter from the production of metals and machinery. The remaining 11 percent is divided between petroleum refining (3 percent) and a hundred other smaller categories. According to the United Nations Environment Programme, more than 7 million chemicals have now been discovered or created by humankind, and several thousand new ones are added each year. Of the 80,000 now in common use in significant quantities, most are produced in a manner that also creates chemical waste, much of it hazardous. While many kinds of hazardous chemical waste can be managed fairly easily, other kinds can be extremely dangerous to large numbers of people in even minute quantities. Unfortunately, there is such a wide range of waste labeled "hazardous" that the public is often misled about what is really dangerous and what is not. Most troubling of all, many of the new chemical waste compounds are never tested for their potential toxicity.

In addition, we now produce significant quantities of heavy metal contaminants, like lead and mercury, and medical waste, including infectious waste. Nuclear waste, of course, is the most dangerous of all, since it is highly toxic and remains so for thousands of years. Indeed, the most serious waste problems appear to be those created by federal facilities involved in nuclear weapons production. These problems may have received less attention in the past because most federal facilities are somewhat isolated from their communities. In contrast, the public has become outraged by the dumping of hazardous waste into landfills, because numerous studies and disastrous events have shown that the practice is simply not safe. Basically, the technology for disposing of waste hasn't caught up with the technology of producing it.

Few communities want to serve as a dumping ground for toxic waste; studies have noted the disproportionate number of landfills and hazardous waste facilities in poor and minority areas. For example, a major study, *Toxic Wastes and Race in the United States,* by the United Church of Christ, came to the following conclusion:

> Race proved to be the most significant among variables tested in association with the location of commercial hazardous waste facilities. This represented a consistent national pattern. Communities with the greatest number of commercial hazardous waste facilities had the highest composition of racial and ethnic residents. In communities with two or more facilities or one of the nation's five largest landfills, the average minority percentage of the population was more than three times that of communities without facilities (38% vs. 12%).

It's practically an American tradition: waste has long been dumped on the cheapest, least desirable land in areas surrounded by less fortunate citizens. But the volume of hazardous waste being generated is now so enormous that it is being transported all over the country by haulers who are taking it wherever they can. A few years ago, some were actually dumping it on the roads themselves, opening a faucet underneath the truck and letting the waste slowly drain out as they crossed the countryside. In other cases, hazardous waste was being turned over to unethical haulers controlled by organized crime who dumped the waste on the side of the road in rural areas or into rivers in the middle of the night. There is some evidence that we have made progress in addressing these parts of the problem.

However, the danger we face as a result of improper waste hauling is nothing compared to what happens in most older cities in America every time it rains heavily: huge quantities of raw, untreated sewage are dumped directly into the nearest river, creek, or lake. Since the so-called storm water sewers in these cities were built to connect to the sewer system (before the combined pipes reach the processing plant), the total volume of water during a hard rain is such that the processing plant would be overwhelmed if it didn't simply open the gates, forget about treating the raw sewage, and just dump it directly into the nearest large body of water. This practice is being allowed to continue indefinitely because local officials throughout the country have convinced Congress that the cost of separating the sewers that carry human waste from the sewers that carry rainwater would be greater than the cost of continuing to poison the rivers and oceans. But no effort has been made to calculate the cost of the growing contamination. Could it be because Congress, and indeed this generation of voters, seem to feel that this practice is acceptable because the cost of handling the waste properly will be borne by us, and much of the cost for fouling the environment can be shunted off on our children and their children?

Though federal law purports to prohibit the dumping of municipal sewage and industrial waste into the oceans by 1991, it is obvious that the increasing volumes being generated and the enormous cost of the steps required to prevent ocean dumping will make that deadline laughably irrelevant. Currently, our coastal waters receive 2.3 trillion gallons of municipal effluent and 4.9 billion gallons of industrial wastewater each year, most of which fails to pass muster under the law. Nor are we the only nation guilty of this practice. Germany's river system carries huge quantities of waste toward the sea each day. Most rivers throughout Asia and Europe, Africa and Latin America, are treated as open sewers, especially for industrial waste and sewage. And, as previously noted, the first major tragedy involving chemical waste in the water was in Japan in the 1950s, at Minimata. International cooperative efforts have focused on regional ocean pollution problems, such as the Mediterranean, the North Sea, and the Caribbean.

The disposal of hazardous waste has received a good deal of attention in recent years, though there is still much to be done. For one thing, how do we know which waste is truly hazardous and which isn't? We produce more industrial waste than any other kind, but do we really know enough about it? Most industrial waste is disposed of on sites owned by the generator, often next to the facility that creates the waste. The landfills and dumps used by industry are therefore often far from public view and — especially because these companies create jobs — their waste is usually noticed only when it escapes from the site by means of underground water flow or dispersal by the winds.

* * *

Much more difficult to hide are the landfills used for municipal solid waste. Many of us grew up assuming that although every town and city needed a dump, there would always be a hole wide enough and deep enough to take care of all our trash. But like so many other assumptions about the earth's infinite capacity to absorb the impact of human civilization, this one too was wrong. Which brings us to the second major change concerning our production of waste: the volume of garbage is now so high that we are running out of places to put it. Out of 20,000 landfills in the United States in 1979, more than 15,000 have since reached their permanent capacity and closed. Although the problem is most acute in older cities, especially in the Northeast, virtually every metropolitan area is either facing or will soon face the urgent need to find new landfills or dispose of their garbage by some other means.

Those landfills still in operation feature mountains of garbage that are reaching heroic proportions: Fresh Kills Landfill on Staten Island, for instance, receives 44 million pounds of New York City garbage every single day. According to a study by a *Newsday* investigative team, it will soon become "the highest point on the Eastern Seaboard south of Maine." It will soon legally require a Federal Aviation Administration permit as a threat to aircraft.

Dr. W. L. Rathje, a professor of anthropology at the University of Arizona and perhaps the leading "garbologist" in the world, testified to the epic scale of these modern landfills before one of my subcommittee hearings: "When I was a graduate student, I was told that the largest monument ever built by a New World civilization was the Temple of the Sun, constructed in Mexico around the time of Christ and occupying thirty million cubic feet of space. Durham Road Landfill near San Francisco is two mounds compiled since 1977 solely out of cover dirt and the municipal solid waste from three California cities. I can still remember my shock when my students calculated that each mound was seventy million cubic feet in volume, a total of nearly five Sun Temples. Landfills are clearly the largest garbage middens (i.e., refuse heaps) in the history of the world."

What is in these mountains? Various forms of paper, mostly newspapers and packaging, take up approximately half the space. Another 20 percent or so is made up of yard waste, construction wood, and assorted organic waste, especially food. (Rathje found that 15 percent of all the solid food purchased by Americans ends up in landfills.) An unbelievable conglomeration of odds and ends accounts for the rest, with almost 10 percent made up of plastic, including the so-called biodegradable plastic. (Starch is added to the plastic compound as an appetizer for microorganisms, who will theoretically disassemble the plastic as they consume the starch.) Rathje dryly noted that he was skeptical of such claims: "In our landfill refuse from decades past we have uncovered corncobs with all their kernels still intact. If microorganisms won't eat corn-on-the-cob, I doubt whether they will dig cornstarch out of plastics."

But much organic waste does ultimately decompose, in the process generating a great deal of methane, which poses a threat of explosions and underground fires in older dumps that do not have proper venting or control. More significant, it contributes to the increased amount of methane entering the atmosphere. As we now know, rising levels of methane are one reason that the greenhouse effect has become so dangerous.

As existing landfills close, cities throughout the United States are desperately searching for new ones. And they are not easy to find. In fact, in my home state of Tennessee, to take one example, the single hottest political issue in the majority of our ninety-five counties is where to locate a new landfill or incinerator. Since these problems have customarily been addressed at the local level, they have not been defined as national issues, even though they generate more political controversy nationwide than many other issues. Now, however, the accumulation of waste has gotten so out of hand that cities and states have begun shipping large quantities across state lines. The Congressional Research Service has estimated that more than 12 million tons of municipal solid waste were shipped across state lines in 1989. Although some of this volume is due to the fact that some major cities are next to a state line and some is due to formal interstate compacts for regional disposal facilities (which can be among the more responsible alternatives), there is an enormous growth in shipments by private haulers to landowners in poorer areas of the country who are ready to make a dollar by having garbage dumped on their property.

I remember the day that citizens from the small Tennessee town of Mitchellville (pop. 500) called me to complain about four smelly boxcars dripping with garbage from New York City that had been sitting in the hot sun for a week on a railroad siding in their town. "What worries me," said one resident to a reporter from the *Nashville Banner*, "is that so many germs are carried through the air, viruses and this type of thing. When that wind is blowing that stuff all over town, them little germs are not saying, 'Now, we can't leave this boxcar, you know we've got to stay here.' " Mitchellville's vice mayor, Bill Rogers, said, "A lot of the time you can see water, or some kind of liquid, dripping out the bottom of the cars, and some of them contain pure New York garbage." As it turned out, the mayor had agreed to let the hauling company, Tuckasee Inc., bring trash from New York, New Jersey, and Pennsylvania to a landfill thirty-five miles from the railroad siding for a fee of $5 per boxcar, which looked like a good deal for a city whose annual budget is less than $50,000.

Small communities like Mitchellville throughout the Southeast and Midwest are being deluged with shipments of garbage from the Northeast. Rural areas of the western United States are receiving garbage from large cities on the Pacific coast. No wonder that bands of vigilantes have formed to patrol the highways and backroads in areas besieged by trucks of garbage from larger population centers. One of my favorite spoofs on *Saturday Night Live* was a mock commercial for a product called the Yard-a-Pult, a scaled-down model of a medieval catapult, just large enough for the backyard patio, suitable for the launching of garbage bags into your neighbor's property. No need for recycling, incineration, or landfills. The Yard-a-Pult is the ultimate in "out of sight, out of mind" convenience. Unfortunately, the fiction is disturbingly like the reality of our policy for dealing with waste.

Sometimes truth is even stranger than fiction. One of the most bizarre and disturbing consequences of this considerable shipment of waste is the appearance of a new environmental threat called backhauling. Truckers take loads of chemical waste and garbage in one direction and food and bulk liquids (like fruit juice) in the opposite direction — in the same containers. In a lengthy report, the *Seattle Post-Intelligencer* found hundreds of examples of food being carried in containers that had been filled with hazardous waste on the first leg of the journey. Although the trucks were typically washed between loads, the drivers (at some threat to their jobs) described lax inspections, totally inadequate washouts, and the use of liquid deodorizers, themselves dangerous, to mask left-over chemical smells. In 1990, Senators Jim Exon, Slade Gorton, and I joined with Congressman Bill Clinger to pass legislation prohibiting this practice.

But no legislation, by itself, can stop the underlying problem. When one means of disposal is prohibited, the practice continues underground or a new method is found. And what used to be considered unthinkable becomes commonplace because of the incredible pressure from the mounting volumes of waste.

One especially disquieting example is the idea of shipping waste across national borders. Probably the most famous example of this was the so-called garbage barge, which left Islip, Long Island, in early 1987 and wandered for six months in search of a port that would accept its 3,186 tons of commercial garbage. Before returning to Long Island, the barge was ordered out of ports in North Carolina, Louisiana, Florida, Mexico, Belize, and the Bahamas as well as other New York ports. For many, its mock-epic journey became a symbol of the crisis created when older landfills were filled up by the rapidly increasing amounts of waste.

More important, however, is that we are now trying to export garbage; many of the disposal sites proposed for the contents of the garbage barge were, after all, in foreign countries. For more than a year after the media had us all chuckling over the garbage barge, a cargo ship called the *Khian Sea,* which carried 15,000 tons of toxic ash from Philadelphia's incinerators, sailed from the Caribbean to West Africa to Southeast Asia, searching for a port. After a two-year journey, the ship finally dumped its load in an unknown location, according to officials in Singapore quoted by *Newsday.*

On the West Coast, some municipal officials in California have begun negotiating with the Marshall Islands in the South Pacific to receive regular shipments of solid waste. The residents of these islands, many of whom are suffering from the lingering effects of the U.S. government's atmospheric nuclear testing program in the 1950s, would not ordinarily consider receiving such an unsavory and even dangerous import, but their poverty forces them to do so.

Meanwhile, Greenpeace recently disclosed that officials in Baltimore were negotiating with authorities in China for permission to dump tens of thousands of tons of municipal solid waste in Tibet. Nothing could be more cynical. The Tibetan people are powerless to prevent Chinese officials from destroying the ecology of their homeland because of China's armed subjugation of Tibet for the last forty years. But the shipment has not taken place, and the United States has not yet become heavily involved in cross-border waste trafficking.

Nevertheless, the growing problems associated with international waste shipments have led to much debate, and one African leader recently denounced "garbage imperialism," a sentiment widely shared by others in the Organization of African Unity, which once condemned a rash of dumping incidents as "a crime against Africa." Such concerns finally prompted an international treaty in 1989 called the Basle Convention, which — if ratified by a sufficient number of countries — will limit the dumping of waste from industrial countries in the Third World.

Meanwhile, developing countries already have their own problems with waste, especially in the large and growing cities. In Cairo, for example, it is not uncommon to see garbage taken to the roofs of the ramshackle homes to decompose in the sun. In many Third World cities, untreated sewage flows freely in the gutters and streets, even as piles of garbage are picked over by legions of impoverished men, women, and children. Early in 1991, these conditions led to a massive outbreak of cholera in Peru and nearby areas of surrounding countries. By fall, this waste-borne disease had spread to Mexico — with a handful of cases as far north as the Gulf Coast of Texas.

In the Philippines, a growing mountain of garbage — called Smokey Mountain — in a suburb of Manila has become a kind of waste city, with 25,000 people living in cardboard huts perched on top of stilts stuck into the giant heap of garbage. According to Uli Schmetzer of the *Chicago Tribune,* they stake out territories in the midst of the waste, even though they and their children are choking on smoke from the fires fueled by decomposition: "Ten people squeeze into a hut the size of a bathroom. There is no shrub, no tree, just the stink of rotting refuse, day and night. And the methane gas produced by the compost." And these waste mountains are rising in the Third World not only because of the pressures of population growth; equally responsible is a pattern of conspicuous consumption that has been exported to these countries along with Western culture and its consumer products.

The latest scheme masquerading as a rational and responsible alternative to landfills is a nationwide — and worldwide — move to drastically increase the use of incineration. In the United States, the percentage of municipal waste incinerated more than doubled — from 7 percent in 1985 to over 15 percent in just four years — and investments in new incineration capacity are expected to double that percentage again in the next several years. In some of these projects, the heat generated by the incineration process is used as a source of energy to make steam, which is then sold to help offset the cost. In still other designs, the waste is molded into burnable pellets of "refuse-derived fuel." But even though the virtue of converting waste to energy is widely touted, the actual amount of energy produced is small and the principal and overwhelming reason for building such plants is that something has to be done with the massive amounts of garbage we create.

The huge new investment in new incinerators — almost $20 billion worth — is being made even though major health and environmental concerns have never been adequately answered. According to congressional investigators, the air pollution from waste incinerators typically includes dioxins, furans, and pollutants like arsenic, cadmium, chlorobenzenes, chlorophenols, chromium, cobalt, lead, mercury, PCBs, and sulfur dioxide. In the case of mercury emissions, a lengthy study by the Clean Water Fund found that "municipal waste incinerators are now the most rapidly growing source of mercury emissions to the atmosphere. Mercury emissions from incinerators [have] surpassed the industrial sector as a major source of atmospheric mercury [and] are likely to double over the next five years. If the incinerators under construction and planning come on line, with currently required control technology, mercury emissions from this source are likely to double. This growth will add millions of pounds of mercury to the ecosystem in the next few decades unless action is taken now." Mercury, of course, does not break down in the environment but rather accumulates, especially in the food chain, by means of a process called bio-accumulation, which concentrates progressively larger amounts in animals at the top of the food chain, such as the fish we catch in lakes and rivers.

The principal consequence of incineration is thus the transporting of the community's garbage — in gaseous form, through the air — to neighboring communities, across state lines, and, indeed, to the atmosphere of the entire globe, where it will linger for many years to come. In effect, we have discovered yet another group of powerless people upon whom we can dump the consequences of our own waste: those who live in the future and cannot hold us accountable. It is still basically a Yard-a-Pult approach.

But toxic air pollution is not the only problem. Incineration also creates a new solid waste problem that is in some ways worse than the one we have now. While 90 percent of the solid waste volume is reduced by incineration, the 10 percent that remains as ash is highly toxic, much more hazardous than the larger volume of waste before incineration. The burning concentrates some of the most

The Yard-A-Pult, invented for a "commercial" on *Saturday Night Live,* invites disposal of waste by catapulting it over the back fence into the yards of nearby neighbors. Our real waste disposal practices are not as different from this spoof as we would like to believe.

toxic ingredients, such as heavy metals, and complicates the task of finding a place to dump them. And 10 percent of a whole lot is still a lot.

Most communities do not even treat this toxic ash as hazardous waste. And because of political pressures from communities increasingly desperate to find a way to deal with their garbage, Congress and the EPA have been reluctant to require that the ash be handled as the hazardous waste that it is, since this would make its disposal much more expensive and significantly change the economics of incineration. Municipal officials also like incineration because it doesn't require a new way of thinking about waste. A single garbage truck can still collect all of the neighborhood garbage and not worry about sorting or recycling. Instead of dumping it in the landfill, it just dumps it in the incinerator.

The underlying problem remains that we are simply generating too much garbage and waste of all kinds. As long as we continue this habit, we will be under growing pressure to use even unsafe disposal methods. As the former New York State sanitation commissioner Brendan Sexton has bluntly put it, "People can complain about these incinerators all they want. They can argue against them, they can write to editors, but in the end, the garbage is going to win."

Many communities in the United States have decided that the real answer is recycling, the reintroduction of what used to be considered useless waste back into the stream of commerce. And some recycling projects have been remarkably successful. The states of Washington and New Jersey have achieved high rates of recycling; Seattle and Newark, along with San Francisco and San Jose, are among the cities with the best recycling records. But many have found that products manufactured and packaged for the mass marketplace often have features that frustrate recycling efforts. For example, some newspaper supplements and many magazines have glossy surfaces made of substances that cannot be processed by machines that recycle paper. Many plastic containers have built-in components that make recycling prohibitively expensive and complicated. Most packaging is designed exclusively for its usefulness in marketing its product, with no thought to the space it consumes in landfills or the toxic chemicals it releases into the air when burned. As a result, far less municipal waste is recycled today than is burned in incinerators.

Moreover, in order for waste (or "postconsumer resources," as some recyclers have labeled it) to be reintroduced into commerce, there has to be a market for it. Unfortunately, most manufacturers are comfortably locked into a pattern of purchasing virgin raw materials and are not equipped, either by habit or machinery, to use recyclable raw materials even though they might well be cheaper, after an admittedly difficult adjustment period. In addition, there are often public subsidies for the use of virgin materials but no comparable encouragement for using the recyclable substitutes. Take paper, for example. Many of the largest paper consumers and manufacturers have large investments in forests and tree farms, and they are therefore loath to use recycled paper instead of making additional profit by cutting the trees in which they have invested and for which they receive large tax subsidies.

In conducting workshops in Tennessee and hearings in Washington on recycling, I have found overwhelming public enthusiasm for the process. But I have also found tremendous disappointment among individuals and groups who have dutifully collected and sorted those elements of their municipal waste that they knew could be profitably recycled, only to find that it was impossible to find buyers for the material. Most people who have experienced this problem believe strongly that federal legislation is required to even the odds between recyclable and virgin materials, to discourage the sale of nonrecyclable products and packages, and to ensure that claims of recyclability are not misleading. (Such legislation is pending before Congress.) In order for recycling to work, more than individual enthusiasm is necessary. The system has to change and our mass processes have to be modified.

Our thinking has to change too. We cannot simply create larger and larger quantities of waste and dump it into the environment and pretend that it doesn't matter. As with all of our most critical environmental problems, the waste disposal crisis stems from our lost sense of place within the natural world. In nature, all species produce waste, virtually all of which is "recycled" — not by that species itself but by other forms of life with which it has a symbiotic relationship. Especially toxic elements in the waste stream are removed and isolated naturally to allow slower processes to eventually render them nontoxic. Of course, this supposes the maintenance of a balanced and mutually beneficial relationship between the species involved; any one species that oversteps its boundaries in the system is in danger of no longer being able to escape the consequences of its waste.

In a sense, this natural method actually avoids the creation of "waste" at all, because the waste of one species becomes useful raw material for another. Because we humans have grown both in number and in our power to modify the world around us, we have begun to create waste that far outstrips — in quantity and in toxic potential — the capacity of the natural environment to absorb or reuse it at anything approaching the rate at which it is generated. As a result, we have to find effective ways to recycle our own waste rather than relying on other living things to do it for us, and that is turning out to be a Sisyphean challenge. Better yet, we must drastically reduce the amount of waste generated in the first place.

What's required is a new way of thinking about consumer goods, a challenge to the assumption that everything must inevitably wear out or break and be replaced with a new and improved model, itself destined quickly to wear out or break. This will not be easy, however, for our civilization is presently built on a matrix of interlocking economic and social activities that emphasize constant consumption of new "things." Mass production has made it possible for millions of people to possess the much-desired products of industrial civilization. This development has been almost universally regarded as a major step forward; indeed, it has made possible a tremendous advance in the standard of living and the quality of life for hundreds of millions of people. In the process, however, these products themselves have become not only accessible but also "cheap" — in more ways than one. Since they are easily replaced with others identical to them, they do not need to be treasured, protected, and cared for to the same degree as in the past. Since each product is only one among millions, it no longer deserves to be appreciated for its uniqueness, and since the machines that made it typically strip away any sign of individual craftsmanship and creativity, it can be easily devalued. As a result, something shiny and new can be quickly transformed in our minds into something we can throw away.

If the need to rethink our throwaway mentality has become obvious, it is also clear that the effort has to involve more than a search for mechanical solutions. I have come to believe that the waste crisis — like the environmental crisis as a whole — serves as a kind of mirror in which we are able to see ourselves more clearly if we are willing to question more deeply who we are and who we want to be, both as individuals and as a civilization. Indeed, in some ways the waste crisis serves as perhaps the best vehicle for asking some hard questions about ourselves.

For example, if we have come to see the things we use as disposable, have we similarly transformed the way we think about our fellow human beings? Mass civilization has led to the creation of impersonal, almost industrial, processes for educating, employing, sheltering, feeding, clothing, and disposing of billions of people. Have we, in the process, lost an appreciation for the uniqueness of each one? Have we made it easier to give up on someone who needs extra attention or repair? Traditional societies venerate the oldest among them as unique repositories of character and wisdom. We, however, are all too willing to throw them away, to think of them as used up, no longer able to produce new things to consume. We mass-produce information and in the process devalue the wisdom of a lifetime, assuming that it can easily be replaced by skimming a froth of essential data off the floodtide of information rushing through our culture. For similar reasons, we have devalued the importance of education (even as we increase the lip service we pay it). Education is the recycling of knowledge, and since we have emphasized the production and constant consumption of massive quantities of information, we don't feel the same need to respect and reuse the refined accumulation of learning treasured by those who have come before us.

At times, we still marvel at the way another human being manifests the experience of life, but that sense of wonder seems more difficult to sustain now, perhaps because we have devalued the idea of commitment to others — whether to latchkey children, ailing parents, abandoned spouses, neglected friends and neighbors, or indeed any of our fellow citizens. One of the most horrifying examples of our degraded appreciation of the individual is a new category among the homeless called throwaway children, children who are thrown out of their homes because they have become too difficult to handle or because their parent or parents no longer have the extra time for their special needs. And every so often we read about a newborn baby literally thrown into a garbage can or a trash compactor because the mother is for whatever reason overwhelmed by the prospect of raising the child and despairs of finding the understanding and assistance she needs in our society. Throwaway children: nothing could better illustrate my strong belief that *the worst of all forms of pollution is wasted lives.*

By definition, a wasted life is one that is seen as having no value in the context of human society. Similarly, if we see ourselves as separate from the earth, we find it easy to devalue the earth. The two issues — wasting lives and wasting the earth — are intimately connected, because until we see that all life is precious, we will continue to degrade both the human community and the natural world. Consider these words from a homeless eight-year-old boy in New York City in 1990: "When our baby die we start to sit by the window. We just sit and sit all wrapped up quiet in old shirts an' watch pigeons. That pigeon she fly so fast. She move nice. A real pretty flier. She open her mouth and take in the wind. We just spread out crumbs, me and my [four-year-old] brother. And we wait. Sit and wait, there under the window sill. She don't even see us till we slam down the window. And she break. She look with one eye. She don't die right away. We dip her in, over and over, in the water pot we boils on the hot plate. We wanna see how it be to die slow like our baby die."

If we feel no connection to those in our own communities whose lives are being wasted, who are we? Ultimately, as we lose our place in the larger context in which we used to define our purpose, the sense of community disappears, the feeling of belonging dissipates, the meaning of life itself slips from our grasp.

Believing ourselves to be separate from the earth means having no idea how we fit into the natural cycle of life and no understanding of the natural processes of change that affect us and that we in turn are affecting. It means that we attempt to chart the course of our civilization by reference to ourselves alone. No wonder we are lost and confused. No wonder so many people feel their lives are wasted. Our species used to flourish within the intricate and interdependent web of life, but we have chosen to leave the garden. Unless we find a way to dramatically change our civilization and our way of thinking about the relationship between humankind and the earth, our children will inherit a wasteland.

PART II

THE SEARCH FOR BALANCE

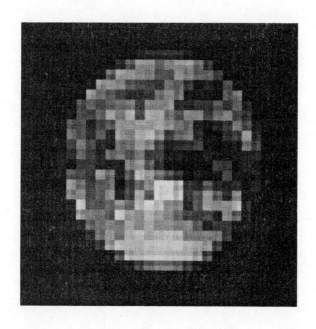

9

Self-Stewardship

The new strategic threats to the global environment are becoming increasingly apparent, but do we understand how and why we created them in the first place? If our relationship to the ecological system is no longer healthy, how did we make so many poor choices along the way?

For part of the explanation we must look to politics. Too often, politics and politicians have not served us well on environmental issues, but there is also a fundamental problem with the political system itself. Aside from its uninspired response to the environmental crisis, our political system itself has now been exploited, manhandled, and abused to the point that we are no longer making consistently intelligent choices about our course as a nation. For one thing, the way we make political choices has been distorted by the awesome power of the new tools and technologies now available for political persuasion. Thirty-second television commercials and sophisticated public opinion polling can now calibrate and target a political message with frightening speed and accuracy, and they can do more to manipulate the opinions of voters in two weeks than all the speeches and debates and political organizations together can accomplish in ten years.

These new technologies are not inherently bad, but they are so much more powerful than the ones we were using when our political system was created that we haven't yet comprehended their consequences for the system as a whole. New technology often magnifies our power to accomplish old objectives, and these new political tools allow politicians to seek the instant gratification of votes and high approval ratings but to ignore the true meaning of what we are doing. Increasingly, we concentrate on form to the exclusion of substance. And since the substance of politics is hard choices, it is precisely the hard choices that are excluded wherever possible. They are hidden, neglected, postponed, and ignored. And the voters are distracted with all manner of clever and extremely powerful manufactured messages. Means become ends. Tactics prevail over principles. Too often, principles themselves become tactics, to be changed as the circumstances warrant.

Indeed, in the age of electronic image-making, reasoned discourse of the sort imagined by our Founders plays less of a role than ever before. Impressions and affect have become the coin of our political realm. Skillful "visual rhetoric" has become as important as logic, knowledge, or experience in determining a candidate's success.

Having been a politician for many years now, I say this from personal experience. At an early age, I learned many political skills simply by observing my parents; I also learned that these skills are valuable only insofar as they serve worthy goals. Later, I learned the visual rhetoric of my own television generation and found myself unconsciously practicing a new set of "personality skills." But I am increasingly struck by how easy it is for every politician — myself included — to get lost in the forms of personality traits designed to please and rhetoric designed to convey a tactical impression. Voice modulation, ten-second "sound bites," catchy slogans, quotable quotes, newsworthy angles, interest group buzzwords, priorities copied from pollsters' reports, relaxation for effect, emotion on cue — these are the forms of modern politics, and together they can distract even the best politician from the real work at hand.

What does it say about our culture that personality is now considered a technology, a tool of the trade, not only in politics but in business and the professions? Has everyone been forced to become an actor? In sixteenth-century England, actors were not allowed to be buried in the same cemeteries as "God-fearing folk," because anyone willing to manipulate his personality for the sake of artifice, even to entertain, was considered spiritually suspect. Actors today, of course, are honored and respected for their skill at manipulating personality. And in politics, as well, these skills are now highly valued. The technology of politics and the technology of personality have been fused together by the technology of television.

At least where the presidency is concerned, there is still some symmetry between the skills needed to be elected and the skills needed to govern. After all, the ability of a president to communicate effectively on television is essential. But there is this problem: while a president elected primarily because of an appealing image and personality may be able to communicate effectively, that is no guarantee that he or she can deal with the substance of government policies or provide a clear, inspiring vision of our national destiny.

Where Congress is concerned, the ability to communicate a winning personality on television is much less relevant to the skills needed after the election. A keen sense for visual rhetoric is almost totally irrelevant to the task of writing laws — although it is, of course, quite relevant to the nearly constant work of being re-elected.

These maladies did not suddenly arrive with the television age. It was Machiavelli who wrote that "above all, the prince must be an actor." Certainly Mark Twain and Jonathan Swift would recognize the same human impulses today that they described in earlier eras. But the brute force of the new techniques now available for manipulating mass thought and the degree to which they have come to dominate elections mark a dramatic change from anything that happened in American politics before. The principal harm comes not so much from their direct effects on voters, but rather from the indirect way they radically distort the process of democracy as it was understood in the age dominated by print media. The new tools of persuasion frequently crowd out the semblance of dialogue between voters and candidates that used to take place; worse still, they simulate that dialogue and make many people believe that it is still there, when all too often it just isn't.

These techniques inevitably encourage inauthenticity in a politician's message: why present genuine ideas and true character if artificial ones are more effective in the marketplace of power. And nowhere is this lack of authenticity more of a problem than in our political dialogue. "Get it while you can; forget about the future" has been enshrined as the political ethic of the age. It is not so much the easy lies we tell each other as the hard truths that are never told at all. It has become too easy for those of us in public office to evade responsibility for the tough decisions that ought to be made but are instead ignored. As a result, there is astonishing irresponsibility in the face of grave and unprecedented crises — both in the White House and in Congress. It isn't just the environment. Look at the budget, where we're borrowing a billion dollars every twenty-four hours and in the process endangering the future for our children, yet *nobody is doing anything about it*. Why not? Because genuine political dialogue has been almost completely replaced by a high-stakes competition for the ever shorter attention span of the electorate. The future whispers while the present shouts. Somehow, we have convinced ourselves that we care far less about what happens to our children than about avoiding the inconvenience and discomfort of paying our own bills. So instead of accepting responsibility for our choices, we simply dump huge mountains of both debt and pollution on future generations.

It is increasingly difficult to avoid the conclusion that our political system is itself in deep crisis. The shallowness of the political dialogue breeds cynicism in the public at large; voter participation in the United States has been declining with each successive presidential election and is now at an all-time low. Meanwhile, opinion polls show that distaste for the profession of politics as it is currently practiced is now widespread. No wonder: voters are becoming increasingly jaded by the use of techniques that manipulate the appearance of sincerity as a device for cultivating favor. Put simply, most people are just fed up with the artificiality of most political communication today. And the resulting frustration is intensified because so many people feel our civilization's deepening crisis in their bones and want to see it addressed. Of course the polls pick that up too, and the process becomes even more cynical: sweeping change is repeatedly promised but seldom delivered; candidates promise bold leadership but after the election run with the pack. And as voters lose faith in the ability of their elected leaders to make a difference, they inevitably lose faith in their own ability to make a difference. At that point, it becomes clear to everyone that the political system is simply not working.

Often, when a process or machine is not working as intended, it is because we have not yet learned how to operate it. But in this case we resist that conclusion. After all, we Americans have been the architects and pioneers of self-government. How could our system fail us? What could be wrong?

The United States has long been the natural leader of the global community of nations. Ever since the great voyages of discovery five hundred years ago, the political imagination of Western civilization has been focused on the New World, the place where hope has a second chance and where, in the words of F. Scott Fitzgerald, "for the last time man came face to face with something commensurate to his capacity for wonder." The mythic destiny of the New World seemed to be fulfilled by the birth there of modern democracy. And for the last two hundred years, its promise has been embodied in a remarkable republic, empowered to protect the "inalienable" rights of individuals, and in a political system based on a constitutional government in which each center of power is carefully balanced against all the others.

Government, as a tool used to achieve social and political organization, may be considered a technology, and in that sense *self*-government is one of the most sophisticated technologies ever created. Indeed, the language used by the framers of the Constitution suggests an acute awareness of the almost hydraulic forces operating in society; in a way, the Constitution is a blueprint for an ingenious machine that uses pressure valves and compensating forces to achieve a dynamic balance between the needs of the individual and the needs of the community, between freedom and order, between passions and principles. This "machine" is a daring and wonderfully effective invention, and it represents the most important breakthrough in the entire history of the search for better political technologies — a point best illustrated by the fact that despite today's dizzying pace of change, a document written more than two hundred years ago is still universally recognized as the world's most forward-looking charter for self-government.

As succeeding generations have watched this revolutionary experiment not only survive but prosper, America's hold on the imagination of all humankind has grown steadily stronger. More and more people everywhere have come to believe that the United States, for all its mistakes and excesses, holds the key to important truths about the future of human civilization. And one of those truths is that self-government is probably best understood as a journey toward freedom that is still under way. Indeed, it was one of the new leaders of Eastern Europe, Václav Havel, who noted in an address to a joint session of Congress in 1990 that we in the United States have not yet reached our goal and still travel toward "the eternally receding horizon of freedom."

From the beginning, our leadership of the world community has been based on much more than military and economic strength. The American drive to correct injustice — from the abolition of slavery to the granting of women's suffrage — has constantly renewed our moral authority to lead. But we have not always lived up to our potential. By the end of World War I, in which the United States played the pivotal role, the political center of gravity had shifted decisively across the Atlantic. But in the aftermath of that war to "make the world safe for democracy," the United States failed to provide the leadership that the world community desperately needed and wanted. The decision to turn inward after the war — to back out of the infant League of Nations and to choose isolationist and protectionist policies — helped to ensure the chaos and divisiveness that characterized the next two decades and, in the minds of many, nourished the roots of World War II.

The lesson of those years was an important factor in shaping our determination not to repeat the same mistakes after World War II. Indeed, the strong bipartisan support in Congress for both the Marshall Plan in Europe and the careful reconstitution and stewardship of Japan came in part from a widespread awareness of how much the tragedy of World War II was a direct result of the failure after 1918 of the Congress and the people of the United States — not President Woodrow Wilson — to discharge the nation's new obligation of world leadership.

We ought to heed that lesson today. By failing to provide world leadership in the aftermath of the victory over communism and in the face of the assault by civilization on the global environment, the United States is once again inviting a descent toward chaos. History is change, and change is a relentless, driving force. Now that the human community has developed into a truly global civilization, we have a choice: either we search for the means to steer the changes shaping our new common history or we will be steered by them — randomly and chaotically. Either we move toward the light or we move toward the darkness.

Just as in 1918, this choice will be made disproportionately by the United States. Ironically, then the president offered vision and leadership, but the people failed to follow; this time the people appear ready, but the president does not. In the immediate aftermath of the Persian Gulf War, a poll asked the American people for their views about what role the United States should play in the world. The greatest support by far — an incredible 93 percent — supported a proposal for "the U.S. using its position to get other countries to join together to take action against world environmental problems."

No doubt if the American people were asked whether they supported specific measures that might be necessary to actually follow through on this proposition, the results would be quite different. In fact, almost every poll shows Americans decisively rejecting higher taxes on fossil fuels, even though that proposal is one of the logical first steps in changing our policies in a manner consistent with a more responsible approach to the environment. But this pattern is a common one: the American people often give their leaders permission to take action by signaling agreement in principle while reserving the right to object strenuously to each and every specific sacrifice necessary to follow through. A popular idea doesn't always spawn a popular plan: the Marshall Plan is a fine example. Though the people and Congress supported the notion of the United States leading a European recovery program, as soon as President Truman proposed granting large sums of taxpayer dollars to make the plan work, his approval figures in the public opinion polls dropped substantially almost overnight. Similarly, there is no doubt that several measures that would be necessary here in the United States in order to meet these threats would be unpopular and carry enormous political risk. But the American people are, nevertheless, beginning to give their leaders permission to challenge the nation to take bold, visionary, and even difficult steps to confront the environmental crisis forthrightly and responsibly. At open meetings in communities throughout Tennessee, I have found that voters are willing to go much further to meet the crisis than most politicians assume is possible — but they are waiting for leadership. Indeed, I am convinced they are hungry to hear hard truths and are nearly ready to make the all-out effort necessary for an effective response.

Yet President Bush and his advisers continue to oppose suggestions that the United States offer leadership in organizing a global response to the crisis, ostensibly because they are not yet convinced there is a problem. After standing in front of Boston Harbor, pledging to be "the environmental president" and promising to "confront the greenhouse effect with the White House effect," President Bush spent the first two years of his term arguing that no action on global warming was necessary or advisable until the completion of a major international scientific study of the problem then under way. But when the study's long-awaited conclusion called for bold world action on an urgent basis, the president claimed that yet more study was needed before any substantive action was advisable.

Even worse, the president and his administration have concentrated on symbolic actions designed to lull the public into believing that something is being done when actually nothing is. The designation of the Environmental Protection Agency as a cabinet-level department, for example, convinced some casual observers that some substantive progress had taken place. On the other hand, Bush's trip to the Grand Canyon for a "photo op" in the fall of 1991 inspired cynicism as deep as the Canyon itself. The president does deserve credit for helping Congress to enact a compromise version of the Clean Air Act to reduce air pollution — a genuine accomplishment although it was watered down by administration amendments before it was passed and weakened by White House interference with EPA's implementation of it afterward. And even in that bill, the president insisted that all mention of global warming be taken out. Moreover, the administration fought to the end against amendments I wanted to include — provisions dealing with carbon dioxide and other global warming gases and provisions for eliminating more quickly the chemicals destroying the ozone layer.

The president's chief of staff, John Sununu, has openly derided the notion of global warming, campaigning actively to dampen any moves within the government to confront the issue. According to firsthand reports, Mr. Sununu asked for a special program that he could run on his own personal computer to simulate one of the large global climate models. He hoped that the simulation would buttress his disagreement with the scientific community's concern about global warming. Ironically, the program he used clearly confirmed the general scientific consensus. (That didn't matter, of course: his mind appears to have been made up on the subject. And the president allows Sununu not only to set policy in Bush's name but also to stifle disagreement within the administration.)

Secretary of State James Baker began his tenure on a promising note: his first public speech as secretary was about global warming, eloquently establishing it as a top priority in foreign policy. But after two years of inaction by the White House, a series of highly visible embarrassments at international conferences when the rest of the world pointed out the absurdity of constant U.S. arguments for delay, and after, according to all accounts, continued and al-most-obsessive hostility to the issue by the chief of staff, Baker announced at the end of 1990 that it would be a conflict of interest for him to remain involved in the global warming issue since he owned stock in oil companies. Though I count Secretary Baker as a close friend and hold him in the highest regard, one cannot help but wonder whether his recusal on global warming — which hasn't been matched by a withdrawal from discussions about our policy toward OPEC, the Gulf crisis, or other issues with a direct impact on oil companies — had something to do with his keen political sense that he was never going to win the argument with Sununu, and he did not want to be associated with the disastrous and immoral policy on which the White House has been insisting. But with or without the secretary, the State Department is playing a big role in the development of U.S. policy. Disturbingly, the United States has, at times, carefully pursued a coordinated strategy with the world's largest oil producer, Saudi Arabia, in blocking progress in international discussions about global warming.

Why are the president and his chief of staff so hostile to a problem of such urgency? They have, after all, gone to great lengths to combat any who urge an aggressive response. For example, one television network anchorman told me that Ed Rogers, an assistant of Sununu's, called network executives the day a report on the seriousness of global warming was released; he helped persuade the networks to downplay its significance, thereby squelching prominent treatment of it on the evening news. Rogers, who had arranged at least one meeting involving Saudis, left Sununu's staff in 1991 to represent a Saudi sheik implicated in the BCCI banking scandal, until his $600,000 fee caused so much concern among Bush advisers that he was forced to give up his client. While still working as Sununu's right-hand man, Rogers was the White House spokesman most involved in cajoling news organizations to downplay the global warming issue. It certainly isn't unusual for the White House to try and put its "spin" on the presentation of the news. But why such sensitivity on this particular issue? When one of the leading scientists studying global warming, Dr. James Hansen of NASA, described to my subcommittee the connection between warmer temperatures and more frequent droughts in some areas, White House officials censored his testimony and insisted that he describe the phenomenon not as "probable" — which his studies had led him to conclude — but "highly speculative."

Why would the Bush White House go to such lengths to avoid facing the facts about the environment? Is it because the necessary changes would be sufficiently discomfiting to voters and to companies enjoying the status quo as to cause some political risk? Whatever the reason, President Bush's refusal to provide leadership in this crisis is, in my view, a historic failure that will, if not soon remedied, be regarded by future generations as unforgivable.

It is, of course, partisan for me as a Democrat to assess the performance of President Bush; and in his failure to act, he is not alone. Congress is also at fault, as are most other world leaders. But the United States is the only nation truly in a position to lead the world in facing up to a global crisis and organizing an adequate response. British prime minister John Major, a strong ally of President Bush on most issues, broke with the president on this one in 1991 when he condemned the U.S. failure to lead: "The United States accounts for 23 percent [of global CO_2 emissions]. The world looks to them for decisive leadership on this issue as on others." If the history of this century is any guide, it is safe to say that if we do not lead the world on this issue, the chances of accomplishing the massive changes necessary to save the global environment will be negligible. If the United States does choose to lead, however, the possibility of success becomes much greater. Moreover, in spite of the inevitable disruptions that the transition to a new pattern for our civilization will entail, the consequences of not making that transition are unthinkable. Besides, there would almost certainly be substantial economic and geopolitical benefits for the United States, as there have been almost every time we have taken a leadership role. And if the United States can in fact be persuaded to catalyze and coordinate an effective global response, it will once again redeem its promise as the last best hope of humankind on earth.

If inspiration for such leadership is needed, there are precedents. Once before, the world faced a terror which only the United States could confront. In the 1930s, when *Kristallnacht* revealed the nature of Hitler's intentions toward the Jews, there was a profound failure of historical imagination. The United States — and the rest of the world — was slow to act. Few could conceive of the Holocaust to follow, but from a distance the pattern of cruelty and destruction now seems clear. As the prospect of war increased in Europe, many refused to recognize what was about to happen, even when Jews were rounded up and sent to the concentration camps. World leaders waffled and waited, hoping that Hitler was not what he seemed, that world war could be avoided. Later, when aerial photographs revealed the truth of the camps, many pretended not to see. But if it took a long time for the world to respond to Hitler, because of him it took only a short time for Roosevelt to respond to Einstein's letter about building an atomic bomb. A threshold of moral alertness had been crossed.

Now warnings of a different sort signal an environmental holocaust without precedent. But where is the moral alertness that might make us more sensitive to the new pattern of environmental change? Once again, world leaders waffle, hoping the danger will dissipate. Yet today the evidence of an ecological *Kristallnacht* is as clear as the sound of glass shattering in Berlin. We are still reluctant to believe that our worst nightmares of a global ecological collapse could come true; much depends on how quickly we can recognize the danger. How much more evidence is needed by the body politic to justify taking action?

Vigorous response to a crisis often requires a profound shift in thinking, and the recent changes in Eastern Europe and the Soviet Union remind us how quickly it can happen. But the forces that drive dramatic changes are often large, opposing ideas that move slowly and ultimately press against one another with incredible force, like the tectonic plates responsible for continental drift and earthquakes. For forty-five years in Europe, one large idea called democracy pressed against another large idea called communism along a fault line that ran right through Berlin. Although few changes were visible on the surface of the political landscape, great stresses built up in hearts and minds throughout the communist world. The relaxing of geopolitical tensions in the late 1980s eased the friction that had locked the edges of the plates together just enough to allow them to slip, and — suddenly, in one great heaving motion — democracy moved decisively over communism, submerging it and sending out powerful shock waves that knocked down the Berlin Wall, along with almost every political structure in the communist world.

These changes appeared to be impossible before they began, but when people changed their way of thinking about communism, the scope of possible political change expanded. Similarly, as changes in our thinking about the environment take place, we can expand the range of what is politically imaginable. Already, public awareness is changing dramatically throughout the world. In many countries, political leaders are feeling increased pressure to respond to the desire for change. But no one wants cataclysm along with it; we hope that a measured, yet aggressive, response to the crisis will diminish the chance of violent, tectonic change later on. We have a clear choice: we can either wait for change to be imposed on us — and so increase the risk of catastrophe — or we can make some difficult changes on our own terms, and so reclaim control of our destiny.

The decisive factor will be our political system. Enlightened governments — and their leaders — must play a major role in spreading awareness of the problems, in framing practical solutions, in offering a vision of the future we want to create. But the real work must be done by individuals, and politicians need to assist citizens in their efforts to make new and necessary choices.

This last point is crucial: men and women who care must be politically empowered to demand and help effect remedies to ecological problems wherever they live. As the dramatic environmental problems in Eastern Europe show, freedom is a necessary condition for an effective stewardship of the environment. Here in the United States, a hugely disproportionate number of the worst hazardous waste sites are in poor and minority communities that have relatively little political power because of race or poverty or both. Indeed, almost wherever people at the grass-roots level are deprived of a voice in the decisions that affect their lives, they and the environment suffer. I have therefore come to believe that an essential prerequisite for saving the environment is the spread of democratic government to more nations of the world.

But as we attempt to make other governments more accountable to their citizens, we need to pay close attention to the problems that

These young boys, like everything else in Copsa Mica, Romania, are covered with carbon — one of humankind's pollutants in the air and water of Eastern Europe.

currently inhibit the proper functioning of our democracy — and remedy.them. By strengthening our own political system, we will empower new environmental stewards in areas where they are needed the most.

This task is crucial. Because if our basic method of making group decisions is not working properly, it is both an important explanation for why we have plunged headlong down a blind alley and an obstacle to coping with the problems that have resulted. Success in changing our destructive relationship to the global environment will depend on our ability to develop a keener understanding of how to make self-government respond to the environmental concerns shared by millions more people around the world each year. In fact, the agendas of the environmental movement and the democracy movement must become intertwined. The future of human civilization depends on our stewardship of the environment and — just as urgently — our stewardship of freedom.

The powerful forces working against stewardship are the same in both cases: greed, self-involvement, and a focus on short-term exploitation at the expense of the long-term health of the system itself. The current weakness of our political system reflects an emphasis on expediency and a failure to nurture our capacity for self-determination. We have not paid adequate attention to the serious problems undermining the accountability of government and the confidence citizens have in it. Too many people now feel that they have no way to exercise any real influence over the important decisions by government that affect their lives, that large campaign contributors have access to the decision-makers but the average citizen does not, that powerful special interests control the outcomes but a mere voter does not, that self-interested individuals and groups who can benefit from the decisions find a way to hot-wire the process while the broader public interest is ignored.

When the lack of accountability is due to corruption, the damage to democracy is especially severe. And in many countries, corruption is one of the principal causes of environmental destruction. To take only one of literally thousands of examples, concessions to clear-cut the rain forest of Sarawak, in East Malaysia, were sold personally by the minister of environment for Sarawak. Even though he was officially responsible for protecting the integrity of the environment, he enriched himself personally by selling permission to destroy it.

But the moral compromises involved in corruption for personal enrichment, ugly as they are, are not the ones that cause the most damage to our stewardship of freedom. A more subtle and pervasive temptation is the desire to attain and hold on to power, even when doing so means avoiding hard choices and ignoring the truth. In this regard, one of the most deadly threats to the stewardship of democracy is a lack of leadership. Indeed, even though the resiliency of self-government is contrasted with the brittleness of dictatorships that rely on a single "strong man," democracy is actually extremely vulnerable to a lack of leadership. Especially in times of rapid change, the ability of leaders to provide vision and to catalyze appropriate responses to dangers is critical. In my view, President Bush has sought to avoid that kind of leadership, concentrating instead on short-term political concerns. In other circumstances, his failure might be written off as par for the political course — but not in the circumstances we now face.

Perhaps the most serious threat to our stewardship of self-determination — one that may be even more threatening than all of the others put together — is that so many people have come to feel that the process of change in which we are now swept up has gone so far and gained so much momentum that it has outstripped our capacity to affect it. They fear that forces beyond our control now guide our destiny and that our means of responding are simply too cumbersome and unwieldy. The institutions of government and the systems with which we make choices about the future are indeed unwieldy, but in order to redeem the promise of democratic government, we must make all these institutions more accountable. Those still mired in the past must be swept forward and changed — despite their inertia.

And perhaps the most sluggish of all, after our political system, is our system of economics.

10

Eco-nomics: Truth or Consequences

Free market capitalist economics is arguably the most powerful tool ever used by civilization. As a system for allocating resources, labor, finance, and taxation, for determining the production, distribution, and consumption of wealth, and for directing decisions about virtually every aspect of our lives together, classical economics reigns supreme. Its laws are so pervasive that we take them largely for granted, like the laws of motion and gravity — which, incidentally, were codified by Sir Isaac Newton at the beginning of the Scientific Revolution, only a few decades before Adam Smith codified the major principles that still underlie economics today.

Rival systems, like communism, have been unable to compete in the marketplace of ideas. Although communism failed in large part because it suffocated political freedom, its parallel assault on economic freedom was its real undoing. Indeed, the stunning collapse of the Soviet Union and its empire in Eastern Europe has been due in large part to the perception on both sides of the Iron Curtain that capitalism, because it better incorporates classical economic theory, is simply superior to communism in theory and practice.

And indeed it is. But capitalism's recent triumph over communism should lead those of us who believe in it to do more than merely indulge in self-congratulation. We should instead recognize that the victory of the West — precisely because it means the rest of the world is now more likely to adopt our system — imposes upon us a new and even deeper obligation to address the shortcomings of capitalist economics as it is now practiced.

The hard truth is that our economic system is partially blind. It "sees" some things and not others. It carefully measures and keeps track of the value of those things most important to buyers and sellers, such as food, clothing, manufactured goods, work, and, indeed, money itself. But its intricate calculations often completely ignore the value of other things that are harder to buy and sell: fresh water, clean air, the beauty of the mountains, the rich diversity of life in the forest, just to name a few. In fact, the partial blindness of our current economic system is the single most powerful force behind what seem to be irrational decisions about the global environment.

Fortunately, these shortcomings can be fixed — albeit with great difficulty. The first step is recognizing that economics, like any tool, distorts our relationship to the world even as it gives us impressive new powers. Because we come to rely so completely on the capabilities conferred by our economic system, we adapt our thinking to its contours and begin to assume that our economic theory can provide a comprehensive analysis of whatever we wish it to interpret.

However, just as our eyes fail to see all but a narrow portion of the light spectrum, our economics fails to see — let alone measure — the full value of major parts of our world. Indeed, what we do see and measure is a very thin band within the full spectrum of the costs and benefits resulting from our economic choices. And in both cases, what is out of sight is out of mind.

Much of what we don't see with our economics involves the accelerating destruction of the environment. Many popular textbooks on economic theory fail even to address subjects as basic to our economic choices as pollution or the depletion of natural resources. Although these issues have been studied by many microeconomists in specific business contexts, they have generally not been integrated into economic theory. "There is no point of contact between macroeconomics and the environment," says the World Bank economist Herman Daly, a leading student of the problem.

Consider the most basic measure of a nation's economic performance, gross national product (GNP). In calculating GNP, natural resources are not depreciated as they are used up. Buildings and factories are depreciated; so are machinery and equipment, cars and trucks. So why, for instance, isn't the topsoil in Iowa depreciated when it washes down the Mississippi River after careless agricultural methods have lessened its ability to resist wind and rain? Why isn't that loss measured as an economic cost of the process by which our grain was produced last year? If the rate of topsoil loss is high enough in a given year, the nation may end up poorer, even when the value of the grain produced is taken into account. Meanwhile, our economic reports will assure us that, to the contrary, we are richer for having grown the grain, and richer still because we didn't spend the money required to grow it in an ecologically sound manner and thus keep the topsoil from washing away. This is now more than economic theory: largely because we failed to see the value of growing grain in an ecologically sound manner, we have lost more than half of all the topsoil in Iowa.

There are thousands of other examples. Here is one: the heavy use of pesticides may ensure that the grain we grow achieves the highest possible short-term profits, but the careless and excessive use of pesticides poisons the groundwater reservoirs beneath the field. When we add up the costs and benefits of growing the grain, the loss of that freshwater resource will be ignored. And largely because we have failed to measure the economic value of clean, fresh groundwater, we have contaminated more than half of all the underground reservoirs in the United States with pesticide runoff and other poisonous residues that are virtually impossible to remove.

Or take still another situation, one a little farther from home. When an underdeveloped nation cuts down a million acres of tropical rain forest in a single year, the money received from the sale of the logs is counted as part of that country's income for the year. The wear and tear on the chain saws and logging trucks as a result of a year's work in the rain forest will be entered on the expense side of the ledger, but the wear and tear on the forest itself will not. In fact, nowhere in the calculation of that country's GNP will there be an entry reflecting the stark reality that a million acres of rain forest is now gone. This ought to strike anyone as alarming, if not absurd. Yet when the World Bank, the International Monetary Fund, regional development banks, and national lending authorities decide what kinds of loans and monetary assistance to give countries around the world, they base their decisions on how a loan might improve the recipients' economic performance. And for all these institutions, the single most important measure of progress in economic performance is the movement of GNP. For all practical purposes, GNP treats the rapid and reckless destruction of the environment as a good thing!

Robert Repetto, an economist at the World Resources Institute, has led a team that studied the effects of this distortion in national income accounting on the development pattern of Indonesia. That nation's net losses of forest resources now exceed timber harvests: so much topsoil has eroded that the net value of the timber crop has been reduced by approximately 40 percent. Yet while this economic tragedy was unfolding and Indonesia was racing toward the precipice, the official economic reports all showed a rosy picture of steady progress.

Recently I asked the United Nations officials responsible for periodically revising the definition of GNP why this blindness is allowed to remain in our methods of calculation. The definition of GNP and other key yardsticks of economic performance are reviewed by the world community under the aegis of the United Nations every twenty years. And economists like Daly, Repetto, Robert Costanza of the University of Maryland, and others have long urged the changes I was recommending. The officials, who were then beginning their review procedures for this twenty-year cycle, acknowledged the good sense of these changes but claimed it would be difficult and inconvenient to make them now. "Perhaps at the next review," they said — twenty years from now.

What a striking contrast between the awesome power and efficiency our economic system displayed in its philosophical rout of Marxism-Leninism and the abject failure of the very same system to even take note of the poisoning of our water, the fouling of our air, the destruction of tens of thousands of living species every year. We make billions of economic choices every day, and the consequences are bringing us steadily closer to the brink of ecological catastrophe.

Classical economists like to argue that all participants in the struggle between supply and demand have "perfect information" — that everyone who makes an economic choice within this all-powerful, all-encompassing framework of calculation can safely be presumed to know all of the facts surrounding and supporting their choices, even if marginal errors of judgment are allowed. The logical extension of "perfect information" is what classical economists call the market-clearing feature of the economic system, which they also assume to be perfect. This notion is best illustrated by the famous story in which an elderly man is walking down the sidewalk with his young granddaughter when she notices a $20 bill and starts to pick it up. "No, no," says the grandfather, stopping the little girl's hand in midair. "If there was a $20 bill there on the sidewalk, it would already be gone. It can't be real."

Such theories border on intellectual arrogance, especially in light of the inability of classical economics to deal with the idea of accounting for lost natural resources. Just as our current system of economics makes absurd and unrealistic assumptions about the information actually available to real people in the real world, it insists upon equally absurd assumptions that natural resources are limitless "free goods."

This assumption stems in part from the fact that the system of national income accounts was established by John Maynard Keynes before the end of the colonial era, during which supplies of natural resources did indeed seem limitless. In fact, it is not entirely coincidental that much of the worst environmental devastation today is taking place in countries that have emerged from their colonial status only in the last generation. Patterns of abusive exploitation of the environment have a momentum that is difficult to reverse — especially if the prevailing economic assumptions were set in place by those who were primarily interested in removing the natural resources from these countries.

Accounting blindness is not limited to the valuation of products alone, however. According to the First Law of Thermodynamics, neither matter nor energy can be either created or destroyed; natural resources, therefore, are transformed into both useful products, called goods, and harmful by-products, including what we sometimes call pollution. Not surprisingly, our economic system measures the efficiency of production, or "productivity," in a way that keeps better track of the good things we produce than the bad. But every production process creates waste; why isn't it accounted for? If a country produces huge amounts of aluminum, for example, why isn't the calcium fluoride sludge, an inevitable by-product, accounted for?

Indeed, improvements in productivity — the single most significant measure of economic "progress" — are currently calculated by a method that embodies yet another absurd assumption: if a new technique has both good and bad consequences, it is permissible, under some circumstances, to measure only the good and simply ignore the bad. When the number of good things produced with each unit of labor, raw materials, and capital go up — usually because somebody has cleverly figured out a "better" way to perform the task at hand — then productivity is said to increase. But what if the clever new process results not only in the increased production of good things but also in an even larger increase in the number of bad things? Shouldn't that count? After all, it may cost a lot of money to deal with the consequences of the extra bad things.

And the absurdity doesn't stop here. Later, when expenditures are required to clean up the pollution, they are usually included in the national accounts as another positive entry on the ledger. In other words, the more pollution we create, the more productive contributions we can make to national output. The *Exxon Valdez* oil spill in Prince William Sound, and efforts to clean it up, to take one example, actually increased our GNP.

Classical economics also fails to account properly for all the costs associated with what we call consumption. Every time we consume something, some sort of waste is created, but this fact is conveniently forgotten by classical economists. When we consume millions of tons of chlorofluorocarbons (CFCs) each year, are they gone? If so, then what is eating the hole in the ozone layer? When we consume 14 million tons of coal each day and 64 million barrels of oil, are they gone? If so, where is all the extra carbon dioxide (CO_2) in the atmosphere coming from?

None of these hidden costs is accounted for properly; indeed, the way our economic system measures productivity doesn't make sense *even within the logic of the system itself*. It is almost as if the ultrarational "economic man" of classical theory actually believes in magic. If our economic goods are produced from natural resources that never have to be depreciated because their supply is limitless, if the production process leaves no unwanted by-products whatsoever, and if our products disappear without a trace when they are consumed, then we are witnessing powerful magic indeed.

I remember as a young child sitting with my father in his office while a man who appeared perfectly rational explained in detail his plans for a machine that would transform lead into gold. I suspected my father of being kinder and more patient than he might otherwise have been in order to give me the opportunity to hear one of the last alchemists on earth. In fact, however, alchemists are anything but an endangered species, because when we pretend to consume goods and resources, we are actually transforming them into a different chemical and physical substance. It is industrial alchemy of a very dangerous form. And at some point, the hidden costs of this alchemy will have to be paid.

Classical economics defines productivity narrowly and encourages us to equate gains in productivity with economic progress. But the Holy Grail of progress is so alluring that economists tend to overlook the bad side effects that often accompany improvements. The problem is, of course, they almost always go together, and wisdom requires balancing the good against the bad to determine whether the overall result is positive or negative. If we measure the value of what we do and consistently ignore important side effects, we will continue to set ourselves up for nasty surprises. When a "new" environmental catastrophe is discovered, for instance, we can often look back and see an accumulation of thousands of seemingly defensible but poorly thought-out decisions, all made according to criteria that do not themselves make any sense when all of the costs and risks are balanced against the benefits. Why weren't these consequences considered ahead of time? The answer lies in our economic system's ability to conceal the ill effects of many choices by resorting to an intellectual device labeled "externalities."

The bad things economists want to ignore while they measure the good things are often said to be too difficult to integrate into their calculations. After all, the bad things usually cannot be sold to anyone, and the responsibility for dealing with their consequences can often be quietly pushed on to someone else. Therefore, since the effort to keep track of the bad things would complicate the valuation of the good things, the bad things are simply defined away as external to the process and called externalities.

This habit of using an arbitrary definition to exclude inconvenient facts from the calculation of what is good and what is bad is a form of dishonesty. Philosophically, it is similar in some ways to the moral blindness implicit in racism and anti-Semitism — which also use arbitrary definitions to justify exclusions from the calculus of right and wrong. A racist, for example, can be seen as a person who draws a circle of value around himself and those of his own race in order to exclude, by definition, those of other races. The racist then often makes choices that artificially inflate the value of those inside the circle at the expense of those outside. Frequently, there is a direct ratio between the increasing value inside the circle and the decreasing value outside. Both slavery and apartheid are examples of this phenomenon at work.

In much the same way, our current system of economics arbitrarily draws a circle of value around those things in our civilization we have decided to keep track of and measure. Then we discover that one of the easiest ways to artificially increase the value of things inside the circle is to do so at the expense of those things left outside the circle. And here too, a direct and perverse ratio emerges: the more pollution dumped into the river, the higher the short-term profits for the polluter and his shareholders; the faster the rain forest is burned, the quicker more pasture becomes available for cattle and the faster they can be turned into hamburgers. Our failure to measure environmental externalities is a kind of economic blindness, and its consequences can be staggering. A mathematician at the University of British Columbia, Colin Clark, has said, "Much of apparent economic growth may in fact be an illusion based on a failure to account for reduction in natural capital."

Robert Repetto and others have suggested a modest change in the way we calculate productivity as a first step toward taking environmental externalities into consideration. He suggests that we carefully measure both beneficial and damaging products from any process and keep track of changes in both categories before measuring changes in productivity. For example, a coal-fired power station produces both kilowatt-hours of electricity and tons of atmospheric pollution. It is easy to evaluate the economic significance of the electricity because it is sold. But it is also possible to evaluate at least some of the economic significance of the atmospheric emissions. Sulfur oxides cause crop losses downwind from the power station, along with materials damage, visibility losses, and medical bills for the treatment of respiratory distress. A great deal of work has gone into calculating the real cost of the effects associated with each additional ton of sulfur dioxide emissions. So far, these valuations are a lot less precise than the values established by the market for electricity. Still, that difficulty ought not to be used as a convenient excuse for asserting that the cost should be placed at zero; there is a well-accepted and agreed-upon range, and some value within that range could and should be used in calculating the costs and benefits of each ton of coal burned.

Coal-burning power stations also provide a good illustration of a related point. When a new law like the Clean Air Act is passed, requiring a reduction in sulfur dioxides, we are told that the productivity of coal-fired power stations will go down — based on a calculation that completely ignores the savings that will result from lower expenditures to deal with the consequences of pollution every time a ton of coal is burned. Even if we changed the calculation of productivity only enough to include those economic impacts of pollution for which we already have accepted values, we would be that much closer to an accurate definition of true gains or losses.

Past a certain point, however, it is impossible to put a price on the environmental effect of our economic choices. Clean air, fresh water, the sun rising through the mist on a mountain lake, an abundance of life on the land, in the air, and in the sea — the value of these things is incalculable. It would be cynical indeed to conclude that because such treasures have no price, it is reasonable to make decisions based on the assumption that they are worthless. As Oscar Wilde said, "A cynic is one who knows the price of everything and the value of nothing."

In drawing a circle of value around those things we consider important enough to measure in our economic system, we not only exclude a great deal that is important in the environment, we also discriminate against future generations. The accepted formulas of conventional economic analysis contain short-sighted and arguably illogical assumptions about what is valuable in the future as opposed to the present; specifically, the standard "discount rate" that assesses cost and benefit flows resulting from the use or development of natural resources routinely assumes that all resources belong totally to the present generation. As a result, any value that they may have to future generations is heavily "discounted" when compared to the value of using them up now or destroying them to make way for something else. The effect is to magnify the power of one generation to compromise all future generations. In the words of Herman Daly, "There is something fundamentally wrong in treating the earth as if it were a business in liquidation."

In 1972 the Bruntland Commission, established by the United Nations to examine the connection between economic development and protection of the environment, focused our attention on the need for "intergenerational equity" — an insistence that decisions by the present generation be made with an awareness of their impact on future generations. Although this phrase has become a fixture in the rhetoric about the environment, it is not yet reflected in the way our economic system measures the effect of our decisions in the real world. As a result, we continue to act as if it is perfectly all right to use up as many natural resources in our own lifetime as we possibly can.

The current debate over sustainable development is based on the widespread recognition that many investments by major financial institutions, such as the World Bank, have stimulated economic development in the Third World by encouraging the short-term exploitation of natural resources, thus emphasizing short-term cash flow at the expense of longer-term, sustainable growth. This pattern has prevailed both because of a tendency to discount the future value of natural resources and because of a failure to properly depreciate their value as they are used up in the present.

This partial blindness in the way we account for the impact of our decisions on the natural world is also a major obstacle to our efforts to formulate sensible responses to the strategic threats now facing the environment. Typically, we cite hugely inflated estimates of the expense involved in changing our current policies, with no analysis whatsoever of the expense associated with the impact of the changes that will occur if we do nothing.

For example, the loss of 75 percent of California's annual moisture has long been predicted by some climatologists as a consequence of global warming. Yet because the scale of the problem is so large, no one seems to even consider including the cost of the water shortage in California in our calculation of the benefits of an aggressive program to counter global warming. We should also calculate the costs of doing nothing, because the consequences of the seven-year drought are already staggering and may get worse. One of the lesser costs came to my attention as chairman of the Senate subcommittee that oversees NASA, and it illustrates my point perfectly. Early in 1991, NASA announced that the drought in California had dried up deepwater reservoirs far below the dry lake bed on which the space shuttle lands when it returns from orbit. The six-foot fissures unexpectedly appearing in the surface of the lake bed may eventually pose a threat to the landing strip. If a new landing facility becomes necessary, it will cost a lot of money. It seems only fair that this new landing strip be included on the expense side of the ledger in calculating the costs of doing nothing to counter global warming. (When I suggested that this new expense be added to the Office of Management and Budget's analysis, they said, "You've got to be kidding." "Only partly," I responded.)

But the problem goes far beyond our response to the drought in California. In many ways, the Bush administration's entire cost-benefit analysis is misguided and reflects an apparent inability to see the magnitude of the environmental crisis. Thus far, the administration has been blind to the true value of preserving the environment while keenly aware — like Oscar Wilde's cynic — of the price. When President Bush welcomed an international conference on the global environment in the spring of 1990, his staff prepared materials for the visiting negotiators that contained a graphic illustration of the administration's approach to balancing short-term monetary gains against long-term environmental destruction. In the illustration, several bars of gold rested on one tray of a scale; on the other tray perched the entire earth and all its natural systems, seemingly with a weight and value roughly equivalent to the six bars of gold. A scientist, or perhaps an economist, is noting the careful balance on her clipboard. Although several delegates from other countries commented privately that it seemed to be an ironic symbol of Bush's approach to the crisis, the president and his staff seemed wholly oblivious of the absurdity of their willingness to place the entire earth in the balance.

. Some of America's best corporations are doing a much better job of responding creatively to the crisis. Those that have made a strong commitment to environmental responsibility have found, to their surprise, that when they start to "see" their pollution and look for ways to minimize it, they begin to "see" new ways to cut down on their use of expensive raw materials and new ways to improve efficiency in virtually every part of the production process. Some of these companies have also reported that this new attentiveness to each stage of production has also resulted in a sharp reduction in product defects. For example, the 3M company credits its Pollution Prevention Pays program with major improvements in profits; Xerox and several other companies have reported the same experience.

Some companies are trying to guess whether the new public awareness of the environment is temporary or permanent. Major

President Bush presented a brochure including this drawing to delegates at his 1990 White House Conference on the global environment. The administration has tried to convince the world that the environment faces no serious dangers and that the wisdom of any effort to rescue it is outweighed by the cost — represented here by six bars of gold.

paper mills, for example, facing a round of investment in new capacity, must decide whether the current interest in recycled paper is here to stay. If so, then large investments in recycling plants will be profitable; if not, they may face serious risks in making such investments. Such prophecies often tend to be self-fulfilling, of course. But here is where the government can play an important role — and too often has failed to do so. The Bush administration talks loudly about the tendency of a free marketplace to solve all problems. But many of our markets are highly regulated, often in hidden ways. In the case of the paper industry, for instance, tax-payers currently subsidize the manufacture of paper products made from virgin timber, both as the largest single purchaser and by further subsidizing the construction of logging roads into national forests. In addition, the federal government pays the entire cost of managing the forest system, including many activities that ex-clusively benefit the timber industry. All of these policies encourage further destruction of a critical natural resource.

The Bush administration and the entire U.S. government ought to understand the economic significance of a healthy environment as a kind of infrastructure supporting future productivity. If it is destroyed, many jobs now at risk will be lost. A case in point is the heated dispute between the timber industry in the Pacific North-west and conservationists eager to protect the endangered spotted owl. This issue has been billed as a conflict between jobs and the environment. But if the remaining 10 percent of old-growth forest is logged out, as the timber industry prefers, the jobs will be lost anyway. The only question is whether the effort to create new jobs will begin now or later, after the forest is completely gone.

The current administration also ought to do a much better job of encouraging appropriate technologies, since they can be an impor-tant benefit to set against all the costs of environmental degrada-tion. Japan, for example, is already implementing an ambitious plan to cultivate what it believes will be a massive global market for new technologies for renewable energy and environmentally be-nign processes. Tragically, however, after having developed the first products using wind and solar energy, the United States is now a net importer of both technologies.

There is an Alice-in-Wonderland quality to much of our current approach to economic analysis. Even as we have ignored the consequences for the environment of our present economic decisions, attention has been focused on increasingly frenetic speculation, merger mania, asset shuffling, and a range of other activities largely unrelated to the creation of competitive goods and services. The result is not only a diminished competitive position for the United States in the world economy, but also an acceleration of the trend toward the kind of short-term thinking that will make it harder to formulate a creative and effective response to the environmental crisis.

But it is not yet too late to avert the worst effects of this crisis, and the United States ought to lead the way. Victory in our epic struggle with communism, whose features were infinitely worse — for individuals and for the environment — than anything our economic system has produced, should give us confidence, as well as a sense of obligation, to address the challenges now before us. We must correct the shortcomings in the rules and procedures that guide the millions of daily decisions that are the nerves and sinews of Adam Smith's invisible hand: we must address the deficiencies of our current methods for defining what is progress and what is absurdity.

Some of the changes needed will be relatively simple to implement. Others will be more difficult. But all will require the courage to see things as they are, to avoid deceiving ourselves, to train ourselves to recognize when sophisticated imbecilities are substituted for serious analysis. In 1989, for example, the president's Council of Economic Advisers concluded in its annual report that "there is no justification for imposing major costs on the economy in order to slow the growth of greenhouse gas emissions." Part of the reasoning used to support this conclusion was that "the average temperature differential between New York City and Atlanta is as large as the most extreme predictions of warming yet there is no evidence that Atlanta's warmer climate produces a greater health risk than New York's." But if New York will be as hot as Atlanta, what will Atlanta be like? What will southern California be like? What will the midwestern droughts be like? What changes will occur in the global climate pattern? These questions and others were of course ignored as the political equivalent of externalities.

Years from now, if their policies are followed and global warming causes terrible destruction to the global environment with no serious effort having been made to stop it, members of the Bush administration will no doubt be properly humble and contrite. And it will hardly be the first time that devotion to the convenience of the present has blinded decision-makers to their obligation to prepare for the future. But the time for action is now, and for inspiration we might look back to one of history's most visionary leaders.

On November 12, 1936, Winston Churchill grew so exasperated with the continuing failure of Britain to prepare for Hitler's onslaught that he charged in a speech to the House of Commons: "The Government simply cannot make up their minds, or they cannot get the Prime Minister to make up his mind. So they go on in strange paradox, decided only to be undecided, resolved to be irresolute, adamant for drift, solid for fluidity, all-powerful to be impotent. . . . The era of procrastination, of half-measures, of soothing and baffling expedients, of delays, is coming to its close. In its place we are entering a period of consequences."

11

We Are What We Use

One of the characteristics that distinguish human beings from all other living creatures is our ability to use information to create symbolic representations of the world around us. By manipulating information about the world, or sharing it with others, we learn how to manipulate the world itself.

This way of relating to the world has been so successful for us that by now it is second nature. We not only take it for granted, we incorporate it into every other strategy we devise for gaining power over the world around us. It is hardly surprising, therefore, that over the course of history we have become increasingly dependent on information in all its forms. But this dependence has for the most part been unquestioned; rarely do we examine the negative impact of information on our lives.

We have always placed a high value on knowledge. When faced with a problem, our first instinct is to seek more and more information that will help us understand it. For most of history, a great deal of what we call culture has consisted of beguiling ways to share especially valuable information about our world and how we can relate to it productively: how to make an arrowhead with a groove in the side to allow the animal's blood to drain out, how to weave a basket that will catch the grain but not the dust, how to dance to the hunt and the harvest while singing the secrets of the moon and the seasons, how to tell stories that captivate children and teach them important lessons about life.

In ancient cultures, the storehouse of accumulated information was invariably embedded in a larger story of life communicated to

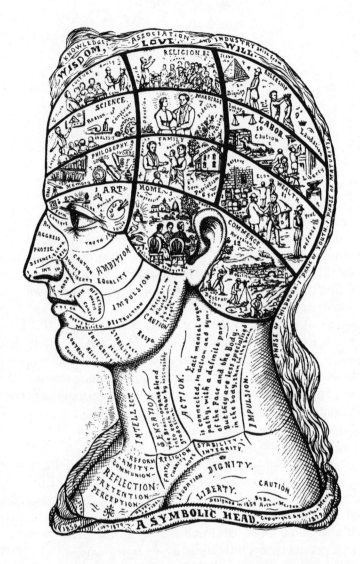

The inner structures of human thought have changed as we have adapted to new technologies. Of necessity, we reconfigure our mental "software" to accommodate the distinctive pattern of information we receive in large quantities from conduits like books, television, and computer terminals. This 1879 drawing is from a manual on the discredited practice known as phrenology, the study of bumps on the head and their alleged relationship to localized thoughts in the human brain.

successive generations. The social, cultural, and ecological context in which information was acquired and used remained fresh in the minds of those wielding it. But a story is the simplest technology, and as more sophisticated technologies for collecting, storing, and communicating information were invented — such as legal codes and financial accounting — they received special attention because of the new power they conferred. In the Middle Ages, for example, guilds and crafts devoted to the specialized knowledge of important skills became a primary source of identity for their members. Inevitably, as the amount of information communicated through these new techniques grew into denser and more valuable messages to successive generations, we had to reconfigure our minds to receive this stream of information, and to remember it and use it.

But something was lost along the way, because the amount of attention necessary for this mental work detracted from the attention we paid to the context of the communication, within which the power it conferred was wielded. Those who took such pride and care in building the bridge over the River Kwai, for example, almost forgot the context in which their skill was applied. One of the things we began to ignore was the way these new information technologies changed us and the context of our lives. The more information we consumed, the more our mental lives were dominated by direct experience with information representing the world rather than direct experience of the world itself. And the more we grew accustomed to experiencing the world indirectly, by means of ever more complex representations, the more we hungered for information of all kinds — and the more we turned our attention to inventing new ways to create it.

This cycle accelerated dramatically when civilization discovered the scientific method. Acquiring knowledge about the natural world had long been a focus for human enterprise, and the scientific method gave us a powerful new way to investigate natural phenomena and reduce them to a collection of smaller bits of information, each one susceptible to explanation, repetition — and manipulation.

Soon, the amount of raw information generated began to increase rapidly, and our power to manipulate nature grew explosively. So too did our respect for the awesome productivity of this new way of relating to the world. Impressed with our own cleverness, we made heroes of our inventors and, later, our industrialists. We grew to believe that no matter what problem we confronted, all we had to do was to apply the scientific method, reduce the problem to its component bits of information, and then experiment with them until they yielded a technological solution.

But as the industrial age gave way to the information age, the production of information began to far outstrip our capacity to use it. John Stuart Mill has been described as "the last man to know everything"; now none of us can hope to be fully conversant with the knowledge of our time. Indeed, no one can even hope to know "everything" in his or her own field of study anymore.

We now face a crisis entirely of our own making: we are drowning in information. We have generated more data, statistics, words, formulas, images, documents, and declarations than we can possibly absorb. And rather than create new ways to understand and assimilate the information we already have, we simply create more, and at an increasingly rapid pace.

Our current approach to information resembles our old agricultural policy. We used to store mountains of excess grain in silos throughout the Midwest and let it rot, while millions around the world died of starvation. It was easier to subsidize growing more corn than to create a system for feeding those who were hungry. Now we have silos of excess data rotting (sometimes literally) while millions hunger for the solutions to unprecedented problems.

It is interesting to note the similarity between this crisis in our relationship to information and the crisis in our relationship to the natural world. Just as we automated the process for converting oxygen into carbon dioxide (CO_2) — with inventions like the steam engine and automobile — without taking into account the limited ability of the earth to absorb CO_2, we have also automated the process of generating data — with inventions like the printing press and computer — without taking into account our limited ability to absorb the new knowledge thus created.

In fact, we have now generated vast mountains of data that never enter a single human mind as thought. The Landsat satellite photography program, for example, is capable of taking a complete photograph of every square inch of the earth's surface every eighteen days, and it has done so for much of the last twenty years. However, in spite of our critical need to understand what has been happening to the surface of the earth during that time, more than 95 percent of all those photographs have never been seen by anyone. Instead, the images are collected and stored on magnetic tape in the equivalent of digital silos and left to gather dust and rot.

Perhaps this sort of data should be called "exformation" instead of information, since it exists completely outside the brain of any living person. But however it's labeled, the problem is growing rapidly worse. Within a few years, the new Mission to Planet Earth program will, as currently designed, beam down from orbit more information every hour than currently exists in all the earth sciences combined. Why? To help us decide whether the environmental crisis is real fifteen years from now. No doubt, the information will be valuable. But waiting for it will be dangerous, especially since many of us believe that we already have more than enough information to make the decision to act. And coping with all that data will be extremely difficult, not least because most all of it will never enter a single human brain.

Vast amounts of unused information ultimately become a kind of pollution. The Library of Congress, for instance, receives more than ten thousand periodicals each year — from India alone! And given that some of our accumulated information and knowledge is dangerous — such as the blueprint for an atomic bomb — keeping track of all the data can become as important as it is difficult. What if this toxic information leaks into the wrong places? The credit bureau's file on your personal life, to take a less deadly example, should not be available to anyone who wants it.

It is not a coincidence that we have a crisis in education coinciding with our surfeit of information. Education is the recycling of knowledge, but we find it easier to generate new facts than to conserve and use the knowledge we already have. So when faced with the problem of ignorance, we immediately create more and more information without seeming to realize that while it may be valuable, it is no substitute for knowledge — much less wisdom. Indeed, by generating raw data in much larger quantities than ever before, we have begun to interfere with the process by which information eventually becomes knowledge. When it is allowed to run its normal course, the process actually resembles fermentation: information is first distilled into knowledge, which is then — sometimes — fermented into wisdom. Now, however, so much more information is collected each day than ever before that the slow process by which it is converted to knowledge has been overwhelmed by an avalanche of new data.

If we need to do a better job of processing these enormous quantities of information, we also need to better understand the powers — both good and bad — inherent in how we communicate information. The first of our information technologies, after all, was spoken language, and its power has always been revered. My own religious tradition teaches me: "In the Beginning was the Word." Indeed, in the Judeo-Christian creation story it was through speaking that God accomplished his purpose: " 'Let there be light.' And there was light." Likewise, the emergence of the second information technology — written language — is generally credited with the real beginning of what we consider organized civilization.

What gets less attention, however, is the fact that how we communicate information can change us as we use it. Information technology, like any technology, mediates our relationship to whatever we describe with it, because in the process of trying to capture the full meaning of a real phenomenon in a symbolic representation, we leave some features out and, by selective inclusion, distort the significance of others. Of necessity, we configure our minds to the contours of the symbolic representation. All information technologies — words chiseled in stone, beautiful manuscripts copied by monks, the printing press, satellite television, and computer graphics sent by optical fiber — have expanded our ability to understand the world around us. But these technologies have also created distinctive patterns of distortion and have thus changed the way our minds receive, remember, and understand the world.

We normally adapt so completely to the communication technology we are using that we become oblivious of its distorting effects. Spoken words, for example, are levelers of experience. Even as they convey distinctions, contrasts, and subtleties, they also homogenize and routinize — simply because one's direct experience is often with the words that convey the meaning rather than the meaning itself. The commandment against using the name of God in vain, for instance, is premised in part on the assumption that "God" has such significance that the name should virtually implode in the user's mind. Once it is reduced into an easily manipulated symbol, however, "God" can be repeated over and over, and in the process become drained of much of its capacity to trigger reverence and awe. Moreover, it can be used out of context to artificially magnify the significance of even the most trivial and mundane concerns.

Reproduced images have the same leveling effect. In his classic essay, "The Work of Art in the Age of Mechanical Reproduction," Walter Benjamin describes how a work of art that is technologically replicated loses its "aura" or sanctity. Anyone who has seen a print of the Mona Lisa or *The Luncheon of the Boating Party* is familiar with this effect: no matter how faithful to the original, the print has inevitably lost the impact of the original. If we see the print in several different locations, each new encounter will diminish the impact of what remains of the experience that it seeks to replicate over and over again. It is a trade-off: many more people can experience something of what the original painting conveyed — in fact, a great deal of it — but the experience of looking at the print simply cannot be compared to the experience of seeing the original.

Whenever any technology is used to mediate our experience of the world, we gain power but we also lose something in the process. The increased productivity of assembly lines in factories, for example, requires many employees to repeat the identical task over and over until they lose any feeling of connection to the creative process — and with it their sense of purpose.

Something like this has happened in our relationship to nature. The more we rely on technology to mediate our relationship to nature, the more we encounter the same trade-off: we have more power to process what we need from nature more conveniently for more people, but the sense of awe and reverence that used to be present in our relationship to nature is often left behind. This is a primary reason that so many people now view the natural world merely as a collection of resources; indeed, to some people nature is like a giant data bank that they can manipulate at will. But the cost of such perceptions is high, and much of our success in rescuing the global ecological system will depend upon whether we can find a new reverence for the environment as a whole — not just its parts.

Too many of us, however, display a reverence only for information and analysis. The environmental crisis is a case in point: many refuse to take it seriously simply because they have supreme confidence in our ability to cope with any challenge by defining it, gathering reams of information about it, breaking it down into manageable parts, and finally solving it. But how can we possibly hope to accomplish such a task? The amount of information — and exformation — about the crisis is now so overwhelming that conventional approaches to problem-solving simply won't work. Furthermore, we have encouraged our best thinkers to concentrate their talents not on understanding the whole but on analyzing smaller and smaller parts.

Despite — or perhaps because of — the so-called information age, what's required now is a Jeffersonian approach to the environment. Thomas Jefferson, like other leading thinkers of his time, aspired to a catholic understanding of the whole of knowledge, and when he and his colleagues in Philadelphia turned to the task of creating the world's first constitutional self-government, they combined an impressive understanding of human nature with a full command of jurisprudence, politics, history, philosophy, and Newtonian physics. The world as a whole has now arrived at a watershed comparable in some ways to the challenge that confronted the founders two hundred years ago. Just as the thirteen colonies faced the task of defining a framework to unite their common interests and identity, the people of all nations have begun to feel that they are part of a truly global civilization, united by common interests and concerns — among the most important of which is the rescue of our environment. If we are to succeed, we must resist being overwhelmed by the flood of information and refuse to consider the natural world as merely a convenient bank of resources and coded information. We must be bold enough to use Jefferson's formula and seek to combine a catholic understanding of the nature of civilization with a comprehensive command of the way in which the environment functions.

* * *

The impact of technology in our lives goes far beyond its effect on our methods of information processing, of course. Indeed, the scientific and technological revolution has almost completely changed the physical realities of our relationship to the earth. With a dizzying array of new tools, technologies, and processes, we have extended our senses and magnified our ability to work our will on the world around us. We can now see the rings of Saturn, the atoms of a molecule, the valves inside the human heart, and the entire earth rising above the moon's horizon. We can hear the recorded voices of speakers who have long since died, the music of whales at the bottom of the sea, and the cries of a baby trapped in an abandoned well a thousand miles away. We can walk down the aisle of a plane traveling twice the speed of sound, leave Europe at lunchtime and arrive the same day in New York for a late breakfast. We can grasp the levers of a giant crane and, like Atlas, lift the weight of a thousand men.

The still-accelerating scientific and technological revolution is magnifying the power of every one of the 5.5 billion people on earth to recreate physical reality in the image of his or her own intention. Every ambition, every hunger, every desire, every fear, and every hope now resonate in the human heart with more powerful implications for the world around us. Ancient habits of thought now carry new significance because of our power to transform even the boldest thought into action. But like the Sorcerer's Apprentice, who learned how to command inanimate objects to serve his whims, we too have set in motion forces more powerful than we anticipated and that are harder to stop than to start.

Among all the problems arising out of the scientific revolution, the effect of nuclear weaponry on the perception of war has been singled out for special and intense scrutiny. Nuclear weapons pose an obvious and deadly threat, and for the past forty-five years, millions have protested that our world is unsafe as long as this technology is available for use during a war. But nuclear weapons have dramatically changed our perception of war, which over the long term may prove beneficial. After all, the long and protracted Cold War between the United States and the Soviet Union never led to direct armed conflict partly because of the new awareness in both countries of the unthinkably horrible consequences of war in the nuclear age. The subsequent shift by the Soviet Union and all of Eastern Europe from communism to democracy and capitalism — largely without violence — might never have been possible before this shift in thinking about the acceptability of war began.

Just as war has been a part of civilization for thousands of years, so too has our age-old practice of exploiting the earth for sustenance, for food, water, shelter, clothing, and our other basic needs. Science and technology have given us, especially in this century, many thousands of new tools that magnify our power to exploit the earth for what we need — as well as for what we merely want. None of these new technologies by itself even approaches the significance of nuclear weapons, but taken together, their combined impact on the earth's natural systems makes the consequences of unrestrained exploitation every bit as unthinkable as the consequences of all-out nuclear war.

It was relatively easy to recognize the dramatic qualitative difference between Dr. Oppenheimer's atomic bomb and Dr. Nobel's dynamite, partly because our attention was focused on a single prominent technology. By contrast, it is extremely difficult to aggregate all of the powerful new technologies affecting our relationship to the earth and the range of needs and desires we seek to satisfy with them. The cumulative impact of these technologies is qualitatively different from the cumulative impact of their predecessors, but since there are so many of them, most performing services that have become integrated into our lives, it is very difficult to recognize this dramatic change in circumstances as a historical event that has transformed our relationship to the earth.

We have also fallen victim to a kind of technological hubris, which tempts us to believe that our new powers may be unlimited. We dare to imagine that we will find technological solutions for every technologically induced problem. It is as if civilization stands in awe of its own technological prowess, entranced by the wondrous and unfamiliar power it never dreamed would be accessible to mortal man. In a modern version of the Greek myth, our hubris tempts us to appropriate for ourselves — not from the gods but from science and technology — awesome powers and to demand from nature godlike privileges to indulge our Olympian appetite for more. Technological hubris tempts us to lose sight of our place in the natural order and believe that we can achieve whatever we want.

And, far too often, our fascination with technology displaces what used to be a fascination with the wonder of nature. Like the young child who thinks bread originates on a store shelf, we begin to forget that technology acts *upon* nature to meet our needs. As the population increases and our desire for higher rates of consumption continues to grow, we ask civilization for more of everything we want while ignoring the stress and strain tearing at the fabric of every natural system. Because we feel closer to the supermarket than to the wheatfield, we pay far more attention to the bright colors of the plastic in which the bread is wrapped than we do to the strip mining of the topsoil in which the wheat was grown. Thus, as we focus our attention more and more on using technological processes to meet our needs, we numb the ability to feel our connections to the natural world.

Often, when we seek to artificially enhance our capacity to acquire what we need from the earth, we do so at the direct expense of the earth's ability to provide naturally what we are seeking. For example, when we increase agricultural production by using technologies that increase topsoil erosion, we damage the land's ability to grow more food in the future. And we frequently ignore the impact of our technological alchemy on natural processes. Thus, when we manufacture millions of internal combustion engines and in the process automate the conversion of oxygen into CO_2 and other gases, we interfere with the earth's ability to cleanse itself of the impurities that are normally removed from the atmosphere.

In order to change the destructive pattern of our current relationship to the environment, we have to develop a new understanding of technology's role in magnifying the harmful effects of once-benign impulses and activities. In many cases, the technologies themselves need to be changed. For example, it makes little sense to continue manufacturing cars and trucks that get twenty miles per gallon and pump nineteen pounds of CO_2 into the atmosphere per gallon. And in fact we need to make a strategic decision to accelerate the development of new technologies, like solar electricity production, which have less harmful effects on the environment. But in every case, success will require careful attention to the way we relate through technology to the environment, and a much greater awareness of the profound effect any powerful technology can have on that relationship.

Sometimes a shift from one technology to another will transform established patterns. To take one example, the invention of the printing press led to a completely new approach to government and politics. Some modern nations were formed only after the printing press made it possible to widely distribute the collection of shared ideas and values (usually in a shared language) on which a nation could be founded. Many historians argue that the American Revolution might never have occurred without the many printed pamphlets and treatises such as Thomas Paine's *Common Sense,* which spread the idea of a new American nation.

The dominant technologies of any age shape our often unspoken assumptions about what is possible and what is likely. The Constitution, for example, establishes a system of checks and balances through three branches of government, each designed to be equal with the other two. However, the authors of the Constitution assumed that each branch would communicate with the people primarily through the printed page. When, in the middle of the twentieth century, electronic broadcasting replaced newspapers as the dominant means of mass communication, the relative prominence of the three branches of government changed, at least as they are perceived by the people. The president, unlike Congress and the courts, spoke with one voice on radio and, with the emergence of television, projected one face and personality into the living room of virtually every American home. The members of the House and Senate and the Supreme Court were never seen or heard all together through this medium — except when they applauded the president during his State of the Union address. Thus, since real political power in a democracy flows from the people, the new prominence of the president compared to that of the other branches of government soon amounted to a kind of constitutional amendment by technology.

Consider another technology that affects our system of government. When the technology of war is no longer large armies and navies that require months to assemble and move into action but intercontinental ballistic missiles that reach their targets in less time than it takes for Congress to establish a quorum, doesn't that threaten to undercut Congress's power to declare war? Here again, it is almost as if the Constitution has been amended by technology. And here too, a new technology had a very different technological coefficient than did older technology in its interaction with our system of government. (The phrase "technological coefficient" refers to the unique way any technology affects that part of the world upon which it is used.)

The shift from one technology to another, even if the new one is used for the same basic purpose, can profoundly alter the relationship between different elements in a system. Moreover, new generations of technology now appear so rapidly that the shift from one to the other is sometimes disconcertingly abrupt. This too can cause problems in our relationship to the environment.

Consider, for example, the genesis of the best-known hazardous chemical waste site in the United States, Love Canal. Early in this century, not long after Thomas Edison learned how to harness the power of electrical energy, the new chemical-processing industry — made possible by and dependent on large amounts of electricity — built its factories as close as possible to sources of hydroelectric power like Niagara Falls. Edison had chosen to sell what we now call direct current, which loses most of its electricity when transmitted over long distances. So, not surprisingly, good industrial building sites next to Niagara Falls were soon in short supply.

An entrepreneur named Colonel William Love hit upon the idea of digging a canal a few miles upstream, where the Niagara River bends sharply back toward itself as it winds down the side of a hill toward the falls. Love recognized that a canal connecting two arms of the river bend would create an artificial waterfall capable of generating electricity for new chemical plants he hoped to entice to building sites along the new canal. But shortly after Love began digging his canal, he learned that a Russian émigré named Nikola Tesla had invented a new way to use the form of electricity called alternating current, which could cover relatively long distances with only modest losses of power.

Suddenly, chemical plants no longer needed to be next to the generator, and the electricity from Niagara Falls was transmitted to new plants that began to spring up miles away. When they looked for a place to dump their chemical waste, they found an obsolete, partly completed canal. After it was filled, it was covered with a layer of dirt. Years later, a residential subdivision was built along both sides of the canal. Right in the center was a new elementary school for the children, who never knew why it was called Love Canal until the chemicals oozed up into the playground.

We can see something like this same pattern in the way our society has allowed the inner city to become a toxic dumping ground for crime, drug abuse, poverty, ignorance, and desperation. In this case too there was a shift, not from a single technology but from the industrial age itself — which had put a premium on the clustering of factories and housing near ports, where a critical mass of coal, raw materials, and workers could be maintained. It was followed by the post-industrial age, in which families moved into suburbs and new jobs and established new patterns in their lives. The inner cities they left behind, no longer essential to efficient production, became — in part — repositories for wasted lives.

Sometimes it is not the technology that changes but the setting in which it is used. In Kenya, for example, a tribe with a successful technique for cultivating land and growing crops in the highland areas of the Rift Valley was pressured by population growth to migrate into the lowlands. But the agricultural technology that had served them well for generations caused catastrophic soil erosion in their new area, with its greater rainfall and different soil. Similarly, it may be quite inappropriate to transplant the industrial culture that works in a wealthy developed nation into a poor developing country where all the social conditions are different.

Our relationship to technology can also be complicated by the way two or more powerful technologies interact with one another. We are all familiar with the occasional warnings on prescription medicine to alert us about potential drug interactions: two perfectly good medications, each useful and effective when taken alone, can in combination cause an extremely harmful reaction. The same thing can occur with technologies. I have often wondered whether a similarly harmful reaction in the political culture of the United States might be caused by the coexistence of television and print technology as rival systems for communicating — and in the process organizing — political thought. Frequently, people who read about an event or idea in the newspaper come away with wildly different impressions from those of people who watched the same event or idea on the evening news. Each medium tends to create its own way of thinking, and each tends to frustrate the other. In the process, the country as a whole seems unable to define our objectives, much less move coherently toward them.

In a different context, the writer Octavio Paz once observed that, in his opinion, India's seeming social paralysis results in part from the coexistence of the world's most rigidly monotheistic religious system, Islam, and the world's most elaborately pantheistic religion, Hinduism. In the same way, I wonder whether America's political paralysis might stem in part from the coexistence of two powerful but clashing media for communicating political thought.

In examining how science and technology have transformed our relationship to the natural world, it seems useful to refine our definition of "technology." In addition to tools and devices, we should include systems and methods of organization that enhance our ability to impose our will on the world. Any collection of processes that together make up a new way to magnify our power or facilitate the performance of some task can be understood as a technology. Even a large new system of thought like market economics or democracy can be recognized as a device for producing certain results; like other devices, they sometimes have consequences that are difficult to anticipate.

Given this broader definition, the human body can be viewed as a kind of technology. The way we think about the environment begins, of course, with the way we experience the earth, and our principal contact with the earth is through our five senses. But although most of us take them for granted, our senses are actually quite limited in their ability to provide information about the world. Even as they give us our primary sense of what the world is like, they limit our experience, channeling it into patterns that reflect only the information they can receive and process. As a result, we begin to believe that the limited information we receive represents the entirety of what exists, so we are usually surprised to find that something invisible to us is an important part of our world, especially if it poses a serious threat to which we must respond.

For example, the chemicals destroying the ozone layer, chloro-fluorocarbons (CFCs), are odorless, tasteless, and colorless. In other words, as far as our unaided senses are concerned, they don't exist. Similarly, the extra concentrations of CO_2 in the atmosphere that have accumulated during the last several decades are invisible unless we use sophisticated means of measuring them. Moreover, the infrared waves that are the particular kind of solar radiation trapped by the extra CO_2 and the CFCs, are in that part of the spectrum that is invisible to the human eye. Indeed, part of our difficulty in responding to the ecological crisis is that its symptoms are not yet setting off alarms that we can sense directly through hearing, tasting, seeing, smelling, or touching. Over the past few years, many people have noticed that the summers seem hotter and the droughts longer; if this apparently direct evidence of global warming is causing people to take the problem more seriously, how much more urgent the crisis would seem if we could taste CFCs or see CO_2!

Our bodies and minds, then, are hardly perfect technologies. And what complicates things further is the role of gender in how we experience the world. A famous experiment by the psycho-analyst Erik Erikson illustrates the point. Forty years ago, Erikson gave blocks to a group of children and carefully noted the shapes and structures they built. The girls were far more likely to build structures that seemed to define a space that was contained within the structure. The boys, by contrast, were more likely to build structures that extended from a base outward and upward to penetrate the space around them.

It certainly seems as if the way our civilization as a whole relates to the environment has been characterized by a determined exten-sion outward into nature, with far too little emphasis on patterns that might contain, protect, and nurture the environment. Accord-ing to this view, for the last few thousand years, Western civiliza-tion has emphasized a distinctly male way of relating to the world and has organized itself around philosophical structures that de-value the distinctly female approach to life. For example, as the scientific and technological revolution has picked up speed, we have seemed to place a good deal more emphasis on technologies that extend and magnify abilities — such as fighting wars — historically associated more with males than females. At the same time, new ways to reduce our scandalously high rate of infant mortality have received far less attention. Indeed, our approach to technology itself has been shaped by this same perspective: devices take precedence over systems, ways to dominate nature receive more attention than ways to work with nature. Ultimately, part of the solution for the environmental crisis may well lie in our ability to achieve a better balance between the sexes, leavening the dominant male perspective with a healthier respect for female ways of experiencing the world.

As is the case with gender, stage of life has a profound effect on the way an individual relates to the world. Adolescents, for example, have a sense of immortality that dulls their perception of some physical dangers. During middle age, on the other hand, emotionally mature adults naturally experience a desire to spend more time and effort on what Erikson has called generativity: the work of bringing forth and nurturing possibilities for the future. The metaphor is irresistible: a civilization that has, like an adolescent, acquired new powers but not the maturity to use them wisely also runs the risk of an unrealistic sense of immortality and a dulled perception of serious danger. Likewise, our hope as a civilization may well lie in our potential for adjusting to a healthy sense of ourselves as a truly global civilization, one with a mature sense of responsibility for creating a new and generative relationship between ourselves and the earth.

Our experience of life is also shaped by another aspect of our physical being that is so taken for granted, we almost never notice it. Every person shares the basic architecture of the human body, which has two virtually identical halves on either side of a plane that bisects our bodies like a mirror. Known as bilateral symmetry, this mirroring feature of our bodies has extensive implications for how we experience the world. In almost everything we do to or with the world, we divide the task into two conceptual halves — consolidation and manipulation — and assign each half to opposite sides of the machine our body resembles. At breakfast this morning, I consolidated my grapefruit with my left hand to keep it from moving on the plate and then manipulated it with my right hand, first by cutting portions away from the whole with a knife, then by eating them with a spoon. When I play catch with my children, I consolidate the baseball in my glove with one hand, then reach to grasp it with the other and throw it to one of them.

We also use the two halves of our brain to relate to the world in two very different ways: one half is more adept at maintaining a sense of context and spatial proportion while the other is more adept at the manipulation of thought called logic. Some linguists have noted that the only feature common to virtually all languages is a reliance on the subject-verb dichotomy. Indeed, each sentence on this page begins with a noun ("sentence") and moves toward the period by means of a verb ("moves"). We have emphasized action upon the world; yet in the words of Father Thomas Berry, "The Universe is a communion of subjects, not a complex of objects."

This point about bilateral symmetry may seem obscure, but in my view it suggests what is perhaps the single most dangerous way in which modern technology has distorted our relationship to the earth, for technology has vastly magnified our various abilities to manipulate nature far beyond the extent to which it has thus far magnified our abilities to conserve and protect nature. We now have a thousand incredibly powerful new ways to manipulate and transform the natural systems of our fragile earth, but our notions of how to consolidate and protect the environment against unintended consequences are still rudimentary. And our reckless manipulations of nature are far more likely to have catastrophic collateral damage precisely because we have failed to think of how to safeguard the stability and continuity of their context.

Just as technology, when used thoughtlessly, can disrupt the ecological balance of the world, so too can some technologies disrupt the ecological balance in the way we experience the world. By enhancing some senses more than others, by magnifying some abilities more than others, and by heightening some potentials more than others, technologies can profoundly alter how we perceive, experience, and then relate to the world. In the latter half of this century, for example, we have manipulated nature in unheard-of ways and then, as problems have resulted, we have reflexively looked for still more ways to manipulate nature in hopes of fixing the damage from the initial intervention.

In discussions of the greenhouse effect, I have actually heard adult scientists suggest placing billions of strips of tin foil in orbit to reflect enough incoming sunlight away from the earth to offset the larger amount of heat now being trapped in the atmosphere. I have heard still others seriously propose a massive program to fertilize the oceans with iron to stimulate the photosynthesis by plankton that might absorb some of the excess greenhouse gases we are producing. Both of these proposals spring from the impulse to manipulate nature in an effort to counteract the harmful results of an earlier manipulation of nature. We seem to make it easier to consider even harebrained schemes like these than to consider the seemingly more difficult task of revisiting the wisdom of those earlier manipulations, which don't seem to have a healthy relationship to their context, for they are in the process of destroying it.

In its deepest sense, the environmentalism that concerns itself with the ecology of the whole earth is rising powerfully from the part of our being that knows better, that knows to consolidate, protect, and conserve those things we care about before we manipulate and change them, perhaps irrevocably.

12

Dysfunctional Civilization

At the heart of every human society is a web of stories that attempt to answer our most basic questions: Who are we, and why are we here? But as the destructive pattern of our relationship to the natural world becomes increasingly clear, we begin to wonder if our old stories still make sense and sometimes have gone so far as to devise entirely new stories about the meaning and purpose of human civilization.

One increasingly prominent group known as Deep Ecologists makes what I believe is the deep mistake of defining our relationship to the earth using the metaphor of disease. According to this story, we humans play the role of pathogens, a kind of virus giving the earth a rash and a fever, threatening the planet's vital life functions. Deep Ecologists assign our species the role of a global cancer, spreading uncontrollably, metastasizing in our cities and taking for our own nourishment and expansion the resources needed by the planet to maintain its health. Alternatively, the Deep Ecology story considers human civilization a kind of planetary HIV virus, giving the earth a "Gaian" form of AIDS, rendering it incapable of maintaining its resistance and immunity to our many insults to its health and equilibrium. Global warming is, in this metaphor, the fever that accompanies a victim's desperate effort to fight the invading virus whose waste products have begun to contaminate the normal metabolic processes of its host organism. As the virus rapidly multiplies, the sufferer's fever signals the beginning of the "body's" struggle to mobilize antigens that will attack the invading pathogens in order to destroy them and save the host.

The obvious problem with this metaphor is that it defines human beings as inherently and contagiously destructive, the deadly carriers of a plague upon the earth. And the internal logic of the metaphor points toward only one possible cure: eliminate people from the face of the earth. As Mike Roselle, one of the leaders of Earth First!, a group espousing Deep Ecology, has said, "You hear about the death of nature and it's true, but nature will be able to reconstitute itself once the top of the food chain is lopped off — meaning us."

Some of those who adopt this story as their controlling metaphor are actually advocating a kind of war on the human race as a means of protecting the planet. They assume the role of antigens, to slow the spread of the disease, give the earth time to gather its forces to fight off and, if necessary, eliminate the intruders. In the words of Dave Foreman, a cofounder of Earth First!, "It's time for a warrior society to rise up out of the earth and throw itself in front of the juggernaut of destruction, to be antibodies against the human pox that's ravaging this precious, beautiful planet." (Some Deep Ecologists, it should be added, are more thoughtful.)

Beyond its moral unacceptability, another problem with this metaphor is its inability to explain — in a way that is either accurate or believable — who we are and how we can create solutions for the crisis it describes. Ironically, just as René Descartes, Francis Bacon, and the other architects of the scientific revolution defined human beings as disembodied intellects separate from the physical world, Arne Naess, the Norwegian philosopher who coined the term Deep Ecology in 1973, and many Deep Ecologists of today seem to define human beings as an alien presence on the earth. In a modern version of the Cartesian dénouement of a philosophical divorce between human beings and the earth, Deep Ecologists idealize a condition in which there is no connection between the two, but they arrive at their conclusion by means of a story that is curiously opposite to that of Descartes. Instead of seeing people as creatures of abstract thought relating to the earth only through logic and theory, the Deep Ecologists make the opposite mistake, of defining the relationship between human beings and the earth almost solely in physical terms — as if we were nothing more than humanoid bodies genetically programmed to play out our bubonic destiny, having no intellect or free will with which to understand and change the script we are following.

The Cartesian approach to the human story allows us to believe that we are separate from the earth, entitled to view it as nothing more than an inanimate collection of resources that we can exploit however we like; and this fundamental misperception has led us to our current crisis. But if the new story of the Deep Ecologists is dangerously wrong, it does at least provoke an essential question: What new story can explain the relationship between human civilization and the earth — and how we have come to a moment of such crisis? One part of the answer is clear: our new story must describe and foster the basis for a natural and healthy relationship between human beings and the earth. The old story of God's covenant with both the earth and humankind, and its assignment to human beings of the role of good stewards and faithful servants, was — before it was misinterpreted and twisted in the service of the Cartesian world view — a powerful, noble, and just explanation of who we are in relation to God's earth. What we need today is a fresh telling of our story with the distortions removed.

But a new story cannot be told until we understand how this crisis between human beings and the earth developed and how it can be resolved. To achieve such an understanding, we must consider the full implications of the Cartesian model of the disembodied intellect.

Feelings represent the essential link between mind and body or, to put it another way, the link between our intellect and the physical world. Because modern civilization assumes a profound separation between the two, we have found it necessary to create an elaborate set of cultural rules designed to encourage the fullest expression of thought while simultaneously stifling the expression of feelings and emotions.

Many of these cultural rules are now finally being recognized as badly out of balance with what we are learning about the foundations of human nature. One such foundation is, of course, the brain, which is layered with our evolutionary heritage. Between the most basic and primitive part of our brain, responsible for bodily functions and instinct, and the last major structure within the brain to evolve, the part responsible for abstract thought and known as the neocortex, is the huge portion of our brain that governs emotion, called the limbic system. In a very real sense, the idea that human beings can function as disembodied intellects translates into the absurd notion that the functions of the neocortex are the only workings of our brains that matter.

Yet abstract thought is but one dimension of awareness. Our feelings and emotions, our sensations, our awareness of our own bodies and of nature — all these are indispensable to the way we experience life, mentally and physically. To define the essence of who we are in terms that correspond with the analytical activity of the neocortex is to create an intolerable dilemma: How can we concentrate purely on abstract thinking when the rest of our brain floods our awareness with feelings, emotions, and instincts?

Insisting on the supremacy of the neocortex exacts a high price, because the unnatural task of a disembodied mind is to somehow ignore the intense psychic pain that comes from the constant nagging awareness of what is missing: the experience of living in one's body as a fully integrated physical and mental being. Life confronts everyone with personal or circumstantial problems, of course, and there are many varieties of psychic pain from which we wish to escape. But the cleavage between mind and body, intellect and nature, has created a kind of psychic pain at the very root of the modern mind, making it harder for anyone who is suffering from other psychological wounds to be healed.

Indeed, it is not unreasonable to suppose that members of a civilization that allows or encourages this cleavage will be relatively more vulnerable to those mental disorders characterized by a skewed relationship between thinking and feeling. This notion may seem improbable, since we are not used to looking for the cause of psychological problems in the broad patterns of modern civilization. But it is quite common for epidemiologists to trace the cause of physical disorders to patterns adopted by societies that place extra stress on especially vulnerable individuals. Consider, for example, how the pattern of modern civilization almost certainly explains the epidemic level of high blood pressure in those countries — like the United States — that have a diet very high in sodium. Although the precise causal relationship is still a mystery, epidemiologists conclude that the nearly ubiquitous tendency of modern civilization to add lots of salt to the food supply is responsible for a very high background level of hypertension. In the remaining pre-industrial cultures where the food supply is not processed and sodium consumption is low, hypertension is virtually unknown, and it is considered normal for an elderly man's blood pressure to be the same as that of an infant. In our society, we assume that it is natural for blood pressure to increase with age.

Resolving high blood pressure is much easier than resolving deep psychological conflicts, however. Most people respond to psychic pain the way they respond to any pain: rather than confront its source, they recoil from it, looking immediately for ways to escape or ignore it. One of the most effective strategies for ignoring psychic pain is to distract oneself from it, to do something so pleasurable or intense or otherwise absorbing that the pain is forgotten. As a temporary strategy, this kind of distraction isn't necessarily destructive, but dependence on it over the long term becomes dangerous and, finally, some sort of addiction. Indeed, it can be argued that every addiction is caused by an intense and continuing need for distraction from psychic pain. Addiction *is* distraction.

We are used to thinking of addiction in terms of drugs or alcohol. But new studies of addiction have deepened our understanding of the problem, and now we know that people can become addicted to many different patterns of behavior — such as gambling compulsively or working obsessively or even watching television constantly — that distract them from having to experience directly whatever they are trying to avoid. Anyone who is unusually fearful of something — intimacy, failure, loneliness — is potentially vulnerable to addiction, because psychic pain causes a feverish hunger for distraction.

The cleavage in the modern world between mind and body, man and nature, has created a new kind of addiction: I believe that our civilization is, in effect, addicted to the consumption of the earth itself. This addictive relationship distracts us from the pain of what we have lost: a direct experience of our connection to the vividness, vibrancy, and aliveness of the rest of the natural world. The froth and frenzy of industrial civilization mask our deep loneliness for that communion with the world that can lift our spirits and fill our senses with the richness and immediacy of life itself.

We may pretend not to notice the emptiness we feel, but its effects may be seen in the unnatural volatility with which we react to those things we touch. I can best illustrate this point with a metaphor drawn from electrical engineering. A machine using lots of electrical energy must be grounded to the earth in order to stabilize the flow of electricity through the machine and to prevent a volatile current from jumping to whatever might touch it. A machine that is not grounded poses a serious threat; similarly, a person who is not "grounded" in body as well as mind, in feelings as well as thoughts, can pose a threat to whatever he or she touches. We tend to think of the powerful currents of creative energy circulating through every one of us as benign, but they can be volatile and dangerous if not properly grounded. This is especially true of those suffering from a serious addiction. No longer grounded to the deeper meaning of their lives, addicts are like someone who cannot release a 600-volt cable because the electric current is just too strong: they hold tightly to their addiction even as the life force courses out of their veins.

In a similar way, our civilization is holding ever more tightly to its habit of consuming larger and larger quantities every year of coal, oil, fresh air and water, trees, topsoil, and the thousand other substances we rip from the crust of the earth, transforming them into not just the sustenance and shelter we need but much more that we don't need: huge quantities of pollution, products for which we spend billions on advertising to convince ourselves we want, massive surpluses of products that depress prices while the products themselves go to waste, and diversions and distractions of every kind. We seem increasingly eager to lose ourselves in the forms of culture, society, technology, the media, and the rituals of production and consumption, but the price we pay is the loss of our spiritual lives.

Evidence of this spiritual loss abounds. Mental illness in its many forms is at epidemic levels, especially among children. The three leading causes of death among adolescents are drug- and alcohol-related accidents, suicide, and homicide. Shopping is now recognized as a recreational activity. The accumulation of material goods is at an all-time high, but so is the number of people who feel an emptiness in their lives.

Industrial civilization's great engines of distraction still seduce us with a promise of fulfillment. Our new power to work our will upon the world can bring with it a sudden rush of exhilaration, not unlike the momentary "rush" experienced by drug addicts when a drug injected into their bloodstream triggers changes in the chemistry of the brain. But that exhilaration is fleeting; it is not true fulfillment. And the metaphor of drug addiction applies in another way too. Over time, a drug user needs a progressively larger dose to produce an equivalent level of exhilaration; similarly, our civilization seems to require an ever-increasing level of consumption. But why do we assume that it's natural and normal for our per capita consumption of most natural resources to increase every year? Do we need higher levels of consumption to achieve the same distracting effect once produced by a small amount of consumption? In our public debates about efforts to acquire a new and awesome power through science, technology, or industry, are we sometimes less interested in a careful balancing of the pros and cons than in the great thrill sure to accompany the first use of the new enhancement of human power over the earth?

The false promise at the core of addiction is the possibility of experiencing the vividness and immediacy of real life without having to face the fear and pain that are also part of it. Our industrial civilization makes us a similar promise: the pursuit of happiness and comfort is paramount, and the consumption of an endless stream of shiny new products is encouraged as the best way to succeed in that pursuit. The glittering promise of easy fulfillment is so seductive that we become willing, even relieved, to forget what we really feel and abandon the search for authentic purpose and meaning in our lives.

But the promise is always false because the hunger for authenticity remains. In a healthy, balanced life, the noisy chatter of our discourse with the artificial world of our creation may distract us from the deeper rhythms of life, but it does not interrupt them. In the pathology of addiction, this dialogue becomes more than a noisy diversion; as their lives move further out of balance, addicts invest increasing amounts of energy in their relationship to the objects of their addiction. And once addicts focus on false communion with substitutes for life, the rhythm of their dull and deadening routine becomes increasingly incompatible, discordant, and dissonant with the natural harmony that entrains the music of life. As the dissonance grows more violent and the clashes more frequent, peaks of disharmony become manifest in successive crises, each one more destructive than the last.

The disharmony in our relationship to the earth, which stems in part from our addiction to a pattern of consuming ever-larger quantities of the resources of the earth, is now manifest in successive crises, each marking a more destructive clash between our civilization and the natural world: whereas all threats to the environment used to be local and regional, several are now strategic. The loss of one and a half acres of rain forest every second, the sudden, thousandfold acceleration of the natural extinction rate for living species, the ozone hole above Antarctica and the thinning of the ozone layer at all latitudes, the possible destruction of the climate balance that makes our earth livable — all these suggest the increasingly violent collision between human civilization and the natural world.

Many people seem to be largely oblivious of this collision and the addictive nature of our unhealthy relationship to the earth. But education is a cure for those who lack knowledge; much more worrisome are those who will not acknowledge these destructive patterns. Indeed, many political, business, and intellectual leaders deny the existence of any such patterns in aggressive and dismissive tones. They serve as "enablers," removing inconvenient obstacles and helping to ensure that the addictive behavior continues.

The psychological mechanism of denial is complex, but again addiction serves as a model. Denial is the strategy used by those who wish to believe that they can continue their addicted lives with no ill effects for themselves or others. Alcoholics, for example, aggressively dismiss suggestions that their relationship to alcohol is wreaking havoc in their lives; repeated automobile crashes involving the same drunk driver are explained away in an alcoholic's mind as isolated accidents, each with a separate, unrelated cause.

Thus the essence of denial is the inner need of addicts not to allow themselves to perceive a connection between their addictive behavior and its destructive consequences. This need to deny is often very powerful. If addicts recognize their addiction, they might be forced to become aware of the feelings and thoughts from which they so desperately need distraction; abandoning their addiction altogether would threaten them with the loss of their principal shield against the fear of confronting whatever they are urgently trying to hold at bay.

Some theorists argue that what many addicts are trying to hold at bay is a profound sense of powerlessness. Addicts often display an obsessive need for absolute control over those few things that satisfy their craving. This need derives from, and is inversely proportional to, the sense of helplessness they feel toward the real world — whose spontaneity and resistance to their efforts at control are threatening beyond their capacity to endure.

It is important to recognize that this psychological drama takes place at the border of conscious awareness. Indeed, it is precisely that border which is being defended against the insistent intrusions of reality. Meanwhile, the dishonesty required to ensure that reality doesn't breach the ramparts often assumes such proportions that friends find it hard to believe that addicts don't know what they are doing to themselves and those around them. But the inauthenticity of addicts is, in one sense, easy to explain: they are so obsessed with the need to satisfy their craving that they subordinate all other values to it. Since a true understanding of their behavior might prove inhibiting, they insist they have no problem.

We are insensitive to our destructive impact on the earth for much the same reason, and we consequently have a similar and very powerful need for denial. Denial can take frightening and bizarre forms. For example, in southern California in 1991, the worsening five-year drought led some homeowners to actually spray-paint their dead lawns green, just as some undertakers apply cosmetics to make a corpse look natural to viewers who are emotionally vulnerable to the realization of death. As Joseph Conrad said in *The Heart of Darkness*, "The conquest of the earth is not a pretty thing when you look into it too much." But we are addicted to that conquest, and so we deny it is ugly and destructive. We elaborately justify what we are doing while turning a blind eye to the consequences. We are hostile to the messengers who warn us that we have to change, suspecting them of subversive intent and accusing them of harboring some hidden agenda — Marxism, or statism, or anarchism. ("Killing the messenger," in fact, is a well-established form of denial.) We see no relationship between the increasingly dangerous crises we are causing in the natural world; they are all accidents with separate, distinct causes. Those dead lawns, for example — could they be related to the fast-burning fires that made thousands homeless late in 1991? No matter; we are certain that we can adapt to whatever damage is done, even though the increasingly frequent manifestations of catastrophe are beginning to resemble what the humorist A. Whitney Brown describes as "a nature hike through the Book of Revelations."

The bulwark of denial isn't always impenetrable, however. In the advanced stages of addiction, when the destructive nature of the pattern becomes so overwhelmingly obvious that addicts find it increasingly difficult to ignore the need for change, a sense of resignation sets in. The addiction has by then so thoroughly defined the pattern of their lives that there seems to be no way out. Similarly, some people are finding it increasingly difficult to deny the destructive nature of our relationship to the earth, yet the response is not action but resignation. It's too late, we think; there's no way out.

But that way spells disaster, and recovery is possible. With addiction, an essential element in recovery is a willingness on the part of addicts to honestly confront the real pain they have sought to avoid. Rather than distracting their inner awareness through behavior, addicts must learn to face their pain — feel it, think it, absorb it, own it. Only then can they begin to deal honestly with it instead of running away.

So too our relationship to the earth may never be healed until we are willing to stop denying the destructive nature of the current pattern. Our seemingly compulsive need to control the natural world may have derived from a feeling of helplessness in the face of our deep and ancient fear of "Nature red in tooth and claw," but this compulsion has driven us to the edge of disaster, for we have become so successful at controlling nature that we have lost our connection to it. And we must also recognize that a new fear is now deepening our addiction: even as we revel in our success at controlling nature, we have become increasingly frightened of the consequences, and that fear only drives us to ride this destructive cycle harder and faster.

What I have called our addictive pattern of behavior is only part of the story, however, because it cannot explain the full complexity and ferocity of our assault on the earth. Nor does it explain how so many thinking and caring people can unwittingly cooperate in doing such enormous damage to the global environment and how they can continue to live by the same set of false assumptions about what their civilization is actually doing and why. Clearly, the problem involves more than the way each of us as an individual relates to the earth. It involves something that has gone terribly wrong in the way we collectively determine our mutual relationship to the earth.

A metaphor can be a valuable aid to understanding, and several metaphors have helped me understand what is wrong with the way we relate to the earth. One that has proved especially illuminating comes out of a relatively new theory about ailing families; a synthesis by psychologists and sociologists of research in addiction theory, family therapy, and systems analysis, this theory attempts to explain the workings of what has come to be called the dysfunctional family.

The idea of the dysfunctional family was first developed by theorists such as R. D. Laing, Virginia Satir, Gregory Bateson, Milton Erickson, Murray Bowen, Nathan Ackerman, and Alice Miller, and more recently it has been refined and brought to a popular audience by writers like John Bradshaw. The problem they have all sought to explain is how families made up of well-meaning, seemingly normal individuals can engender destructive relationships among themselves, driving individual family members as well as their family system into crisis.

According to the theory of dysfunctionality, unwritten rules governing how to raise children and purporting to determine what it means to be a human being are passed down from one generation of a family to the next. The modern version of these rules was shaped by the same philosophical world view that led to the scientific and technological revolution: it defines human beings as primarily intellectual entities detached from the physical world. And this definition led in turn to an assumption that feelings and emotions should be suppressed and subordinated to pure thought.

One consequence of this scientific view was a changed understanding of God. Once it became clear that science — instead of divine provenance — might explain many of nature's mysteries, it seemed safe to assume that the creator, having set the natural world in motion within discernible and predictable patterns, was somewhat removed and detached from the world, out there above us looking down. Perhaps as a consequence, the perception of families changed too. Families came to be seen as Ptolemaic systems, with the father as the patriarch and source of authority and all the other family members orbiting around him. This change had a dramatic effect on children. Before the scientific era, children almost certainly found it easier to locate and understand their place in the world· because they could define themselves in relation both to their parents and to a God who was clearly present in nature. With these two firm points of reference, children were less likely to lose their direction in life. But with God receding from the natural world to an abstract place, the patriarchal figure in the family (almost always the father) effectively became God's viceroy, entitled to exercise godlike authority when enforcing the family's rules. As some fathers inevitably began to insist on being the sole source of authority, their children became confused about their own roles in a family system that was severely stressed by the demands of the dominant, all-powerful father.

Parents were accorded godlike authority to enforce the rules, and, as Bradshaw and others argue, one of the most basic rules that emerged is that the rules themselves cannot be questioned. One of the ways dysfunctional families enforce adherence to the rules and foster the psychic numbness on which they depend is by teaching the separation between mind and body and suppressing the feelings and emotions that might otherwise undermine the rules. Similarly, one of the ways our civilization secures adherence to its rules is by teaching the separation of people from the natural world and suppressing the emotions that might allow us to feel the absence of our connection to the earth.

The rules of both perpetuate the separation of thought from

feeling and require full acceptance of the shared, unspoken lies that all agree to live. Both encourage people to accept that it is normal not to know their feelings and to feel helpless when it comes to any thought of challenging or attempting to change the assumptions and rules upon which the divorce from feeling is based. As a result, these rules frequently encouraged psychological dramas and role playing. Rules that are simultaneously unreasonable and immune to questioning can perpetuate disorders like addiction, child abuse, and some forms of depression. This is the paradigm of the dysfunctional family.

It is not uncommon for one member of a dysfunctional family to exhibit symptoms of a serious psychological disorder that will be found, upon scrutiny, to be the outward manifestation of a pattern of dysfunctionality that includes the entire family. In order to heal the patient, therapists concentrate not on the pathology of the individual but on the web of family relationships — and the unwritten rules and understandings that guide his approach to those relationships.

For example, it has long been known that the vast majority of child abusers were themselves abused as children. In analyzing this phenomenon, theorists have found the blueprint for an archetypal intergenerational pattern: the child who is a victim remembers the intensity of the experience with his body but suppresses the memory of the pain in his mind. In a vain effort to resolve his deep confusion about what happened, he is driven to repeat or "recapitulate" the drama in which a powerful older person abuses a powerless child, only this time he plays the abuser's role.

To take a more subtle example, discussed in Alice Miller's seminal work on dysfunctionality, *The Drama of the Gifted Child,* children in some families are deprived of the unconditional love essential for normal development and made to feel that something inside them is missing. Consequently these children develop a low opinion of themselves and begin to look constantly to others for the approval and validation they so desperately need. The new term "codependency" describes the reliance on another for validation and positive feelings about oneself. The energy fueling this insatiable search continues into adulthood, frequently causing addictive behavior and an approach to relationships that might be described, in the words of the popular song, as "looking for love in all the wrong places." Sadly but almost inevitably, when they themselves have children, they find in the emotional hunger of their own infant a source of intense and undiluted attention that they use to satisfy their still insatiable desire for validation and approval, in a pattern that emphasizes taking rather than giving love. In the process, they neglect to give their own child the unconditional love the child needs to feel emotionally whole and complete. The child therefore develops the same sense that something is missing inside and seeks it in the faces and emotions of others, often insatiably. Thus the cycle continues.

The theory of how families become dysfunctional usually does not require identifying any particular family member as bad or as someone intent upon consciously harming the others. Rather, it is usually the learned pattern of family rules that represents the real source of the pain and tragedy the family members experience in each generation. As a diagnosis, dysfunctionality offers a powerful source of hope, because it identifies the roots of the problems in relationships rather than in individuals, in a shared way of thinking based on inherited assumptions rather than a shared human nature based on inherited destiny. It is therefore subject to healing and transformation.

That's the good news. The bad news is that many dysfunctional rules internalized during infancy and early childhood are extremely difficult to displace. Human evolution, of course, is responsible for our very long period of childhood, during much of which we are almost completely dependent on our parents. As Ashley Montagu first pointed out decades ago, evolution encouraged the development of larger and larger human brains, but our origins in the primate family placed a limit on the ability of the birth canal to accommodate babies with ever-larger heads. Nature's solution was to encourage an extremely long period of dependence on the nurturing parent during infancy and childhood, allowing both mind and body to continue developing in an almost gestational way long after birth. But as a result of this long period of social and psychological development, children are extremely vulnerable to both good and bad influences, and in a dysfunctional family, that means they will absorb and integrate the dysfunctional rules and warped assumptions about life that are being transmitted by the parents. And since much of what parents transmit are the lessons learned during their own childhood, these rules can persist through many generations.

Every culture is like a huge extended family, and perhaps nothing more determines a culture's distinct character than the rules and assumptions about life. In the modern culture of the West, the assumptions about life we are taught as infants are heavily influenced by our Cartesian world view — namely, that human beings should be separate from the earth, just as the mind should be separate from the body, and that nature is to be subdued, just as feelings are to be suppressed. To a greater or lesser degree, these rules are conveyed to all of us, and they have powerful effects on our perception of who we are.

The model of the dysfunctional family has a direct bearing on our ways of thinking about the environment. But this model also helps describe how we have managed to create such a profound and dangerous crisis in our relationship to the environment, why this crisis is not due to our inherently evil or pathogenic qualities, and how we can heal this relationship. As the use of this metaphor suggests, however, the environmental crisis is now so serious that I believe our civilization must be considered in some basic way dysfunctional.

Like the rules of a dysfunctional family, the unwritten rules that govern our relationship to the environment have been passed down from one generation to the next since the time of Descartes, Bacon, and the other pioneers of the scientific revolution some 375 years ago. We have absorbed these rules and lived by them for centuries without seriously questioning them. As in a dysfunctional family, one of the rules in a dysfunctional civilization is that you don't question the rules.

There is a powerful psychological reason that the rules go unquestioned in a dysfunctional family. Infants or developing children are so completely dependent that they cannot afford even to think there is something wrong with the parent, even if the rules do not feel right or make sense. Since children cannot bear to identify the all-powerful parent as the source of dysfunctionality, they assume that the problem is within themselves. This is the crucial moment

when the inner psychological wound is inflicted — and it is a self-inflicted wound, a fundamental loss of faith by the children in themselves. The pain of that wound often lasts an entire lifetime, and the emptiness and alienation that result can give rise to enormous amounts of psychological energy, expended during the critical period when the psyche is formed in an insatiable search for what, sadly, can never be found: unconditional love and acceptance.

Just as children cannot reject their parents, each new generation in our civilization now feels utterly dependent on the civilization itself. The food on the supermarket shelves, the water in the faucets in our homes, the shelter and sustenance, the clothing and purposeful work, our entertainment, even our identity — all these our civilization provides, and we dare not even think about separating ourselves from such beneficence.

To carry the metaphor further: just as children blame themselves as the cause of the family's dysfunction in their relationship with it, so we quietly internalize the blame for our civilization's failure to provide a feeling of community and a shared sense of purpose in life. Many who feel their lives have no meaning and feel an inexplicable emptiness and alienation simply assume that they themselves are to blame, and that something is wrong with them.

Ironically, it is our very separation from the physical world that creates much of this pain, and it is because we are taught to live so separately from nature that we feel so utterly dependent upon our civilization, which has seemingly taken nature's place in meeting all our needs. Just as the children in a dysfunctional family experience pain when their parent leads them to believe that something important is missing from their psyches, we surely experience a painful loss when we are led to believe that the connection to the natural world that is part of our birthright as a species is something unnatural, something to be rejected as a rite of passage into the civilized world. As a result, we internalize the pain of our lost sense of connection to the natural world, we consume the earth and its resources as a way to distract ourselves from the pain, and we search insatiably for artificial substitutes to replace the experience of communion with the world that has been taken from us.

Children in dysfunctional families who feel shame often construct a false self through which they relate to others. This false self can be quite elaborate as the children constantly refine the impression it makes on others by carefully gauging their reactions, to make the inauthentic appear authentic. Similarly, we have constructed in our civilization a false world of plastic flowers and AstroTurf, air conditioning and fluorescent lights, windows that don't open and background music that never stops, days when we don't know whether it has rained or not, nights when the sky never stops glowing, Walkman and Watchman, entertainment cocoons, frozen food for the microwave oven, sleepy hearts jump-started with caffeine, alcohol, drugs, and illusions.

In our frenzied destruction of the natural world and our apparent obsession with inauthentic substitutes for direct experience with real life, we are playing out a script passed on to us by our forebears. However, just as the unwritten rules in a dysfunctional family create and maintain a conspiracy of silence about the rules themselves, even as the family is driven toward successive crises, many of the unwritten rules of our dysfunctional civilization encourage silent acquiescence in our patterns of destructive behavior toward the natural world.

The idea of a dysfunctional civilization is by no means merely a theoretical construct. In this terrible century, after all, we have witnessed some especially malignant examples of dysfunctional civilization: the totalitarian societies of Nazi Germany under Hitler, fascist Italy under Mussolini, Soviet communism under Stalin and his heirs, and the Chinese communism of Mao Zedong and Deng Xaoping, as well as many less infamous versions of the same phenomenon. Indeed, only recently the world community mobilized a coalition of armies to face down the Baathist totalitarianism of Iraq under Saddam Hussein.

Each of these dysfunctional societies has lacked the internal validation that can only come from the freely expressed consent of the governed. Each has demonstrated an insatiable need to thrust itself and its political philosophy onto neighboring societies. Each has been oriented toward expansion through the forceful takeover of other countries. Moreover, each has fostered in its society a seamless web of shared assumptions that most people know are false but that no one dares to question. These societies reflect in macrocosm the pathology of dysfunctionality as it has been observed in families. A developing child in a dysfunctional family searches his parent's face for signals that he is whole and all is right with the world; when he finds no such approval, he begins to feel that something is wrong inside. And because he doubts his worth and authenticity, he begins controlling his inner experience — smothering spontaneity, masking emotion, diverting creativity into robotic routine, and distracting an awareness of all he is missing with an unconvincing replica of what he might have been. Similarly, when the leadership in a totalitarian society dares to look in the faces of its people for signals of what they really feel, it is seldom reassured that all is right with the world. On the contrary, the leadership begins to fear that something is wrong because its people do not — cannot — freely express the consent of the governed. They stare back, trancelike, their vacant sullenness suggesting the uneasiness and apprehension that is so pervasive among oppressed populations everywhere. Denied validation in the countenance of its citizens, the totalitarian leadership feels no choice but to try to expand, out of an insatiable ambition to find — by imposing itself on others —conclusive evidence of its inner value.

Typically, the totalitarian expansion begins with the takeover of a weak and relatively defenseless neighboring society. Hoping that this initial conquest will satiate the aggressor, other societies frequently mute their response, some because they fear they might be the next targets, others because they are sure they will not be. But if the totalitarian society is deeply dysfunctional, it will not be satisfied for long and will continue to feel a need to expand. Alas, this horrifying pattern is all too familiar: totalitarian expansions have directly caused the deaths of more than 100 million human beings in this century.

The phenomenon of modern totalitarianism is, of course, extremely complex and involves political, economic, and historical factors unique to each of its incarnations. But whatever its specific causes, the psychology of totalitarianism has always been characterized by a fear of disorientation within and a search for legitimacy without. The pathology of expansion so evident in modern totalitarian societies results from this dysfunctional pattern, and the sense of wholeness they seek cannot be restored as long as they refuse to confront the dishonesty, fear, and violence eating away at the heart of their national identity.

The unprecedented assault on the natural world by our global civilization is also extremely complex, and many of its causes are related specifically to the geographic and historical context of its many points of attack. But in psychological terms, our rapid and aggressive expansion into what remains of the wildness of the earth represents an effort to plunder from outside civilization what we cannot find inside. Our insatiable drive to rummage deep beneath the surface of the earth, remove all of the coal, petroleum, and other fossil fuels we can find, then burn them as quickly as they are found — in the process filling the atmosphere with carbon dioxide and other pollutants — is a willful expansion of our dysfunctional civilization into vulnerable parts of the natural world. And the destruction by industrial civilization of most of the rain forests and old-growth forests is a particularly frightening example of our aggressive expansion beyond proper boundaries, an insatiable drive to find outside solutions to problems arising from a dysfunctional pattern within.

Ironically, Ethiopia, the first victim of modern totalitarian expansion, has also been an early victim of the dysfunctional pattern that has led to our assault on the natural world. At the end of World War II, after the Italian fascists had been forced out, 40 percent of Ethiopia's land was covered with, and protected by, trees. Less than a half century later, after decades marked by the most rapid population growth in the world, a relentless search for fuelwood, overgrazing, and the export of wood to pay interest on debts, *less than 1 percent* of Ethiopia is covered by trees. First, much of the topsoil washed away; then the droughts came — and stayed. The millions who have starved to death are, in a real sense, victims of our dysfunctional civilization's expansionist tendencies.

In studying the prospects for halting our destructive expansion, one is almost awestruck by our relentless and seemingly compulsive drive to dominate every part of the earth. Always, the unmet needs of civilization fuel the engine of aggression; never can these needs be truly satisfied. The invaded area is laid waste, its natural productivity is eviscerated, its resources are looted and quickly consumed

Starving children like this one in Bangladesh often have distended stomachs because severe deprivation of protein causes fluids to be produced as organs deteriorate. An average of 37,000 children under the age of five die every day from starvation, diarrhea, and easily preventable diseases.

— and all this destruction merely stokes our appetite for still more.

The weakest and most helpless members of the dysfunctional family become the victims of abuse at the hands of those responsible for providing nurture. In a similar fashion, we systematically abuse the most vulnerable and least defended areas of the natural world: the wetlands, the rain forests, the oceans. We also abuse other members of the human family, especially those who cannot speak for themselves. We tolerate the theft of land from indigenous peoples, the exploitation of areas inhabited by the poorest populations, and — worst of all — the violation of the rights of those who will come after us. As we strip-mine the earth at a completely unsustainable rate, we are making it impossible for our children's children to have a standard of living even remotely similar to ours.

In philosophical terms, the future is, after all, a vulnerable and developing present, and unsustainable development is therefore what might be called a form of "future abuse." Like a parent violating the personal boundaries of a vulnerable child, we violate the temporal boundaries of our rightful place in the chain of human generations. After all, the men and women of every generation must share the same earth — the only earth we have — and so we also share a responsibility to ensure that what one generation calls the future will be able to mature safely into what another generation will call the present. We are now, in effect, corruptly imposing our own dysfunctional design and discordant rhythms on future generations, and these persistent burdens will be terribly difficult to carry.

Police officers, doctors, and psychologists who deal with the victims of child sexual abuse often wonder how any adult — especially a parent — could commit such a crime. How could anyone be deaf to the screams, blind to the grief, and numb to the pain their actions cause? The answer, we now know, is that a kind of psychic numbness, induced by the adults' own adaptation to the dysfunctional pattern in which they were themselves raised as children, serves to anesthetize their conscience and awareness in order to facilitate their compulsive repetition of the crime that was visited upon them.

Just as the members of a dysfunctional family emotionally anesthetize themselves against the pain they would otherwise feel, our dysfunctional civilization has developed a numbness that prevents us from feeling the pain of our alienation from our world. Both the dysfunctional family and our dysfunctional civilization abhor direct contact with the full and honest experience of life. Both keep individuals in a seamless web of abstract, unfeeling thought, focused always on others, what others are assumed to be experiencing and what others might say or do to provide the wholeness and validation so desperately sought.

But there is a way out. A pattern of dysfunctionality need not persist indefinitely, and the key to change is the harsh light of truth. Just as an addict can confront his addiction, just as a dysfunctional family can confront the unwritten rules that govern their lives, our civilization can change — must change — by confronting the unwritten rules that are driving us to destroy the earth. And, as Alice Miller and other experts have shown, the act of mourning the original loss while fully and consciously feeling the pain it has caused can heal the wound and free the victim from further enslavement. Likewise, if the global environmental crisis is rooted in the dysfunctional pattern of our civilization's relationship to the natural world, confronting and fully understanding that pattern, and recognizing its destructive impact on the environment and on us, is the first step toward mourning what we have lost, healing the damage we have done to the earth and to our civilization, and coming to terms with the new story of what it means to be a steward of the earth.

13

Environmentalism of the Spirit

Twenty years ago, E. F. Schumacher defined an important new issue arising from the relationship between a technology and the context — social, cultural, political, and ecological — in which it is used. For example, a nuclear power plant can certainly generate a lot of electricity, but it may not be an "appropriate" technology for an underdeveloped nation with an unstable government, a shortage of trained engineers, an absence of any power grid to distribute the electricity generated, and a megalomaniacal ruler anxious to acquire fissionable material with which to construct nuclear weapons. The appropriateness of a technology becomes increasingly important as its power grows and its potential for destroying the environment expands.

It is time we asked a similar question about ourselves and our relationship to the global environment: When giving us dominion over the earth, did God choose an appropriate technology?

Knowing what we do about our new power as a species to interfere with and even overwhelm the earth's natural systems and recognizing that we are now doing so with reckless abandon, one is tempted to answer, the jury is still out.

Whether we believe that our dominion derives from God or from our own ambition, there is little doubt that the way we currently relate to the environment is wildly inappropriate. But in order to change, we have to address some fundamental questions about our purpose in life, our capacity to direct the powerful inner forces that have created this crisis, and who we are. These questions go beyond any discussion of whether the human species is an appropriate technology; these questions are not for the mind or the body but the spirit.

A change in our essential character is not possible without a realistic hope that we can make change happen. But hope itself is threatened by the realization that we are now capable of destroying ourselves and the earth's environment. Moreover, the stress of coping with the complicated artificial patterns of our lives and the flood of manufactured information creates a pervasive feeling of exhaustion just when we have an urgent need for creativity. Our economy is described as post-industrial; our architecture is called post-modern; our geopolitics are labeled post–Cold War. We know what we are not, but we don't seem to know what we are. The forces that shape and reshape our lives seem to have an immutable logic of their own; they seem so powerful that any effort to define ourselves creatively will probably be wasted, its results quickly erased by successive tidal waves of change. Inevitably, we resign ourselves to whatever fate these powerful forces are propelling us toward, a fate we have little role in choosing.

Perhaps because it is unprecedented, the environmental crisis seems completely beyond our understanding and outside of what we call common sense. We consign it to some seldom visited attic in our minds where we place ideas that we vaguely understand but rarely explore. We tag it with the same mental labels we might use for Antarctica: remote, alien, hopelessly distorted by the maps of the world we inhabit, too hard to get to and too unforgiving to stay very long. When we do visit this attic, when we learn about how intricately the causes of the crisis are woven into the fabric of industrial civilization, our hope of solving it seems chimerical. It seems so forbidding that we resist taking even the first steps toward positive change.

We turn by default to an imprudent hope that we can adapt to whatever changes are in store. We have grown accustomed to adapting; we are good at it. After all, we have long since adapted, with the help of technology, to every climate extreme on the surface of the earth, at the bottom of the sea and even in the vacuum of space. It is by adapting, in fact, that we have extended our dominion into every corner of the earth. And so it is tempting to conclude that this familiar strategy is the obvious response to our rapidly emerging dilemma.

But the magnitude of the change to which we must now consider adapting is so large that the proposals quickly trend toward the absurd. A study sponsored by the National Academy of Sciences, for example, suggested that as the earth warms, we might create huge corridors of wilderness as pathways to accommodate all of the species trying to migrate from south to north in search of a familiar climate. (Meanwhile, of course, we are laying siege to many of the wilderness areas that already exist — in the Pacific Northwest, for example — in search of timber and other resources.) Some even imagine that genetic engineering will soon magnify our power to adapt even our physical form. We might decide to extend our dominion of nature into the human gene pool, not just to cure terrible diseases, but to take from God and nature the selection of genetic variety and robustness that gives our species its resilience and aligns us with the natural rhythms in the web of life. Once again, we might dare to exercise godlike powers unaccompanied by godlike wisdom.

But our willingness to adapt is an important part of the underlying problem. Do we have so much faith in our own adaptability that we will risk destroying the integrity of the entire global ecological system? If we try to adapt to the changes we are causing rather than prevent them in the first place, have we made an appropriate choice? Can we understand how much destruction this choice might finally cause?

Believing that we can adapt to just about anything is ultimately a kind of laziness, an arrogant faith in our ability to react in time to save our skin. But in my view this confidence in our quick reflexes is badly misplaced; indeed, a laziness in our spirit has estranged us from our true selves and from the quickness and vitality of the world at large. We have been so seduced by industrial civilization's promise to make our lives comfortable that we allow the synthetic routines of modern life to soothe us in an inauthentic world of our own making. Life can be easy, we assure ourselves. We need not suffer the heat or the cold; we need not sow or reap or hunt and gather. We can heal the sick, fly through the air, light up the darkness, and be entertained in our living rooms by orchestras and clowns whenever we like. And as our needs and whims are sated, we watch electronic images of nature's destruction, distant famine, and apocalyptic warnings, all with the bone-weariness of the damned. "What can we do?" we ask ourselves, already convinced that the realistic answer is nothing.

With the future so open to doubt, we routinely choose to indulge our own generation at the expense of all who will follow. We enshrine the self as the unit of ethical account, separate and distinct not just from the natural world but even from a sense of obligation to others — not just others in future generations, but increasingly even to others in the same generation; and not just those in distant lands, but increasingly even in our own communities. We do this not because we don't care but because we don't really live in our lives. We are monumentally distracted by a pervasive technological culture that appears to have a life of its own, one that insists on our full attention, continually seducing us and pulling us away from the opportunity to experience directly the true meaning of our own lives.

How can we shake loose this distraction? How can we direct our attention to more important matters when our attention has become a commodity to be bought and sold? Whenever a new source of human interest and desire is found, prospectors flock to stake their claim. Using every available tool — newspapers, movies, television, magazines, billboards, blimps, buttons, designer labels, junk faxes — they assault our attention from every side. Advertisers strip-mine it; politicians covet it; pollsters measure it; terrorists steal it as a weapon of war. As the amounts close to the surface are exhausted, the search for fresh supplies leads onto primal paths that run deep into our being, back through our evolutionary heritage, past thought and beyond emotion, to instinct — and a rich vein of primal fears and passions that are also now exploited as raw material in the colossal enterprise of mass distraction. The prospectors of attention fragment our experience of the world, carry away the spoils, and then, in an ultimate irony, accuse us of having short attention spans.

The way we experience the world is governed by a kind of inner ecology that relates perception, emotions, thinking, and choices to forces outside ourselves. We interpret our experience through multiple lenses that focus — and distort — the information we receive through our senses. But this ecology now threatens to fall badly out of balance because the cumulative impact of the changes brought by the scientific and technological revolution are potentially devastating to our sense of who we are and what our purpose in life might be. Indeed, it may now be necessary to foster a new "environmentalism of the spirit." How do we, for example, conserve hope and minimize the quantity of corrosive fear we spill into our lives? How do we recycle the sense of wonder we felt as children, when the world was new? How do we use the power of technology without adapting to it so completely that we ourselves behave like machines, lost in the levers and cogs, lonesome for the love of life, hungry for the thrill of directly experiencing the vivid intensity of the ever-changing moment?

No wonder we have become disconnected from the natural world — indeed, it's remarkable we feel any connection to ourselves. And no wonder we have become resigned to the idea of a world without a future. The engines of distraction are gradually destroying the inner ecology of the human experience. Essential to that ecology is the balance between respect for the past and faith in the future, between a belief in the individual and a commitment to the community, between our love for the world and our fear of losing it — the balance, in other words, on which an environmentalism of the spirit depends.

To some, the global environmental crisis is primarily a crisis of values. In this view, the basic cause of the problem is that we as a civilization base our decisions about how to relate to the environment on premises that are fundamentally unethical. And since religion has traditionally been the most powerful source of ethical guidance for our civilization, the search for villains has led to the doorstep of the major religious systems.

Here in the West, some have charged — inaccurately, I believe — that the Judeo-Christian tradition chartered the relentless march of civilization to dominate nature, beginning with the creation story of Genesis, in which humankind is granted "dominion" over the earth. In its basic form, the charge is that our tradition assigns divine purpose to our exercise of virtually complete power to work our will over nature. It is alleged that by endowing human beings with a completely unique relationship to God and then delegating God's authority over nature to human beings, the tradition sanctions as ethical all choices that put a higher priority on human needs and desires than on the rest of nature. Simply put, according to this view, it is "ethical" to make sure that whenever nature gets in the way of what we want, nature loses.

But this is a cartoon version of the Judeo-Christian tradition, one that bears little resemblance to the reality. Critics attack religion for inspiring an arrogant and reckless attitude toward nature, but they have not always read the relevant texts carefully enough. Although it is certainly true that our civilization is built on the premise that we can use nature for our own ends without regard to the impact we have on it, it is not fair to charge any of the major world religions with promoting this dangerous attitude. Indeed, all of them mandate an ethical responsibility to protect and care for the well-being of the natural world.

In the Judeo-Christian tradition, the biblical concept of dominion is quite different from the concept of domination, and the difference is crucial. Specifically, followers of this tradition are charged with the duty of stewardship, because the same biblical passage that grants them "dominion" also requires them to "care for" the earth even as they "work" it. The requirement of stewardship and its grant of dominion are not in conflict; in recognizing the sacredness of creation, believers are called upon to remember that even as they "till" the earth they must also "keep" it.

This has long been clear to those who have dedicated their lives to these duties. Richard Cartwright Austin, for example, a Presbyterian minister working among the poor in Appalachia, reports on his experience in trying to stop irresponsible strip mining: "I learned early on from my years as a pastor in Appalachia and from the days when I started fighting strip mining in southwest Virginia that the only defense those mountains have from exploitation by the energy conglomerates' bulldozers is the poor, isolated people who live in those hollows, who care so deeply that they would fight for that land. Take those people away and the mountains are totally defenseless. . . . From the biblical point of view, nature is only safe from pollution and brought into a secure moral relationship when it is united with people who love it and care for it."

All around the world, the efforts to stop the destruction of the environment have come mainly from people who recognize the damage being done in that part of the world in which they themselves have "dominion." Lois Gibbs and the other homeowners at Love Canal, Christine and Woodrow Sterling and their family, whose well water was poisoned in West Tennessee, "Harrison" Gnau and the indigenous peoples of the Sarawak rain forest in East Malaysia, Chico Mendes and his rubber tappers in the Amazon, the unemployed fishermen of the Aral Sea — all began their battles to save the environment because of the marriage of dominion and stewardship in their hearts. This is precisely the relationship between humankind and the earth called for in the Judeo-Christian ethic.

In my own religious experience and training — I am a Baptist — the duty to care for the earth is rooted in the fundamental relationship between God, creation, and humankind. In the Book of Genesis, Judaism first taught that after God created the earth, He "saw that it was good." In the Twenty-fourth Psalm, we learn that "the earth is the Lord's and the fullness thereof." In other words, God is pleased with his creation, and "dominion" does not mean that the earth belongs to humankind; on the contrary, whatever is done to the earth must be done with an awareness that it belongs to God.

My tradition also teaches that the purpose of life is "to glorify God." And there is a shared conviction within the Judeo-Christian tradition that believers are expected to "do justice, love mercy, and walk humbly with your God." But whatever verses are selected in an effort to lend precision to the Judeo-Christian definition of life's purpose, that purpose is clearly inconsistent with the reckless destruction of that which belongs to God and which God has seen as "good." How can one glorify the Creator while heaping contempt on the creation? How can one walk humbly with nature's God while wreaking havoc on nature?

The story of Noah and the ark offers further evidence of Judaism's concern for stewardship. Noah is commanded by God to take into his ark at least two of every living species in order to save them from the Flood — a commandment that might appear in modern form as: Thou shalt preserve biodiversity. Indeed, does God's instruction have new relevance for those who share Noah's faith in this time of another worldwide catastrophe, this time one of our own creation? Noah heeded this commandment, and after he and his family and a remnant of every living species on earth survived the Flood, God made a new covenant with him which affirmed His commitment to humankind. Often overlooked, however, is the second half of God's covenant, made not only with Noah but with "all living creatures," again affirming the sacredness of creation, which He pledged to safeguard in "seed time and harvest, cold and heat, summer and winter." It was the promise never again to destroy the earth by floods, which, according to Genesis, is the symbolic message of every rainbow.

In spite of the clear message from a careful reading of these and other Scriptures, critics have gained currency in part because of the prevailing silence with which most denominations have reacted to the growing evidence of an ecological holocaust. Nor does it help that some religious leaders have seemed to encourage environmental recklessness. I remember listening with closed eyes and bowed head to the invocation at the groundbreaking for a new construction project as the minister cited our "dominion over the earth" and then immediately went on to list with great relish every instrument of environmental mayhem he could name, from bulldozers and backhoes to chain saws and steamrollers, as though they were divinely furnished tools we should use with abandon in reshaping the earth for the sheer joy of doing so. Both behaviors — silence in the face of disaster and unthinking enthusiasm for further degradation — do nothing to counter the cartoon image of a faith bent on the domination of nature.

Happily, it has recently become clear that a great movement to protect the earth is stirring deeply within the faith, and many religious leaders are now sounding the alarm. But until now they have seemed reluctant to lend their moral authority to the effort to rescue the earth. Why?

In their defense, it should be said that religious leaders have faced the same difficulties as the rest of us in recognizing this unprecedented pattern of destruction, in comprehending the strategic nature of the threat, and in realizing the profound and sudden change in the relationship between the human species and the rest of the environment. But their failure to act is especially disturbing because the Christian Scriptures carry such a strong activist message. To me, it is best expressed in one of Jesus' parables, recounted in three of the four gospels, the Parable of the Unfaithful Servant. The master of a house, preparing to depart on a journey, leaves his servant in charge of the home and gives him strict instructions to remain alert in case vandals or thieves attempt to ransack the house while the master is away. The servant is explicitly warned that if the vandals come while he is asleep, he still has a duty to protect the house against them — and the fact that he was asleep will not be an acceptable excuse. A question raised by the parable is clear: If the earth is the Lord's and His servants are given the responsibility to care for it, then how are we to respond to the global vandalism now wreaking such unprecedented destruction on the earth? Are we asleep? Is that now an acceptable excuse?

But there is something else at work in organized religion as well. Many of those who might otherwise find themselves in the vanguard of the resistance to this onslaught are preoccupied with other serious matters. For example, Christian theologians and clergy who have traditionally supported a liberal political agenda have inherited a specific set of concerns defined early in this century as the Social Gospel. According to this humane view of the Church's role, followers of Christ should assign priority to the needs of the poor, the powerless, the sick and frail, the victims of discrimination and hatred, the forgotten human fodder chewed up by the cogs of industrial civilization. The moral imperative attached to this set of priorities leads many advocates of the Social Gospel to vigorously resist the introduction of competing concerns which they see as distractions from their appointed task, diluting their already overtaxed resources of money, time, moral authority, and emotional labor. After all, as an issue, "the environment" sometimes seems far from the more palpable sins of social injustice.

On the other hand, politically conservative theologians and clergy have inherited a different agenda, also defined early in this century. The "atheistic communism" against which they have properly inveighed for decades is, for them, only the most extreme manifestation of a statist impulse to divert precious resources —

money, time, moral authority, and emotional labor — away from the mission of spiritual redemption and toward an idolatrous alternative: the search for salvation through a grand reordering of the material world. As a result, they are deeply suspicious of any effort to focus their moral attention on a crisis in the material world that might require as part of its remedy a new exercise of something resembling moral authority by the state. And the prospect of coordinated action by governments all over the world understandably heightens their fears and suspicions.

Thus, with activists of both the left and the right resisting the inclusion of the environment on their list of concerns, the issue has not received the attention from religious leaders one may have expected. This is unfortunate, because the underlying concern is theologically consistent with the perspectives of both sides; equally important, the issue provides a rare opportunity for them to meet on common ground.

As it happens, the idea of social justice is inextricably linked in the Scriptures with ecology. In passage after passage, environmental degradation and social injustice go hand in hand. Indeed, the first instance of "pollution" in the Bible occurs when Cain slays Abel and his blood falls on the ground, rendering it fallow. According to Genesis, after the murder, when Cain asks, "Am I my brother's keeper?" the Lord replies, "Your brother's blood calls out to me from the ground. What have you done?" God then tells Cain that his brother's blood has defiled the ground and that as a result, "no longer will it yield crops for you, even if you toil on it forever!"

In today's world, the links between social injustice and environmental degradation can be seen everywhere: the placement of toxic waste dumps in poor neighborhoods, the devastation of indigenous peoples and the extinction of their cultures when the rain forests are destroyed, disproportionate levels of lead and toxic air pollution in inner-city ghettos, the corruption of many government officials by people who seek to profit from the unsustainable exploitation of resources.

Meanwhile, religious conservatives might be surprised to find that many deeply committed environmentalists have become, if anything, even more hostile to overreaching statism than they are. The most serious examples of environmental degradation in the world today are tragedies that were created or actively encouraged by governments — usually in pursuit of some notion that a dramatic reordering of the material world would enhance the greater good. And it is no accident that the very worst environmental tragedies were created by communist governments, in which the power of the state completely overwhelms the capabilities of the individual steward. Chernobyl, the Aral Sea, the Yangtze River, the "black town" of Copsa Mica in Romania — these and many other disasters testify to the severe environmental threats posed by statist governments.

Both conservative and liberal theologians have every reason, scriptural as well as ideological, to define their spiritual mission in a way that prominently includes the defense of God's creation. Slowly and haltingly, both camps are beginning to do so. But most clergy are still reluctant to consider this cause worthy of their sustained attention; in my opinion, an important source of this reluctance is a philosophical assumption that humankind is separate from the rest of nature, an assumption shared by both liberals and conservatives. The basis for it deserves further attention, especially since the tendency to see human needs as essentially detached from the well-being of the earth's natural systems is not fundamentally Christian in origin. Even so, this tendency reflects a view of the world that was absorbed into the Christian tradition early on; specifically, it was part of the heritage of Greek philosophy, a heritage that had a powerful influence on early Christian thinking and behavior.

A little more than three hundred years before the birth of Christ, Greek culture and philosophy were introduced throughout the lands conquered by Alexander the Great. The inherent power of Greek philosophy as an analytic tool ensured its continued popularity, even as it was adapted to scores of different religious and cultural traditions. It served, of course, as the foundation for the fiercely logical and systematic way of thinking that enabled Rome to conquer all of the "known world," including not only Palestine, where Christianity originated, but also every city in which Christ's disciples preached. It was natural, then, for early Christians to use some of the dominant language and concepts as they spread the Word.

As the world discovered, the greatest Greek philosophers were, first, Plato and, second, Aristotle. The most significant difference between them concerned the relationship between the intellect and physical reality or, in other words, between humankind and nature. Plato believed that the soul exists in a realm quite apart from the body and that the thinker is separate from the world he thinks about. But Aristotle felt that everything in our intellect comes from the senses, and thus the thinker is powerfully connected to the world he thinks about. This dispute began in ancient Greece and continued throughout the early history of Christian thought, through the Middle Ages, and up to the seventeenth century.

One of the most influential thinkers in the early Church, Saint Augustine, recounts how attracted he was, early in the fifth century, to Plato's view of the physical world and how he struggled to overcome his love of Platonic theory before he could "rationalize" his acceptance of Christ's true message. Indeed, this tension — which still exists — has been described by the theologian Michael Novak as the "great temptation of the West." For example, throughout the first five centuries of Christianity, the heresy of Gnosticism — which portrayed physical reality as an illusion — drew powerfully upon Plato's conception of a disembodied spiritual intellect hovering above the material world. But even after the Gnostic view was formally rejected, it periodically resurfaced in various guises, and the Platonic assumption upon which it was based — that man is separate from the world of nature — continued to flourish as a major strain of Christian thought. It may have been overemphasized because of early struggles with paganism.

The heritage of Aristotle's thought, on the other hand, was kept alive principally in the Arabic-speaking world. Alexander, who was tutored by Aristotle, established his thought throughout the lands he conquered, and the city he chose as his capital, Alexandria, became the greatest center of learning in the ancient world. But for many centuries the West was isolated from this intellectual tradition; only after the returning crusaders brought new ideas back to Europe did the West rediscover the other half of its Greek heritage. As the thirteenth century began, Europeans impressed with the intellectual achievements of Arab civilization discovered and translated several works by Aristotle — *Ethics, Politics, Logic,* among others — which had disappeared from Western thought but had been preserved in Arabic. Influenced by the powerful work of Maimonides, the Jewish scholar writing in Arabic (in Alexandria) who reinterpreted Judaism in Aristotelian terms, Saint Thomas Aquinas undertook the same reinterpretation of Christian thought and antagonized the Church establishment with his assertions of an Aristotelian view of the relationship between the spirit and the flesh, between humankind and the world. He saw a philosophical closeness between the soul and physical reality that discomfited the Church. Although his books were banned, burned, and not widely read until almost three centuries later, his powerful thinking eventually played a role in the Church's acquiescence to the impulses that led to the Renaissance, including the impulse to reconnect with the earth. A classic painting by Raphael in 1510 portrays this same philosophical tension at the beginning of the Renaissance: Plato appears with one finger pointing toward the heavens; next to him, Aristotle is gesturing toward the earth.

But just a century later, the reemerging Aristotelian view was dealt a severe blow. On November 10, 1619, René Descartes, soon to become one of the founders of modern philosophy, was a twenty-three-year-old mathematician lying on the banks of the Danube. That day he had a startling vision of a mechanistic world filled with inanimate matter, moving predictably in mathematically determined patterns — patterns that could be discerned and mastered by analytical minds through sustained inquiry and detached observation. In a real sense, Descartes' vision initiated the scientific revolution. It is often said that "all Western philosophy is a footnote to Plato," and much of the credit for this should go to the work of Descartes, who broke through the tension in the seventeenth century between the ideas of Aristotle and Plato with his famous dictum, *Cogito ergo sum,* "I think therefore I am."

By the time Descartes had finished his life's work, Raphael's picture was obsolete as a representation of Western thought. The new modern person pointed decisively upward — away from nature, away from the earth — toward an ethereal realm from which the detached human intellect could observe the movement of mat-

This 1510 painting by Raphael, a detail from the center of a giant Vatican fresco called *The School of Athens,* portrays Plato, on the left, pointing upward to the realm of abstract thought and intellectual idealism; on the right is Aristotle, his right hand gesturing toward the earth, which he argued was the ultimate source — through our senses — of all our thoughts.

ter everywhere in the universe. Floating somewhere above it all, this new disembodied mind could systematically and relentlessly decipher the scientific laws that would eventually enable us to understand nature — and control it. This strange relationship between spirit and nature would later be called that of the "ghost in the machine."

The Church, meanwhile, was supposed to be on guard against any Faustian effort of the people to gain unseemly power to alter God's world, but it fell victim once again to the Platonic vision by reducing its spiritual mission to an effort to guide the inner life of the mind while discounting the moral significance of humankind's manipulations of the natural world. Sir Francis Bacon, lord chancellor of England, author of *The New Atlantis* (1624) and one of the principal founders of the scientific method, undertook to ease any doubts the Church might have about allowing humankind to acquire and exercise the vast new powers of science. Taking "Cartesian dualism" one step further, Bacon argued that not only were humans separate from nature; science, he said, could safely be regarded as separate from religion. In his view, "facts" derived through the scientific method had no moral significance in and of themselves; only "moral knowledge" of matters concerning the distinction between good and evil had religious significance. This facile distinction carried a profound implication: the new power derived from scientific knowledge could be used to dominate nature with moral impunity.

Thus began the long, 350-year separation of science and religion. The astronomical discoveries of Copernicus and Galileo had earlier upset the Church's peaceful coexistence with science, but neither man had intentionally challenged the primacy of the Church's moral teachings as the basis for interpreting the new facts discovered from observing the universe. However, Bacon suggested a moral detour: facts need not be considered in light of their implications. Not long afterward, the Church came to consider science as an adversary, as it posed challenge after challenge to the Church's authority to explain the meaning of existence.

This fundamental shift in Western thinking — which in a very real sense marks the beginning of modern history — gave humankind increasing dominance in the world, as a flood of scientific discoveries began unlocking the secrets of God's blueprint for the universe. But how could this new power be used wisely? Descartes and Bacon ensured the gradual abandonment of the philosophy that humankind is one vibrant strand in an elaborate web of life, matter, and meaning. And ironically, major scientific discoveries have often undermined the Church's tendency to exaggerate our uniqueness as a species and defend our separation from the rest of nature. Charles Darwin's *Origin of Species* claimed jurisdiction for science over the human physical form by placing our evolution in the context of the animal world. Half a century later, Sigmund Freud's explanation of the unconscious claimed part of the mind for nature as well. Thanks to the revolution in thinking they helped to start, it seemed to many that the rational portion of the intellect — that part that created science — became the only remaining province for the moral authority of the Church.

Yet science itself offers a new way to understand — and perhaps begin healing — the long schism between science and religion. Earlier in this century, the Heisenberg Principle established that the very act of observing a natural phenomenon can change what is being observed. Although the initial theory was limited in practice to special cases in subatomic physics, the philosophical implications were and are staggering. It is now apparent that since Descartes reestablished the Platonic notion and began the scientific revolution, human civilization has been experiencing a kind of Heisenberg Principle writ large. The very act of intellectually separating oneself from the world in order to observe it changes the world that is being observed — simply because it is no longer connected to the observer in the same way. This is not a mere word game; the consequences are all too real. The detached observer feels free to engage in a range of experiments and manipulations that might never spring to mind except for the intellectual separation. In the final analysis, all discussions of morality and ethics in science are practically pointless as long as the world of the intellect is assumed to be separate from the physical world. That first separation led inevitably to the separation of mind and body, thinking and feeling, power and wisdom; as a consequence, the scientific method changed our relationship to nature and is now, perhaps irrevocably, changing nature itself.

Although many scientists resist the notion that science can ever be reunited with religion, there is now a powerful impulse in some parts of the scientific community to heal the breach. While Plato tended to emphasize the eternity of existence rather than the concept of creation, and while Descartes' mechanistic explanation also implied an eternal world, many scientists who have had no use for religion in the past now believe that the evidence from recent breakthroughs in astronomy and cosmology points toward a definite beginning for the universe. Some have, as a result, softened their resistance to the notion that the universe, and humankind as a part of it, were "created." Arno Penzias, for example, who shared a Nobel Prize for his discovery of the measurable echo of the Big Bang that accompanied the beginning of time, was asked on a radio call-in show what existed *before* the Big Bang. He said he didn't know but that the answer most consistent with the mathematical evidence was "nothing." When the next caller, infuriated by his answer, accused him of being an atheist, he replied, "Ma'am, I don't think you listened carefully to the implications of what I just said." Those implications — including the notion that some kind of Creator might be responsible for making "something" where there was once only "nothing" — suggest a potential for healing the hostility of science toward religion. And if science and religion are one day reunited, we may recapture a deeper curiosity about not only the nature of existence but its meaning as well, a deeper understanding of not only the universe but of our role and purpose as part of it.

Indeed, there is even, in this emerging scientific view, a palpable "physical" role for human thought in the shaping of reality. Erwin Schrodinger, a pioneer in quantum physics, first offered the astounding view that consciousness is one of the building blocks of the physical universe and that a shift in the "attention" of an observer can have tangible consequences in the location and physical properties of subatomic particles. When he tried to explain one of the enduring puzzles of biology, how a pattern of life can emerge from a formless cluster of molecules, Schrodinger speculated that living organisms are endowed with "an astonishing gift of concentrating a 'stream of order' on [themselves] and thus escaping the decay into atomic chaos." If the mental activity necessary to focus one's "attention" turns out to have tangible consequences of the kind we now associate with a form of physical energy, then, ironically, science may one day definitively disprove Bacon's assertion that there can be a separation between facts and values, between the thoughts of a scientist and the moral duties of a human being.

My own curiosity leads me to this speculation: that the original scientific impulse — before Descartes and before Plato — was made possible by the conception (or revelation) of a single Creator. When Akhnaton first conceptualized a single God and when Judaism introduced the idea of monotheism, it became possible for humankind to develop a new understanding of the nature of all the things they beheld in the world around them. For those who came to believe in a single Creator, there was no longer any reason to imagine that each object and living thing had a unique spiritual force and that each was imbued with mysterious meaning and motivated by unknown powers. Monotheism was a profoundly empowering idea: just as a navigator can — through the technique of triangulation — locate his position anywhere at sea by identifying any other two points with known locations, like familiar stars or constellations, those who came to believe in a single God gained the intellectual power to navigate skillfully through the ocean of superstition and bewilderment that engulfed the ancient world. Whatever these monotheists beheld could be philosophically located with reference to two known points: the Creator, philosophically equidistant from everything He had created, and themselves.

This process of spiritual triangulation identified the natural world as sacred, not because each rock and tree was animated by a mysterious spirit, but because each rock and tree was created by God. Moreover, the physical world was understood, investigated, and ordered in terms of its relationship to the one God who had created it. And the very process by which this inquiry into the nature of the world took place reinforced the assumption that humankind is part of the world, because each inquiry relied on an understanding of our relationship to both God and the physical world in which we live. All three elements — God, human beings, and nature — were understood in relation to one another and were essential to this process of triangulation.

Many centuries after Akhnaton, Plato's intellectual inquiry followed an entirely different path. Although he searched for a single cause behind all things, he attempted to discern their nature by locating them in relation to only one point of reference — human intellect — rather than through a process of philosophical triangulation, which would have relied on two points, humankind and the Creator (or what could also be called a single cause). By assuming that the human intellect is not anchored in a context of meaningful relationships, with both the physical world and the Creator, Plato assured that later explanations of the workings of the world would become progressively more abstract.

Francis Bacon is a case in point. His moral confusion — the confusion at the heart of much of modern science — came from his assumption, echoing Plato, that human intellect could safely analyze and understand the natural world without reference to any moral principles defining our relationship and duties to both God and God's creation. Bacon, for example, was able to enthusiastically advocate vivisection for the pure joy of learning without reference to any moral purpose, such as saving human lives, as justification for the act.

And tragically, since the onset of the scientific and technological revolution, it has seemingly become all too easy for ultrarational minds to create an elaborate edifice of clockwork efficiency capable of nightmarish cruelty on an industrial scale. The atrocities of Hitler and Stalin, and the mechanical sins of all who helped them, might have been inconceivable except for the separation of facts from values and knowledge from morality. In her study of Adolf Eichmann, who organized the death camp bureaucracy, Hannah Arendt coined the memorable phrase "the banality of evil" to describe the bizarre contrast between the humdrum and ordinary quality of the acts themselves — the thousands of small, routine tasks committed by workaday bureaucrats — and the horrific and satanic quality of their proximate consequences. It was precisely the machinelike efficiency of the system that carried out the genocide which seemed to make it possible for its functionaries to separate the thinking required in their daily work from the moral sensibility for which, because they were human beings, they must have had some capacity. This mysterious, vacant space in their souls, between thinking and feeling, is the suspected site of the inner crime. This barren of the spirit, rendered fallow by the blood of unkept brothers, is the precinct of the disembodied intellect, which knows the way things work but not the way they are.

It is my view that the underlying moral schism that contributed to these extreme manifestations of evil has also conditioned our civilization to insulate its conscience from any responsibility for the collective endeavors that invisibly link millions of small, silent, banal acts and omissions together in a pattern of terrible cause and effect. Today, we enthusiastically participate in what is in essence a massive and unprecedented experiment with the natural systems of the global environment, with little regard for the moral consequences. But for the separation of science and religion, we might not be pumping so much gaseous chemical waste into the atmosphere and threatening the destruction of the earth's climate balance. But for the separation of useful technological know-how and the moral judgments to guide its use, we might not be slashing and burning one football field's worth of rain forest every second. But for the assumed separation of humankind from nature, we might not be destroying half the living species on earth in the space of a single lifetime. But for the separation of thinking and feeling, we might not tolerate the deaths every day of 37,000 children under the age of five from starvation and preventable diseases made worse by failures of crops and politics.

But we do tolerate — and collectively perpetrate — all these things. They are going on right now. When future generations wonder how we could go along with our daily routines in silent complicity with the collective destruction of the earth, will we, like the Unfaithful Servant, claim that we did not notice these things because we were morally asleep? Or will we try to explain that we were not so much asleep as living in a waking trance, a strange Cartesian spell under whose influence we felt no connection between our routine, banal acts and the moral consequences of what we did, as long as they were far away at the other end of the massive machine of civilization?

And what would future generations say in response to such a pitiful plea? They might remember the ancient words of the psalmist, who condemned a people who, in their fascination with the works of their own civilization, lost their awareness of the sacred and came to resemble the idolatrous artifacts with which they were fatally enchanted: "They have a mouth but they will not speak, they have eyes but they will not see, they have nostrils but they will not smell, they have hands but they will not feel."

Modern philosophy has gone so far in its absurd pretensions about the separateness of human beings from nature as to ask the famous question: "If a tree falls in the forest and no person is there to hear it, does it make a sound?" If robotic chain saws finally destroy all the rain forests on earth, and if the people who set them in motion are far enough away so that they don't hear the crash of the trees on the naked forest floor, does it matter? This rational, detached, scientific intellect, observing a world of which it is no longer a part, is too often arrogant, unfeeling, uncaring. And its consequences can be monstrous.

The strange absence of emotion, the banal face of evil so often manifested by mass technological assaults on the global environment, is surely a consequence of the belief in an underlying separation of intellect from the physical world. At the root of this belief lies a heretical misunderstanding of humankind's place in the world as old as Plato, as seductive in its mythic appeal as Gnosticism, as compelling as the Cartesian promise of Promethean power — and it has led to tragic results. We have misunderstood who we are, how we relate to our place within creation, and why our very existence assigns us a duty of moral alertness to the consequences of what we do. A civilization that believes itself to be separate from the world may pretend not to hear, but there is indeed a sound when a tree falls in the forest.

The richness and diversity of our religious tradition throughout history is a spiritual resource long ignored by people of faith, who are often afraid to open their minds to teachings first offered outside their own system of belief. But the emergence of a civilization in which knowledge moves freely and almost instantaneously throughout the world has led to an intense new interest in the different perspectives on life in other cultures and has spurred a renewed investigation of the wisdom distilled by all faiths. This panreligious perspective may prove especially important where our global civilization's responsibility for the earth is concerned.

Native American religions, for instance, offer a rich tapestry of ideas about our relationship to the earth. One of the most moving and frequently quoted explanations was attributed to Chief Seattle in 1855, when President Franklin Pierce stated that he would buy the land of Chief Seattle's tribe. The power of his response has survived numerous translations and retellings:

> How can you buy or sell the sky? The land? The idea is strange to us. If we do not own the freshness of the air and the sparkle of the water, how can you buy them? Every part of this earth is sacred to my people. Every shining pine needle, every sandy shore, every mist in the dark woods, every meadow, every humming insect. All are holy in the memory and experience of my people. . . .
>
> If we sell you our land, remember that the air is precious to us, that the air shares its spirit with all the life it supports. The wind that gave our grandfather his first breath also received his last sigh. The wind also gives our children the spirit of life. So if we sell you our land, you must keep it apart and sacred, a place where man can go to taste the wind that is sweetened by the meadow flowers.
>
> Will you teach your children what we have taught our children? That the earth is our mother? What befalls the earth befalls all the sons of the earth.
>
> This we know: the earth does not belong to man, man belongs to the earth. All things are connected like the blood that unites us all. Man did not weave the web of life, he is merely a strand in it. Whatever he does to the web, he does to himself.
>
> One thing we know: Our God is also your God. The earth is precious to Him and to harm the earth is to heap contempt on its Creator.

A modern prayer of the Onondaga tribe in upstate New York offers another beautiful expression of our essential connection to the earth:

> O Great Spirit, whose breath gives life to the world and whose voice is heard in the soft breeze . . . make us wise so that we may understand what you have taught us, help us learn the lessons you have hidden in every leaf and rock, make us always ready to come to you with clean hands and straight eyes, so when life fades, as the fading sunset, our spirits may come to you without shame.

The spiritual sense of our place in nature predates Native American cultures; increasingly it can be traced to the origins of human civilization. A growing number of anthropologists and archaeomythologists, such as Marija Gimbutas and Riane Eisler, argue that the prevailing ideology of belief in prehistoric Europe and much of the world was based on the worship of a single earth goddess, who was assumed to be the fount of all life and who radiated harmony among all living things. Much of the evidence for the existence of this primitive religion comes from the many thousands of artifacts uncovered in ceremonial sites. These sites are so widespread that they seem to confirm the notion that a goddess religion was ubiquitous throughout much of the world until the antecedents of today's religions — most of which still have a distinctly masculine orientation — swept out of India and the Near East, almost obliterating belief in the goddess. The last vestige of organized goddess worship was eliminated by Christianity as late as the fifteenth century in Lithuania.

The antiquity of the evidence and the elaborate and imaginative analysis used to interpret the artifacts leave much room for skepticism about our ability to know exactly what this belief system — or collection of related beliefs — taught. Its best-documented tenet seems to have been a reverence for the sacredness of the earth — and a belief in the need for harmony among all living things; other aspects of the faith are less clear, and it is probable that many barbaric practices accompanied the more benign beliefs. Still, the archaeological scholarship is impressive, and it seems obvious that a better understanding of a religious heritage preceding our own by so many thousands of years could offer us new insights into the nature of the human experience.

Moreover, virtually all current world religions have much to say about the relationship between humankind and the earth. Islam, for example, offers familiar themes. The Prophet Mohammed said, "The world is green and beautiful and God has appointed you His stewards over it." The central concepts of Islam taught by the Qu'ran — *tawheed* (unity), *khalifa* (trusteeship), and *akhrah* (accountability) — also serve as the pillars of the Islamic environmental ethic. The earth is the sacred creation of Allah, and among Mohammed's many instructions about it is: "Whoever plants a tree and diligently looks after it until it matures and bears fruit is rewarded." The first Muslim caliph, Abu-Baker, drew upon the Qu'ran and the *hadith* (oral traditions of the Prophet) when he ordered his troops: "Do not cut down a tree, do not abuse a river, do not harm animals, and be always kind and humane to God's creation, even to your enemies."

A common thread in many religions is the sacred quality of water. Christians are baptized in water, as a sign of purification. The Qu'ran declares that "we have created everything from water." In the *Lotus "Sutra,"* Buddha is presented metaphorically as a "rain cloud," covering, permeating, fertilizing, and enriching "all parched living beings, to free them from their misery to attain the joy of peace, joy of the present world and joy of Nirvana . . . everywhere impartially without distinction of persons . . . ever to all beings I preach the Law equally . . . equally I rain the Law — rain untiringly."

The sacredness of water receives perhaps the greatest emphasis in Hinduism. According to its teachings, the "waters of life" are believed to bring to humankind the life force itself. One modern Hindu environmentalist, Dr. Karan Singh, regularly cites the ancient Hindu dictum: "The earth is our mother, and we are all her children." And in the *Atharvaveda,* the prayer for peace emphasizes the links between humankind and all creation: "Supreme Lord, let there be peace in the sky and in the atmosphere, peace in the plant world and in the forests; let the cosmic powers be peaceful; let Brahma be peaceful; let there be undiluted and fulfilling peace everywhere."

Sikhism, the northern Indian monotheistic offshoot of Hinduism that was founded around 1500, places a great deal of spiritual significance on the lessons we can learn directly from nature. Its founder, Guru Nanak, said, "Air is the Vital Force, Water the Progenitor, the Vast Earth the Mother of All: Day and Night are nurses, fondling all creation in their lap." According to the Sikh scripture, *Guru Granth Sahib,* human beings are composed of five elements of nature, which teach lessons and inspire strength in the formulation of our character: "Earth teaches us patience, love; Air teaches us mobility, liberty; Fire teaches us warmth, courage; Sky teaches us equality, broad-mindedness; Water teaches us purity, cleanliness."

One of the newest of the great universalist religions, Baha'i, founded in 1863 in Persia by Mirza Husayn Ali, warns us not only to properly regard the relationship between humankind and nature but also the one between civilization and the environment. Perhaps because its guiding visions were formed during the period of accelerating industrialism, Baha'i seems to dwell on the spiritual implications of the great transformation to which it bore fresh witness: "We cannot segregate the human heart from the environment outside us and say that once one of these is reformed everything will be improved. Man is organic with the world. His inner life molds the environment and is itself deeply affected by it. The one acts upon the other and every abiding change in the life of man is the result of these mutual reactions." And, again, from the Baha'i sacred writings comes this: "Civilization, so often vaunted by the learned exponents of arts and sciences will, if allowed to overleap the bounds of moderation, bring great evil upon men."

This sensitivity to the changes wrought by civilization on the earth is also evident in new statements from the leaders of Western religions. Pope John Paul II, for example, in his message of December 8, 1989, on humankind's responsibility for the ecological crisis, said: "Faced with the widespread destruction of the environment, people everywhere are coming to understand that we cannot continue to use the goods of the earth as we have in the past . . . a new *ecological awareness* is beginning to emerge which rather than being downplayed, ought to be encouraged to develop into concrete programs and initiatives." In concluding, the pope directly addressed his "brothers and sisters in the Catholic church, in order to remind them of their serious obligation to care for all of creation. . . . Respect for life and for the dignity of the human person extends also to the rest of creation, which is called to join man in praising God."

Many environmental theorists who think of the Catholic church only long enough to complain bitterly about its opposition to birth control (which many Catholics, in fact, use) might be surprised to read the pope's powerful and penetrating analysis of the ecological crisis and recognize him as an ally: "Modern society will find no solution to the ecological problem unless it *takes a serious look at its lifestyle*. In many parts of the world, society is given to instant gratification and consumerism while remaining indifferent to the damage which these cause. As I have already stated, the seriousness of the ecological issue lays bare the depth of man's moral crisis."

The Judeo-Christian tradition has always presented a prophetic vision, from Joseph's warnings to Pharaoh about the seven lean years to John's jubilant promise in Revelations: "We will praise the Lamb, Triumphant, with *all* creatures." Many prophecies use the images of environmental destruction to warn of transgressions against God's will. For example, for those who believe in the literal truth of the Bible, it is hard to read about the predictions of hurricanes 50 percent stronger than the worst ones today, due to the accumulation of greenhouse gases that we have fostered, without recalling the prophecy of Hosea: "They have sown the wind, and they shall reap the whirlwind."

For some Christians, the prophetic vision of the apocalypse is used — in my view, unforgivably — as an excuse for abdicating their responsibility to be good stewards of God's creation. Former Secretary of the Interior James Watt, who deserved his reputation as an anti-environmentalist, was once quoted as belittling concerns about environmental protection in part because it would all be destroyed by God in the apocalypse. Not only is this idea heretical in terms of Christian teachings, it is an appallingly self-fulfilling prophecy of doom. It is noteworthy that Watt did not see the need to forgo other obligations, however. He did not say, for example, that there was no point in conducting a bargain basement sale of grazing rights on federal rangelands to wealthy friends because the Four Horsemen are galloping this way.

Nevertheless, there is no doubt that many believers and nonbelievers alike share a deep uneasiness about the future, sensing that our civilization may be running out of time. The religious ethic of stewardship is indeed harder to accept if one believes the world is in danger of being destroyed — by either God or humankind. This point was made by the Catholic theologian Teilhard de Chardin when he said, "The fate of mankind, as well as of religion, depends upon the emergence of a new faith in the future." Armed with such a faith, we might find it possible to resanctify the earth, identify it as God's creation, and accept our responsibility to protect and defend it. We might even begin to contemplate decisions based on long-term considerations, not short-term calculations.

And if we could find a way to understand our own connection to the earth — all the earth — we might recognize the danger of destroying so many living species and disrupting the climate balance. James Lovelock, the originator of the Gaia hypothesis, maintains that the entire complex earth system behaves in a self-regulating manner characteristic of something alive, that it has managed· to maintain critical components of the earth's life support systems in perfect balance over eons of time — until the unprecedented interference of modern civilization: "We now see that the air, the ocean and the soil are much more than a mere environment for life; they are a part of life itself. Thus the air is to life just as is the fur to a cat or the nest for a bird. Not living but something made by living things to protect against an otherwise hostile world. For life on earth, the air is our protection against the cold depths and fierce radiations of space."

Lovelock insists that this view of the relationship between life and the nonliving elements of the earth system does not require a spiritual explanation; even so, it evokes a spiritual response in many of those who hear it. It cannot be accidental, one is tempted to conclude, that the percentage of salt in our bloodstreams is roughly the same as the percentage of salt in the oceans of the world. The long and intricate process by which evolution helped to shape the complex interrelationship of all living and nonliving things may be explicable in purely scientific terms, but the simple fact of the living world and our place on it evokes awe, wonder, a sense of mystery — a spiritual response — when one reflects on its deeper meaning.

We are not used to seeing God in the world because we assume from the scientific and philosophical rules that govern us that the physical world is made up of inanimate matter whirling in accordance with mathematical laws and bearing no relation to life, much less ourselves. Why does it feel faintly heretical to a Christian to suppose that God is in us as human beings? Why do our children believe that the Kingdom of God is *up,* somewhere in the ethereal reaches of space, far removed from this planet? Are we still unconsciously following the direction of Plato's finger, looking for the sacred everywhere except in the real world?

It is my own belief that the image of God can be seen in every corner of creation, even in us, but only faintly. By gathering in the mind's eye all of creation, one can perceive the image of the Creator vividly. Indeed, my understanding of how God is manifest in the world can be best conveyed through the metaphor of the hologram, which I mentioned in the introduction. (Using a technological metaphor to make a spiritual point is not as odd as it may seem. The Bible often uses metaphors based on the technology of the time. For example, God scatters spiritual seeds on the barren land as well as the rich soil, some of them grow and some of them don't; the wheat must be separated from the tares; at the end of time men shall beat their swords into plowshares and their spears into pruning hooks.) When the light of a laser beam shines on a holographic plate, the image it carries is made visible in three dimensions as the light reflects off thousands of microscopic lines that make up a distinctive "resistance pattern" woven into the plastic film covering the glass plate — much the way a phonograph needle picks up music from the "resistance pattern" of tiny bumps in the grooves of a long-playing record. Each tiny portion of the hologram contains a tiny representation of the entire three-dimensional image, but only faintly. However, due to the novel and unusual optical principles on which holography is based, when one looks not at a small portion but at the entire hologram, these thousands of tiny, faint images come together in the eye of the beholder as a single large, vivid image.

Similarly, I believe that the image of the Creator, which sometimes seems so faint in the tiny corner of creation each of us beholds, is nonetheless present in its entirety — and present in us as well. If we are made in the image of God, perhaps it is the myriad slight strands from earth's web of life — woven so distinctively into our essence — that make up the "resistance pattern" that reflects the image of God, faintly. By experiencing nature in its fullest — our own and that of all creation — with our senses and with our spiritual imagination, we can glimpse, "bright shining as the sun," an infinite image of God.

PART III

STRIKING THE BALANCE

14

A New Common Purpose

Modern industrial civilization, as presently organized, is colliding violently with our planet's ecological system. The ferocity of its assault on the earth is breathtaking, and the horrific consequences are occurring so quickly as to defy our capacity to recognize them, comprehend their global implications, and organize an appropriate and timely response. Isolated pockets of resistance fighters who have experienced this juggernaut at first hand have begun to fight back in inspiring but, in the final analysis, woefully inadequate ways. It is not that they lack courage, imagination, or skill; it is simply that what they are up against is nothing less than the current logic of world civilization. As long as civilization as a whole, with its vast technological power, continues to follow a pattern of thinking that encourages the domination and exploitation of the natural world for short-term gains, this juggernaut will continue to devastate the earth no matter what any of us does.

I have come to believe that we must take bold and unequivocal action: we must make the rescue of the environment the central organizing principle for civilization. Whether we realize it or not, we are now engaged in an epic battle to right the balance of our earth, and the tide of this battle will turn only when the majority of people in the world become sufficiently aroused by a shared sense of urgent danger to join an all-out effort. It is time to come to terms with exactly how this can be accomplished. Having attempted in earlier chapters to understand the crisis from the perspectives offered by the earth sciences, economics, sociology, history, information theory, psychology, philosophy, and religion, I now want to examine, from my vantage point as a politician, what I think can be done about it.

Politics, broadly defined, is the means by which we make collective decisions and choices. We now confront a set of choices as difficult as any in human history. The art of politics must be brought to bear in defining these choices, raising public awareness of the imminent danger facing us, and catalyzing decisions in favor of a collective course of action that has a reasonable chance of success.

There is no doubt that with sufficient agreement on our goals, we can achieve the victory we are seeking. Although very difficult changes in established patterns of thought and action will be required, the task of restoring the natural balance of the earth's ecological system is both within our capacity and desirable for other reasons — including our interest in social justice, democratic government, and free market economics. Ultimately, a commitment to healing the environment represents a renewed dedication to what Jefferson believed were not merely American but universal inalienable rights: life, liberty, and the pursuit of happiness.

The hard part, of course, will be securing a sufficient measure of agreement that difficult comprehensive changes are needed. Fortunately, however, there are ample precedents for the kinds of pervasive institutional changes and shared effort that will be necessary. Though it has never yet been accomplished on a global scale, the establishment of a single shared goal as the central organizing principle for every institution in society has been realized by free nations several times in modern history. Most recently, a coalition of free nations committed to democracy and free markets demonstrated a remarkable capacity to persevere for nearly half a century in their effort to prevent the spread of communism by military, political, or economic means. To the surprise of many, this coalition secured a resounding victory for the idea of freedom in the philosophical war that lasted from the time of the Russian Revolution until the jailers of Eastern Europe released their "enemies of the people" — who were then freely elected as democratic leaders of, by, and for the people. And the political earthquake that accompanied that victory has continued to topple statues of Lenin for several years, from Nicaragua to Angola to Ethiopia, until it brought down the former Soviet Union itself.

What made this dramatic victory possible was a conscious and shared decision by men and women in the nations of the "free world" to make the defeat of the communist system the central organizing principle of not only their governments' policies but of society itself. That is not to say that this goal dominated every waking thought or guided every policy decision, but opposition to communism was the principle underlying almost all of the geo-political strategies and social policies designed by the West after World War II. The Marshall Plan, for example, was conceived primarily as a means of strengthening Western Europe's ability to withstand the spread of the communist idea. Similarly, Mac-Arthur's blueprint for reconstituting Japan's society and economy and Truman's decision in 1947 to extend massive aid to Greece and Turkey were principally motivated by the same objective. NATO and the other military alliances organized under U.S. leadership also grew out of the same central principle. U.S. advocacy of free trade and the granting of foreign aid to underdeveloped nations were in part altruistic but mainly motivated by the struggle against communism. Of course, some of the policies were painful, costly, and controversial. Wars in Korea and Vietnam, the nuclear arms race, arms sales to dictators who disagreed with every American principle save opposition to Soviet communism — these and vir-tually every other foreign policy and national security decision were made because they served the same central principle, albeit in ways that sometimes reflected poor judgment. Though mistakes were made, the basic soundness of the underlying principle con-tinued to motivate the citizens and governments of the free world, and the idea of democracy slowly began to win the battle.

The multiple expressions of anticommunism took some unex-pected forms. Here in the United States, when we built the inter-state highway system, the Defense Interstate Highway Bill au-thorized the money, and the legislation was approved by a majority partly because it would serve our overriding objective, the defeat of communism. When the Soviet Union demonstrated its technologi-cal prowess by sending Sputnik into orbit in 1957, the United States implemented the first federal aid-to-education policy — not because the president and a majority in Congress finally recognized the importance of improving education for its own sake, but because of the new importance of training scientists and engineers in service of our struggle with the communist system. We simultaneously launched the American space program, not because a majority in Congress was suddenly motivated by a desire to explore the universe, but because the program became tied to our desire to defeat the communist idea.

Many of these programs made sense on their own merits; many of their proponents pushed them mainly because of those merits. But they gained sufficient support from society as a whole because they served the central organizing principle to which we were wholeheartedly committed. Commitment sometimes led to terrible excesses: the McCarthyist smear campaigns and the exposure of human guinea pigs to the effects of nuclear radiation are only two examples of how overzealousness can have tragic results. But the point is that virtually every policy and program was analyzed and either supported or rejected primarily according to whether it served our basic organizing principle. Even such widely disparate policies as the green revolution to expand food production in Third World countries and the CIA's encouragement of trade unions in Europe were conceived because they were effective in helping us achieve our main objective.

The long struggle between democracy and communism is in many ways the clearest example of how free societies can sustain a shared commitment to a single overarching goal over a long period of time and in the face of daunting obstacles. But it is hardly the only such example. Before the Cold War, there was an even more consuming central organizing principle behind the policies of the United States and other free nations: the defeat of Nazi Germany and imperial Japan. Industry, commerce, agriculture, transportation — all were mobilized for war. Extremely effective recycling programs were widespread during World War II, not for environmental reasons, but because they helped to win the war. Our resources, our people, our art, and even our gardens played a role in the struggle to save civilization as we knew it.

It is worth remembering how long we waited before finally facing the challenge posed by Nazi totalitarianism and Hitler.

Many were reluctant to acknowledge that an effort on the scale of what became World War II was actually necessary, and most wanted to believe that the threat could be wished away with trivial sacrifices. For several years before the awful truth was accepted, one Western leader spoke out forcefully and eloquently about the gathering storm. Winston Churchill was uncompromising in his insistence that every effort be immediately bent to the task of ensuring Hitler's defeat. After Neville Chamberlain concluded the Munich Pact of 1938, which gave Czechoslovakia to Hitler in return for his pledge not to take over still more territory, most Britons were happy and supported the policy that later was condemned as appeasement. Churchill, however, grasped the essence of what had occurred and of the unavoidable conflict that lay ahead: "I do not begrudge our loyal, brave people . . . the natural, spontaneous outburst of joy and relief when they learned that the hard ordeal would no longer be required of them at the moment; but they should know the truth . . . this is only the beginning of the reckoning. This is only the first sip, the first foretaste of a bitter cup which will be proffered to us year by year unless by a supreme recovery of moral health and martial vigor we arise again and take our stand for freedom."

Thus do we meekly acquiesce in the loss of the world's rain forests and their living species, the loss of the Everglades, the Aral Sea, the old-growth forests of the Pacific Northwest, the topsoil of the Midwest, the vegetation and soils of the Himalayas, Lake Baikal, the Sahel, the unnecessary deaths of 37,000 children every day, the thinning of the stratospheric ozone layer, the disruption of the climate balance we have known since the dawn of the human species. Bitter cups all — but only "the beginning of the reckoning," only the first of a steady stream of progressively more serious ecological catastrophes that will be repeatedly proffered to us and will, sooner or later, arouse us to action and convince us to fight back.

What does it mean to make the effort to save the global environment the central organizing principle of our civilization? For one thing, it means securing widespread agreement that it *should* be the organizing principle, and the way such a consensus is formed is especially important because this is when priorities are established and goals are set. Historically, such a consensus has usually been secured only with the emergence of a life-or-death threat to the existence of society itself; this time, however, the crisis could well be irreversible by the time its consequences become sufficiently clear to congeal public opinion — if not panic. This time, the crisis has a long fuse: the natural processes do not immediately display the full extent of the damage we are inflicting. Once set in motion, however, some of the changes we are imposing will be very difficult to reverse. It is essential, therefore, that we refuse to wait for the obvious signs of impending catastrophe, that we begin immediately to catalyze a consensus for this new organizing principle.

Adopting a central organizing principle — one agreed to voluntarily — means embarking on an all-out effort to use every policy and program, every law and institution, every treaty and alliance, every tactic and strategy, every plan and course of action — to use, in short, every means to halt the destruction of the environment and to preserve and nurture our ecological system. Minor shifts in policy, marginal adjustments in ongoing programs, moderate improvements in laws and regulations, rhetoric offered in lieu of genuine change — these are all forms of appeasement, designed to satisfy the public's desire to believe that sacrifice, struggle, and a wrenching transformation of society will not be necessary. The Chamberlains of this crisis carry not umbrellas but "floppy hats and sunglasses" — the palliative allegedly suggested by a former secretary of the interior as an appropriate response to the increased ultraviolet radiation caused by the thinning of the ozone layer.

Some are willing to assume that we can easily adapt to the effects of our assault on the environment — and indeed, some adaptation will be necessary because of the changes that have already been set irrevocably in motion. But those who propose adaptation as our principal response are really advocating just another form of appeasement. And of course the soothing message of reassurance they bring — that all is well and nothing need be done — is almost always welcome and even flattering to those who believe their complacency is justified.

But there are terrible moral consequences to the current policy of delay, just as there were when we tried to postpone World War II. Then, as now, the real enemy was a dysfunctional way of thinking.

In Nazi Germany, dysfunctional thinking was institutionalized in the totalitarian state, its dogma, and its war machine. Today, a different dysfunction takes the form of ravenous, insatiable consumption, its dogma, and the mechanisms by which ever more resources are obtained. Totalitarianism and consumptionism have led to crises peculiar to advanced industrial civilization: both are examples of alienation and technology run amok. Just as totalitarianism collapses individuals into "the state," the new ideology of consumption collapses individuals into the desire for what they consume, even as it fosters the assumption that we are separate from the earth. It is this strange and destructive way of thinking about our relationship to the physical world that is our real enemy.

The struggle to save the global environment is in one way much more difficult than the struggle to vanquish Hitler, for this time the war is with ourselves. We are the enemy, just as we have only ourselves as allies. In a war such as this, then, what is victory and how will we recognize it?

It is not merely in the service of analogy that I have referred so often to the struggles against Nazi and communist totalitarianism, because I believe that the emerging effort to save the environment is a continuation of these struggles, a crucial new phase of the long battle for true freedom and human dignity. My reasoning here is simple: free men and women who feel individual responsibility for a particular part of the earth are, by and large, its most effective protectors, defenders, and stewards. Wherever this sense of responsibility is diluted or compromised by competing imperatives, the likelihood of stewardship and care for the environment diminishes. For example, when a farmer with a short-term lease is under financial pressure to maximize profits, the land is vulnerable to exploitation. When the officers of a timber company are given annual bonuses based on the size of its quarterly profits, they are likely to cut more trees at a younger age and plant fewer seedlings for harvest in future decades — and care less about the soil erosion that often results. When the voters in a democracy are not prepared by knowledge or conviction to hold their politicians accountable for the polluting of public air and water resources by private parties, then the politicians will be loath to assert the people's right to freely enjoy public property.

The fact that these abuses occur in free countries does not in any sense support the argument that the principles of private ownership, capitalism, or democracy are to blame — any more than the existence of slavery during the first seventy-four years of the American republic could be blamed on representative democracy. As we now understand, the genius of the American founders in conceiving liberty and devising the means for guaranteeing it lies not in the eternal perfection of the laws and institutions they crafted in the late eighteenth century but in the truths they enshrined as guiding principles. Referring to those truths, subsequent generations could and did reinterpret the meaning of freedom for themselves in the context of new knowledge, changed circumstances, and accumulated experience.

Most, though not all, of the generation that wrote the Constitution were partially blind when it came to the inalienable rights of the African Americans held as slaves. They felt themselves separate from people of a different color, so they failed to understand that the rights they so passionately defended for themselves and all others to whom they felt connected by "common destiny" were rights held in common by all. Similarly, most were blind when it came to the right of women to vote. But this blindness did not prevent subsequent generations from developing a fuller understanding of the truths embodied in the Constitution, even if they were not fully visible to those who first had the courage to use them as the foundation stones for democratic government.

Today, most — though not all — are partially blind when it comes to our connection with the natural world. The philosophy of life we have inherited, which tells us we are separate from the earth, obscures our understanding of our common destiny and renders us vulnerable to an ecological catastrophe, just as our forebears' assumption that they were morally and spiritually connected to their slaves led to the catastrophe of the Civil War. What we need now is an expanded understanding of what these freedoms involve and how they can be extended once more.

The largest promise of the democratic idea is that, given the right to govern themselves, free men and women will prove to be the best stewards of their own destiny. It is a promise that has been redeemed against the challenge of every competing idea. The assertion that we might be half slave and half free, that only men should vote, that the common resolve of free nations would wither against the singular will of totalitarianism — all these ideas have fallen while ours remains. But now a new challenge — the threat to the global environment — may wrest control of our destiny away from us. Our response to this challenge must become our new central organizing principle.

The service of this principle is consistent in every way with democracy and free markets. But just as the abolition of slavery required a fuller understanding of the nature of both democracy and private property — and the relationship between the two, so this new struggle will require a still larger conception of how democracy and free markets enhance each other. Just as the extension of civil rights to women and African Americans required a deeper insight into the meaning of democratic government and a broader definition of what all human beings have in common, this global challenge will require a fuller understanding of our connection to all people today and our obligations to future generations.

Let there be no doubt: unless we can grow in these understandings, we will lose our ability to redeem the promise of freedom.

Empowered by a new way of thinking, we can without question succeed in an all-out effort to save the environment. But this effort will require an even deeper respect on the part of governments for the political and economic freedom of individuals; it will also require dramatic measures to ensure that individuals are given both the information to comprehend the enormity of the challenge and adequate political and economic power to be true stewards of the places where they live and work. By themselves, well-motivated individuals cannot hope to win this struggle, but as soon as enough people agree to make it our central organizing principle, success will come within our grasp and we can begin to make rapid progress.

But in those countries that already consider themselves free, there is another political prerequisite as well. The emphasis on the rights of the individual must be accompanied by a deeper understanding of the responsibilities to the community that every individual must accept if the community is to have an organizing principle at all.

This notion is itself an ecological question, in the sense that it involves a balance between rights and responsibilities. In fact, what many feel is a deep philosophical crisis in the West has occurred in part because this balance has been disrupted: we have tilted so far toward individual rights and so far away from any sense of obligation that it is now difficult to muster an adequate defense of any rights vested in the community at large or in the nation — much less rights properly vested in all humankind or in posterity. Today, about the only way to mobilize public opinion sufficiently to stop transgressions of what may be called ecological rights is to spotlight individuals who have been victimized by this or that environmentally unsound practice. The harm done to the community, to the world at large, or to future generations, is then treated as incidental to the harm done to these individuals; their rights are sufficiently similar to our own individual rights that we are willing to defend them, since, after all, we might thereby build desirable protections for ourselves.

This separation from community is clearly related to the assumption that we are separate from the earth. It has not only the same philosophical cause — the overriding faith in the power of the individual intellect — but also the same solution: a more balanced way of thinking about our relationship to the world, including our communities. This reaffirmation of our connection to others involves an obligation to join *with others* in adequately defending and protecting those of our rights — such as the right to breathe clean air and drink clean water — that are naturally among the individual rights belonging to others as well as to us, and are vested in the community — or nation, or world — as a whole.

Another threat to this new organizing principle is the pervasiveness of corruption in both the underdeveloped and developed worlds. This too, in a sense, is an ecological problem. Corruption pollutes the healthy patterns of accountability on which democratic government — and our ability to share stewardship of the environment — depend. Indeed, in almost every case of environmental devastation, corruption has played a significant role in deadening the ability of the political system to respond to the early signals of degradation brought to its attention.

But since corruption affects the system, many feel sufficiently detached to acquiesce in the general lethargy and inertia that allow it to continue. In order for this new central organizing principle to be effectively established, however, the political pollution of corruption must be confronted as an evil that is, in essence, similar to that embodied in the physical pollution of the air and water.

Likewise, the continued tolerance of widespread social injustice has the same corrosive effect on our ability to contemplate vigorous and sustained mutual initiatives. The promotion of justice and the protection of the environment must go hand in hand in any society, whether in the context of a nation's domestic policies or in the design of "North-South" agreements between the industrial nations and the Third World. Without such commitments, the world cannot contemplate the all-out effort urgently needed. Already, the dialogue between poor and wealthy nations is poisoned by Third World cynicism about the industrial nations' motives. But recently it has also been enriched by proposals such as "debt for nature swaps," in which debts are canceled in return for cooperation in protecting endangered parts of the environment.

Rapid economic improvements represent a life-or-death imperative throughout the Third World. Its people will not be denied that hope, no matter the environmental costs. As a result, that choice must not be forced upon them. And from their point of view, why should they accept what we, manifestly, will not accept for ourselves? Who is so bold as to say that any developed nation is prepared to abandon industrial and economic growth? Who will proclaim that any wealthy nation will accept serious compromises in comfort levels for the sake of environmental balance?

The industrial world must understand that the Third World does not have a choice of whether to develop economically. And one hopes it will do so according to a more rational pattern than has thus far been urged upon it. If it does not, then poverty, hunger, and disease will consume entire populations. Long before that, whole societies will experience revolutionary political disorder, and it is not inconceivable that some of the resulting wars could be fought with crude nuclear weapons, because nuclear proliferation continues to reflect our general failure to manage technology wisely. Indeed, some of those wars could be fought over natural resources themselves, like fresh water.

Finally, we must come to a deeper understanding of what is meant by development. Many people of good will recognized early on the need to bring some coherence to the efforts of rich and poor nations to build a more just world civilization; what came to be called development is now the chief means by which wealthy nations — often working through multilateral institutions like the World Bank and the regional development banks — can help underdeveloped nations accelerate their transition into modernity. Unfortunately, the international development programs have often been catastrophic for the countries on the receiving end, because so many of the large projects involved have tried to jump-start industrial growth even if it put the environment at risk. The problems so common in international development programs have been ecologically dangerous in another sense too: there has rarely been much balance between the projects financed by the industrial world and the true needs of the Third World. As a result, too many projects have ended up doing more harm than good, disrupting both ecological balance and societal stability. Part of the price paid is in discouragement, cynicism, and a simplistic conclusion by some that development itself is inherently undesirable. A sad example is the aftermath of the flooding in 1991 of large areas in Bangladesh and the enormous loss of life from drowning, disease, and starvation. The mild response of the industrial world seemed to reflect a fatalistic surrender to the idea that such suffering is certainly tragic but essentially unavoidable. Moreover, serious analysts argued that almost any kind of help from the West was unwise because by facilitating the resettlement of low-lying areas vulnerable to flooding and by increasing the population through the feeding of many who would otherwise starve, Western aid would only sow the seeds of even worse tragedies at the time of the next floods.

Unless the industrial world refines its understanding of how it can help effectively and what kind of development is appropriate, there will be a great many more such political and moral surrenders in the face of horrendous tragedy. We in the rich nations will lie to ourselves and pretend that since development didn't work and often made the problems much worse, the best course of action is to do nothing — to become a silent partner with mass death in the cynical culling of the human species.

* * *

The forces of oppression have always relied on silent partners, the large mass of people who quietly accede to the leaders and institutions who are interested only in consolidating their own power. But the free world owes much to those who have resisted the often intimidating governing forces, and I believe we already owe much to those who have refused to remain silent about environmental degradation. Indeed, one of the most powerful stories of such resistance comes from World War II, when resisters often paid the ultimate price.

In the winter of 1942, the city of Leningrad was under siege, surrounded by Nazi tanks. For 900 horrifying days, the citizens endured artillery fire and aerial bombardment, but the worst was starvation. Before the blockade lifted, it had caused more than 600,000 men, women, and children to starve to death. Those who survived ate sawdust, rats, grass — anything they could find.

At Leningrad's Vavilov Institute, a botanical and agricultural research center, thirty-one scientists remained to guard the unique collection of plants and seeds gathered meticulously from their places of genetic origin all over the world under the direction of the legendary biologist, geneticist, and plant explorer Nikolai Ivanovich Vavilov. Vavilov's colleagues — Vavilov himself had been imprisoned by Stalinists for "the sabotage of agriculture" — were less concerned about the expansion of Nazi Germany or Stalin's expanding *gulag* than about the expansion of industrial civilization into those areas of the natural world that contained the unique genetic resources that govern the world's food supply and genetic diversity. Because these areas were under a siege of their own, the specimens at the institute represented for many species of food crops the only remaining link between the past and the future. Yet even during the bombardment of Leningrad, Vavilov's colleagues bravely planted new generations of crops in order to freshen their genetic stock. And when hungry rats learned to knock the metal boxes of seeds off the shelves to get to the contents, the scientists took turns standing watch to protect their genetic treasures.

Surrounded by edible seeds and sacks of plants such as rice and potatoes, fourteen of the scientists died of starvation in December rather than consume their precious specimens. Dr. Dmytry S. Ivanov, the institute's rice specialist, was surrounded by bags of rice when he was found dead at his desk. He was reported to have said shortly before his death, "When all the world is in the flames of war, we will keep this collection for the future of all people."

The courageous efforts made by these scientists were typical of those made by many men and women of conscience who fought behind the lines during this century's wars in organized resistance movements that tried to slow the momentum of the advancing totalitarian juggernaut. All of them recognized the system they faced as an evil force and felt compelled to fight, even though the odds of succeeding were extremely slim. Some were motivated by religious values; others acted out of nationalist outrage; still others were simply compelled by conscience. Virtually all of these resistance fighters and their movements turned out to be no more than marginal obstacles in the actual battles. But most served as valuable sources of intelligence for the armed strategic response when it was eventually organized, and in some cases they delayed the enemy advance enough to allow the opposition to rally. Most significant, at a time when most of the world was looking the other way, they sounded the alarm — not just in words, but in the inspiring language of courage and conscience.

Today, most of the world is looking the other way, pretending not to notice industrial civilization's terrible onslaught against the natural world. But alarms are now being sounded all over the world in the same familiar tones of courage and conscience. Standing bravely against this new juggernaut, a new kind of resistance fighter has appeared: men and women who have recognized the brutal nature of the force now grinding away at the forests and oceans, the atmosphere and fresh water, the wind and the rain, and the rich diversity of life itself.

They fight against the odds, with little hope of prevailing in the larger war but with a surprising record of success in skirmishes that slow the onslaught and sometimes save the particular corner of the ecological system they have been moved to defend. In the process, they have provided not only examples of courage and resourcefulness, but also our single most valuable source of "intelligence" from the front lines on what works and what doesn't. Pending the call to arms and the organization of a massive effort to preserve the environmental equilibrium that is now under siege, these resistance fighters are trying desperately to draw the world's attention to the truth of what is taking place. They inspire all of us who are coming to a better understanding of this crisis, and at least a few of their stories should be told here.

As individuals, today's resistance fighters often share the character traits psychologists found in those from World War II. Whether these new fighters live in Africa, Asia, Latin America, or environmentally stressed areas of the industrial world, they are in most cases ordinary people with a deeply embedded sense of right and wrong — usually imparted by a strong and caring parent during their upbringing — and a stubborn refusal to bend their principles even when the opposing force appears invincible and even deadly. One such person is Tos Barnett, a former Supreme Court justice of Papua New Guinea (PNG) and a constitutional adviser to the prime minister. Barnett narrowly escaped assassination and had to flee PNG in December 1989, after he submitted the results of his lengthy and courageous investigation of the massive deforestation in PNG and the allegations of corruption in the timber marketing business. His twenty-volume, six-thousand-page report described terrible abuses: bribery of high government officials by Japanese corporations, forced labor camps in the forests with indigenous peoples working seven days a week under deplorable conditions, wanton destruction of homes, massive tax avoidance, and pervasive corruption.

Although Barnett had been appointed by the government to conduct the investigation, the authorities were simply not prepared for his response. Documents relating to the corruption he uncovered were burned deliberately, and his final report was officially suppressed in PNG. Among the companies he named were Sanyo and Sumitomo, names already known for the pattern of destruction they left in Indonesia and Malaysia before PNG became their next target.

Two thousand miles northwest of PNG, at the same time that Tos Barnett began his investigation, thousands of indigenous people in Sarawak, Malaysia, linked arms in human barricades to block the logging roads deep inside the tropical rain forest in a desperate effort to stop indiscriminate and destructive logging. In Sarawak, as in PNG, an investigation revealed that the government officials responsible for the forest were allegedly paid money for concessions by companies who wanted to destroy it. Responding to commercial pressures, the government soon passed a law making it illegal to blockade logging roads. The indigenous peoples, including the Penan, the Kenyah, the Kayan, the Kelabit, and the Lun Bawang, Iban, finally took matters into their own hands after erosion had so damaged their lands that their water was unfit for drinking. Those who relied on the rapidly disappearing forest for their survival became especially desperate. Although these resistance fighters had little chance against the powerful forces arrayed against them, their courage inspired international protests that are still continuing.

One of the Sarawak peoples, the Penan, sent a delegation to the United States with the help of an environmental group, the Friends of the Earth. They walked into my office one winter day, looking a little like visitors from another millennium, their straw headgear and wooden bracelets the only remnants of the culture they left behind, wearing borrowed sweaters as protection against the unaccustomed cold. Using a translator who had painstakingly learned their language, the Penans described how the logging companies had set up floodlights to continue their destruction of the forest all through the night as well as the day. Like the shell-shocked inhabitants of a city under siege, they described how not even the monsoon rains slowed the chain saws and logging machinery that were destroying the ancestral home of their people. Before they left, they gave me the following statement, translated into halting English:

> Almost all the forest reserve of the Penan are gone. The river water has become more silted especially during the rainy season like at present. Many of the village people fall sick. The children often get stomach ache. Food is also not enough. We have to walk to far places to look for food. If we are lucky, after one or two days only can we find food . . . medicines are also difficult to find. When we set up blockades from June to October 1987 the situation became a little better. The river water was beginning to be clear. Forest destruction stopped temporarily . . . many police and soldiers come with helicopters and weapons.
>
> We say that the problems of the Penan make the Penan people set up blockades. Penans want the land and forest of their ancestors. The police and soldiers reply that there are now new laws. If we do not

open our blockades, we will be caught and sent to prison. We Penans do not want to fight with force. We do not want families and village people getting hurt. When the police and soldiers opened the blockades, we did not resist. When we seek the help of the police they do not come. When the company asks, the police come and stay near our village for a long time. Why is the new law so harsh? We want laws that help us. But the new law is most disappointing. We are not being killed by weapons, but when our lands are taken, it is the same as killing us.

These are the front lines of the war against nature now raging throughout the world. These words from the Penans are hauntingly similar to the pleas of the Ethiopians invaded by Mussolini's forces in 1935 and the calls for help from Hungary when Soviet tanks rolled through its streets in 1956. The weak and powerless are the early victims, but the relentless and insatiable drive to exploit and plunder the earth will soon awaken the conscience of others who are only now beginning to interpret the alarms and muffled cries for help. In the famous words of Pastor Martin Niemoller, about how the Nazis were able to take over an entire society: "In Germany the Nazis came first for the Communists, and I didn't speak up because I wasn't a Communist. Then they came for the Jews, and I didn't speak up because I wasn't a Jew. Then they came for the trade unionists, and I didn't speak up because I wasn't a trade unionist. Then they came for the Catholics, and I didn't speak up because I was a Protestant. Then they came for me, and by that time there was no one left to speak for me. "

One who did speak up for the new resistance was Chico Mendes. In late 1988, Senators Tim Wirth, John Heinz, and I, along with Congressmen John Bryant and Gerry Sikorski and a delegation of observers, were on our way to Brazil to meet with Mendes, perhaps the best known of these resistance fighters in recent years, when he was assassinated by a group of wealthy landowners. A native of the Acre province in the Amazon, Mendes organized and led the *seringueiros* (rubber tappers), who harvest the renewable bounty of the rain forest — fruits, nuts, and especially rubber — which they make from sap collected by tapping the rubber trees. Their way of life has helped preserve the rain forest, but they now stand in the way of commercial interests that seek to exploit the forest by bulldozing or burning it down to make way for temporary cattle ranches. On numerous occasions, Mendes and his band of rubber tappers attempted to block the bulldozers and refused to allow the exploiters to cross the rain forest to set fire to nearby plots. Going further, Mendes organized alternative — and sustainable — ways to live profitably in the rain forest and promoted a host of imaginative enterprises to encourage the farmers not to destroy their land but live in harmony with it. As his recognition of the complexity of the issues grew and his innate leadership skills developed, he entered politics, but the wealth and power of the landowners ensured his defeat. When he continued to challenge their interests, they murdered him, three days before Christmas, with a shotgun blast as he stood in the doorway of his home.

When we arrived in Acre, we met with Mendes's widow, Ilzamar, and with his colleagues in the *seringueiro* movement who have vowed to continue to resist the destruction of the Amazon. Their battle is far from over: many others in the movement less visible than Mendes have also been killed, and the rain forest cannot be saved without organized support from the rest of the world. But Mendes's violent death was not wasted, for it dramatically focused global attention on the severe threats to one of the most remarkable ecosystems in the world. Though he wanted to live, this is precisely what Mendes prophesied in his final interview: "If an angel came down from the sky and could guarantee that my death could strengthen this fight, it would be a fair exchange."

Mendes would have appreciated a remarkable woman from Kenya named Wangari Matthai, who founded the Greenbelt Movement. Matthai understands the power inherent in the simple act of planting trees, and by organizing women to plant trees and stop soil erosion, the movement has now planted more than 8 million trees in less than a decade. I spent a day with members of the Greenbelt Movement in rural Kenya in the fall of 1990; they explained to me that the tree plantings offer an opportunity to share information — woman to woman — on family planning and to distribute birth control technologies. In addition, their tree nurseries now serve as genetic storehouses of indigenous food plants carefully matched to the microenvironments of different altitudes

Chico Mendes organized the rubber tappers in the Amazon who harvest renewable products of the rain forest — such as rubber and brazil nuts — and fought against the wholesale burning and destruction by large landowners to earn short-term profits at the expense of long-term ecological tragedy. In December 1988, Mendes was assassinated by a gunman. A rancher and his son were later convicted of the crime.

and soil types in various areas of Kenya. Although Matthai was persecuted and imprisoned in the early years of her movement, she is now considered so popular as to be untouchable, and most of the persecution has ceased.

From Wangari Matthai the world can learn yet another lesson about what works and what doesn't. She and her colleagues used economic incentives to encourage the planting of trees, but they discovered an important key that has helped ensure their success: the compensation is not paid when the tree is planted, but only after it has been nurtured through the seedling stage and become tall and strong enough to survive on its own.

One of the most colorful advocates of global family planning is Mechai Viravayda of Thailand, often called "the P. T. Barnum of birth control." Through eye-catching and humorous promotional activities, he has helped cut Thailand's population growth dramatically — from 3.2 percent in 1970 to less than 1.7 percent in 1990. "If one can get people to laugh together about the topic of family planning, the battle is half over," says Viravayda. Alhough some of his showmanship may seem outlandish — balloon-blowing contests with condoms, the distribution of condoms by traffic policemen on New Year's Eve in what he calls a "cops and rubbers" operation, schemes to give car insurance payments to taxi drivers who sell a quota of condoms, to mention only a few — he has also established a network of thousands of family planning centers, which have also been helpful in Thailand's battle against AIDS.

Mendes, Matthai, and Viravayda come from different continents and cultures, but they have one important trait in common. Like the men and women who became part of resistance movements in World War II, they didn't bring any special training or experience to this new cause. In that they are typical: the people who now devote their lives to healing the environment are, for the most part, "ordinary" people who have a keenly developed sense of right and wrong and the courage to stand by their convictions. They are not people who usually go looking for a fight, but they stand up to injustice when they are confronted by it.

Christine and Woodrow Sterling of Toone, Tennessee, didn't go looking for a fight either. And they certainly had no way of knowing that their sense of right and wrong would end up embodied in two sweeping new federal laws governing the disposal of hazardous chemical waste and one of the largest legal judgments ever granted by a jury in a waste disposal case. They just noticed that their well water tasted funny, and they realized that the strange taste originated with whatever it was the trucks coming from Memphis, seventy-five miles away, had dumped into trenches near their property. And they knew that wasn't right.

Sometimes a large community of ordinary people rises up as one. A few years ago a waste disposal company, Browning & Ferris Industries (BFI), operating through front men, began to secretly buy up leases in a huge area of Henderson County, Tennessee, next to Interstate 40. On Christmas Eve 1983, the news broke: BFI was planning a major regional hazardous waste disposal facility in northern Henderson County. On Christmas morning, 20,000 citizens in the county opened their presents a little hurriedly, then went to work fighting back. Several weeks later, every elementary school student had completed a colorful poster describing in personal terms why the environment of Henderson County had to be protected against BFI's proposal. Every square inch of wall space in the courthouse, from floor to ceiling on all three levels, was covered with the posters. To say their parents were distressed would not be exactly accurate; they were too confident of their ability to stop the dump. As one of them, Marilyn Bullock, put it at a hearing with a crowd of men and women behind her nodding in agreement, "You don't understand. This dump is *not* going to be put here."

What distinguished this movement from a knee-jerk "not in my backyard" (NIMBY) reaction was the organizers' insistence on carefully gathering and evaluating the facts, not just about the company's proposal but about all similar proposals to dump the kind of hazardous waste that BFI transported. A few months later, at a congressional hearing in nearby Jackson, Tennessee, some of the leading experts on waste disposal technology testified that the people of Henderson County, far from being the emotional hotheads described by the allegedly responsible advocates of the waste disposal facility fronting for BFI, were exactly right in their reading of chemistry books and scientific studies, which showed that the proposed dump was utterly irresponsible. In the end, the facility was not built in Henderson County or anywhere else, and Marilyn Bullock and her organization, Humans Against Lethal Trash

(HALT), went on to take the lead in developing state and national legislation. Their moral authority came from having the truth on their side. And it had nothing to do with partisan politics: Henderson County has long been one of the most Republican counties in the United States, and Republicans and Democrats alike have joined the resistance in great numbers.

Some backyards, however, are already seriously polluted. Consider Cancer Alley in the Lower Mississippi River Valley between Baton Rouge and New Orleans, where more than a quarter of America's chemicals are produced and where some of the highest cancer rates in the nation are found. Pat Bryant, an African-American political activist who got his start in the early 1980s by organizing public housing tenants in St. Charles Parish, shifted his attention to the constant respiratory and eye problems of the children who lived near the Union Carbide and Monsanto complexes. In Bryant's view — a view shared by many others — Cancer Alley was made possible by ethnic discrimination and political powerlessness.

I met Bryant in Atlanta at the Southern Environmental Assembly, a gathering of mostly white people. As he said later, "A lot of the environmentalists were middle class. We all speak English, but what we say doesn't always mean the same thing. We must put aside foolish customs that divide us and work together, at least for the sake of our children." True to this vision, Bryant organized a coalition of environmental and labor groups to create the Louisiana Toxics Project, which contributed to the passage of the state's first air quality law in 1989.

But the coalition wasn't finished, and Bryant's view of the problem extended beyond Cancer Alley. The following year, during the Senate's consideration of the Clean Air Act, Bryant and one of the national groups linked with his project brought a glaring loophole to my attention; it would have allowed companies emitting toxic air pollutants (the most deadly class of air pollution) to avoid the tougher emissions standards by buying up neighborhoods downwind of their facilities and creating what environmentalists call "dead zones," large areas devoid of people and inevitably bordering poor neighborhoods, whose property values would drop. Of course, whenever the wind shifted, the toxic pollutants that were supposed to fall in the dead zone would fall somewhere else —

most often on impoverished black families. The national coalition was instrumental in passing the amendment to close the loophole.

Bryant's perspective is especially important because of continuing fears on the part of some activists who work with the poor and oppressed that the environmental movement will divert attention from their priorities. As Bryant puts it, "The environment is the number-one problem in this country. As an African American, my hope and aspiration to be free are greatly dimmed by the prospect of environmental destruction. If we're going to make great strides on this problem, we're going to have to build African American–European American coalitions."

Sometimes, of course, the NIMBY phenomenon raises difficult questions about how and where to locate unpopular facilities. Indeed, among the most highly charged and passionately argued political issues today are proposals to place new landfills or garbage dumps in areas where those nearby feel threatened. But I have found that when the merits of a proposal really do make sense, those who are fighting it often temper their opposition, or at least find it harder to attract support outside their immediate area. More commonly, advocates of a facility that raises serious environmental questions try to distract attention from the real issues by accusing their opponents of adopting a knee-jerk NIMBY approach. And while it's true that people are sometimes too self-interested in their opposition to these issues, in my opinion the NIMBY syndrome is the beginning of a healthy trend. In fact, I am convinced that political support for measures to protect the global atmosphere will one day surge when the meaning of the word "backyard" expands to encompass each person's share of the air we all breathe.

The impetus for that change will come from the forefront of science and from the work of scientists like Dr. Sherwood Rowland, who in 1974 discovered a dramatic change in the chemical composition of our atmosphere. Concentrations of chlorine have increased enormously throughout the world because of the widespread use of chlorofluorocarbons (CFCs). But when he and Dr. Mario Molina, both of the University of California, Irvine, announced their disturbing discovery, Rowland suffered a form of scientific persecution. Suddenly he was no longer invited to address as many scientific meetings; in at least two cases, companies profiting from the suspect chemicals threatened to withhold funding for conferences if Rowland was on the program. But Sherwood Rowland has a keenly developed sense of right and wrong; he decided to fight, and he has been fighting for more than seventeen years now. With his wife, Joanne, he has traveled to conferences and symposia in every part of the world, argued his case, and patiently taken on all comers.

In large part because of the steadfast work of Sherwood Rowland and such colleagues as Mario Molina and Robert Watson of NASA, the world was ready to listen when the ozone hole caused by CFCs suddenly appeared over Antarctica in 1987. Dr. Susan Solomon led an emergency scientific expedition to the South Pole and confirmed what Rowland had hypothesized. Many countries have finally begun taking action, yet even now, when the evidence against CFCs is overwhelming, these life-threatening compounds are still being released into the atmosphere, and some countries still refuse to join the global effort to ban them.

Some members of the resistance have taken the fight for the environment from the scientific journals and symposia to their own backyards and from there to corporate boardrooms and the halls of Congress. One remarkable woman, Lynda Draper, joined the fight in her own kitchen. I learned about her courageous battle in early 1989, when she came to my office for help soon after discovering that General Electric (GE) was planning to release a huge quantity of CFCs into the atmosphere; in fact, they had already begun doing so. As she told me the story (later confirmed by GE), a repairman had knocked on her door in Ellicott City, Maryland, and said that her relatively new refrigerator had a defective compressor that needed to be replaced. Indeed, some GE officials felt they were demonstrating foresight and responsiveness in organizing the largest recall program in the history of the industry; they intended to replace an estimated one to two million compressors that might otherwise fail, leaving some customers with spoiled food.

As Draper told it, the repairman came into her kitchen and inspected her refrigerator. "Then he asked me to open the window. I didn't know why he wanted me to do that, but I did. Then all of a sudden I heard this loud *whoosh!*" Draper, who had worked for environmental groups, immediately realized what was happening: the CFCs from the old compressor were being vented from her refrigerator through her window right into the atmosphere. Horrified, she protested to the repairman. When he explained that it was only a few ounces, she didn't buy it. She went to work finding out how many refrigerators were involved in the recall program and multiplied the total by the number of ounces of CFCs in each; she determined that at least 125 and perhaps as many as 312 tons of CFCs would enter the atmosphere over the course of the recall program. She was determined to stop the company, but the challenge she accepted was steeped in irony: both her father and her grandfather had been long-standing GE employees, and her husband had worked for the company for ten years. At first, Draper took the most obvious course of action — she called up the company to tell them what they were doing and why it was wrong. When the company responded that the amounts were too small to worry about, she decided to complain to local and state officials and finally the EPA. Even then, she got nowhere. By the time she arrived at my office, she had contacted the Public Interest Research Group and had made plans for a press conference to call for a nationwide boycott of all GE products.

In response to Draper's persistent efforts, the company completely changed its corporate policy on CFCs and became the industry leader in CFC reductions, setting standards its competitors are still trying to match. GE developed special equipment for scavenging CFCs instead of releasing them and used this equipment to remove CFCs from other parts of the environment as a way of offsetting what they had released during the recall program. The proposed boycott was never called, and Draper, who started as a PTA volunteer, now works full-time to save the environment. "I intend to keep fighting," she says. "If more people were fighting, we'd make more progress."

Sherwood Rowland and Lynda Draper are, in effect, comrades in arms in the same struggle. But the struggle isn't just about CFCs. Ultimately, it's about the entire relationship between human civilization and the global environment. Slowly, people in all walks of life are coming to understand the enormity of the problem; slowly we are awakening to the strategic threats now posed by our rapidly expanding civilization. Although the resistance is growing, becoming more sophisticated, and scoring some dramatic successes, the larger war to save the earth is being lost. That will change only when the rest of humankind, drawing on the lessons learned by these pioneers and inspired by their courage and sacrifice, finally organizes an all-out response to this unprecedented threat.

Again, we must not forget the lessons of World War II. The Resistance slowed the advance of fascism and scored important victories, but fascism continued its relentless march to domination until the rest of the world finally awoke and made the defeat of fascism its central organizing principle from 1941 through 1945. But too many ignored the early warnings; in June 1936, for instance, Haile Selassie, the emperor of Ethiopia, addressed the world through the League of Nations. His was the first country invaded by the Axis, and in describing the atrocities committed by Mussolini's forces, including the use of poison gas, he said, "Soldiers, women, children, cattle, rivers, lakes, and pastures were drenched continually with this deadly rain. In order to kill off systematically all living creatures, in order the more surely to poison waters and pastures, the Italian command made its aircraft pass over and over again." Selassie said that he wanted both to describe the atrocities against his people and to make it plain that the rest of the world would soon face the same aggression. He came, he said, to "give Europe a warning of the doom that awaits it if it should bow before the accomplished fact. . . . God and history will remember your judgment."

The world is once again at a critical juncture. A relentless advance is again claiming victims throughout the world, and again courageous men and women are standing in the path of destruction and calling upon the rest of the world to help stop the invasions. But this time we are invading ourselves and attacking the ecological system of which we are a part. As a result, we now face the prospect of a kind of global civil war between those who refuse to consider the consequences of civilization's relentless advance and those who refuse to be silent partners in the destruction. More and more people of conscience are joining the effort to resist, but the time has come to make this struggle the central organizing principle of world civilization. We have had a warning of the fate that awaits if we "bow before the accomplished fact." God and history will remember our judgment.

15

A Global Marshall Plan

Human civilization is now so complex and diverse, so sprawling and massive, that it is difficult to see how we can respond in a coordinated, collective way to the global environmental crisis. But circumstances are forcing just such a response; if we cannot embrace the preservation of the earth as our new organizing principle, the very survival of our civilization will be in doubt.

That much is clear. But how should we proceed? How can we create practical working relationships that bring together people who live in dramatically different circumstances? How can we focus the energies of a disparate group of nations into a sustained effort, lasting many years, that will translate the organizing principle into concrete changes — changes that will affect almost every aspect of our lives together on this planet?

We find it difficult to imagine a realistic basis for hope that the environment can be saved, not only because we still lack widespread agreement on the need for this task, but also because we have never worked together globally on any problem even approaching this one in degree of difficulty. Even so, we must find a way to join this common cause, because the crisis we face is, in the final analysis, a global problem and can only be solved on a global basis. Merely addressing one dimension or another or trying to implement solutions in only one region of the world or another will, in the end, guarantee frustration, failure, and a weakening of the resolve needed to address the whole of the problem.

While it is true that there are no real precedents for the kind of global response now required, history does provide us with at least one powerful model of cooperative effort: the Marshall Plan. In a brilliant collaboration that was itself unprecedented, several relatively wealthy nations and several relatively poor nations — empowered by a common purpose — joined to reorganize an entire region of the world and change its way of life. The Marshall Plan shows how a large vision can be translated into effective action, and it is worth recalling why the plan was so successful.

Immediately after World War II, Europe was so completely devastated that the resumption of normal economic activity was inconceivable. Then, in the early spring of 1947, the Soviet Union rejected U.S. proposals for aiding the recovery of German industry, convincing General George Marshall and President Harry Truman, among others, that the Soviets hoped to capitalize on the prevailing economic distress — not only in Germany but also in the rest of Europe. After much discussion and study, the United States launched the basis for the Marshall Plan, technically known as the European Recovery Program (ERP).

The commonly held view of the Marshall Plan is that it was a bold strategy for helping the nations of Western Europe rebuild and grow strong enough to fend off the spread of communism. That popular view is correct — as far as it goes. But the historians Charles Maier and Stanley Hoffman, both professors at Harvard, emphasize the strategic nature of the plan, with its emphasis on the structural causes of Europe's inability to lift itself out of its economic, political, and social distress. The plan concentrated on fixing the bottlenecks — such as the damaged infrastructure, flooded coal mines, and senseless trade barriers — that were impeding the potential for growth in each nation's economy. ERP was sufficiently long-term that it could serve as an overall effort to produce fundamental structural reorientation, not just offer more emergency relief or another "development" program. It was consciously designed to change the dynamic of the systems to which it extended aid, thus facilitating the emergence of a healthy economic pattern. And it was brilliantly administered by Averell Harriman.

Historians also note the Marshall Plan's regional focus and its incentives to promote European integration and joint action. Indeed, from the very beginning the plan tried to facilitate the emergence of a larger political framework — unified Europe; to that end, it insisted that every action be coordinated with all the countries in the region. The recent creation of a unified European parliament and the dramatic steps toward a European political community to accompany the European Economic Community (EEC) have all come about in large part because of the groundwork of the Marshall Plan.

But when it was put in place, the idea of a unified Europe seemed even less likely than the collapse of the Berlin Wall did only a few years ago — and every bit as improbable as a unified global response to the environmental crisis seems today. Improbable or not, something like the Marshall Plan — a Global Marshall Plan, if you will — is now urgently needed. The scope and complexity of this plan will far exceed those of the original; what's required now is a plan that combines large-scale, long-term, carefully targeted financial aid to developing nations, massive efforts to design and then transfer to poor nations the new technologies needed for sustained economic progress, a worldwide program to stabilize world population, and binding commitments by the industrial nations to accelerate their own transition to an environmentally responsible pattern of life.

But despite the fundamental differences between the late 1940s and today, the model of the Marshall Plan can be of great help as we begin to grapple with the enormous challenge we now face. For example, a Global Marshall Plan must, like the original, focus on strategic goals and emphasize actions and programs that are likely to remove the bottlenecks presently inhibiting the healthy functioning of the global economy. The new global economy must be an inclusive system that does not leave entire regions behind — as our present system leaves out most of Africa and much of Latin America. In an inclusive economy, for instance, wealthy nations can no longer insist that Third World countries pay huge sums of interest on old debts even when the sacrifices necessary to pay them increase the pressure on their suffering populations so much that revolutionary tensions build uncontrollably. The Marshall Plan took the broadest possible view of Europe's problems and developed strategies to serve human needs and promote sustained economic progress; we must now do the same on a global scale.

But strategic thinking is useless without consensus, and here again the Marshall Plan is instructive. Historians remind us that it would have failed if the countries receiving assistance had not shared a common ideological outlook, or at least a common leaning toward a set of similar ideas and values. Postwar Europe's strong preference for democracy and capitalism made the regional integration of economies possible; likewise, the entire world is far closer to a consensus on basic political and economic principles than it was even a few short years ago, and as the philosophical victory of Western principles becomes increasingly apparent, a Global Marshall Plan will be increasingly feasible.

It is fair to say that in recent years most of the world has made three important choices: first, that democracy will be the preferred form of political organization on this planet; second, that modified free markets will be the preferred form of economic organization; and, third, that most individuals now feel themselves to be part of a truly global civilization — prematurely heralded many times in this century but now finally palpable in the minds and hearts of human beings throughout the earth. Even those nations that still officially oppose democracy and capitalism — such as China — seem to be slowly headed in our philosophical direction, at least in the thinking of younger generations not yet in power.

Another motivation for the Marshall Plan was a keen awareness of the dangerous vacuum created by the end of the Axis nations' totalitarian order and the potential for chaos in the absence of any positive momentum toward democracy and capitalism. Similarly, the resounding philosophical defeat of communism (in which the Marshall Plan itself played a significant role) has left an ideological vacuum that invites either a bold and visionary strategy to facilitate the emergence of democratic government and modified free markets throughout the world — in a truly global system — or growing chaos of the kind that is already all too common from Cambodia to Colombia, Liberia to Lebanon, and Zaire to Azerbaijan.

The Marshall Plan, however, depended in part for its success on some special circumstances that prevailed in postwar Europe yet do not prevail in various parts of the world today. For example, the nations of Europe had developed advanced economies before World War II, and they retained a large number of skilled workers, raw materials, and the shared experience of modernity. They also

This picture of Earth, showing the Antarctic, Africa, Madagascar, and the Arabian peninsula, is perhaps the most famous picture of our planet from space. It was taken in 1972 during the Apollo program from a point in space halfway to the moon.

shared a clear potential for regional cooperation — although it may be clearer in retrospect than it was at the time, when the prospect of warm relations between, say, Germany and England seemed remote.

In contrast, the diversity among nations involved in a Global Marshall Plan is simply fantastic, with all kinds of political entities representing radically different stages of economic and political development — and with the emergence of "post-national" entities, such as Kurdistan, the Balkans, Eritrea, and Kashmir. In fact, some people now define themselves in terms of an ecological criterion rather than a political subdivision. For example, "the Aral Sea region" defines people in parts of several Soviet republics who all suffer the regional ecological catastrophe of the Aral Sea. "Amazonia" is used by peoples of several nationalities in the world's largest rain forest, where national boundaries are often invisible and irrelevant.

The diversity of the world's nations and peoples vastly complicates the model used so successfully in Europe. Even so, another of the Marshall Plan's lessons can still be applied: within this diversity, the plans for catalyzing a transition to a sustainable society should be made with regional groupings in mind and with distinctive strategies for each region. Eastern Europe, for example, has a set of regional characteristics very different from those of the Sahel in sub-Saharan Africa, just as Central America faces challenges very different from those facing, say, the Southeast Asian archipelago.

Many of the impediments to progress lie in the industrial world. Indeed, one of the biggest obstacles to a Global Marshall Plan is the requirement that the advanced economies must undergo a profound transformation themselves. The Marshall Plan placed the burden of change and transition only on the recipient nations. The financing was borne entirely by the United States, which, to be sure, underwent a great deal of change during those same years, but not at the behest of a foreign power and not to discharge any sense of obligation imposed by an international agreement.

The new plan will require the wealthy nations to allocate money for transferring environmentally helpful technologies to the Third World and to help impoverished nations achieve a stable population and a new pattern of sustainable economic progress. To work, however, any such effort will also require wealthy nations to make a transition themselves that will be in some ways more wrenching than that of the Third World, simply because powerful established patterns will be disrupted. Opposition to change is therefore strong, but this transition can and must occur — both in the developed and developing world. And when it does, it will likely be within a framework of global agreements that obligate all nations to act in concert. To succeed, these agreements must be part of an overall design focused on devising a healthier and more balanced pattern in world civilization that integrates the Third World into the global economy. Just as important, the developed nations must be willing to lead by example; otherwise, the Third World is not likely to consider making the required changes — even in return for substantial assistance. Finally, just as the Marshall Plan scrupulously respected the sovereignty of each nation while requiring all of them to work together, this new plan must emphasize cooperation — in the different regions of the world and globally — while carefully respecting the integrity of individual nation-states.

This point is worth special emphasis. The mere mention of any plan that contemplates worldwide cooperation creates instant concern on the part of many — especially conservatives — who have long equated such language with the advocacy of some supranational authority, like a world government. Indeed, some who favor a common global effort tend to assume that a supranational authority of some sort is inevitable. But this notion is both politically impossible and practically unworkable. The political problem is obvious: the idea arouses so much opposition that further debate on the underlying goals comes to a halt — especially in the United States, where we are fiercely protective of our individual freedoms. The fear that our rights might be jeopardized by the delegation of even partial sovereignty to some global authority ensures that it's simply not going to happen. The practical problem can be illustrated with a question: What conceivable system of world governance would be able to compel individual nations to adopt environmentally sound policies? The administrative problems would be gargantuan, not least because the inefficiency of governance often seems to increase geometrically with the distance between the seat of power and the individuals affected by it; and given the chaotic state of some of the governments that would be subject to that global entity, any such institution would most likely have unintended side effects and complications that would interfere with the underlying goal. As Dorothy Parker once said about a book she didn't like, the idea of a world government "should not be tossed aside lightly; it should be thrown with great force."

But if world government is neither feasible nor desirable, how then can we establish a successful cooperative global effort to save the environment? There is only one answer: we must negotiate international agreements that establish global constraints on acceptable behavior but that are entered into voluntarily — albeit with the understanding that they will contain both incentives and legally valid penalties for noncompliance.

The world's most important supranational organization — the United Nations — does have a role to play, though I am skeptical about its ability to do very much. Specifically, to help monitor the evolution of a global agreement, the United Nations might consider the idea of establishing a Stewardship Council to deal with matters relating to the global environment — just as the Security Council now deals with matters of war and peace. Such a forum could be increasingly useful and even necessary as the full extent of the environmental crisis unfolds.

Similarly, it would be wise to establish a tradition of annual environmental summit meetings, similar to the annual economic summits of today, which only rarely find time to consider the environment. The preliminary discussions of a Global Marshall Plan would, in any event, have to take place at the highest level. And, unlike the economic summits, these discussions must involve heads of state from both the developed and developing world.

In any global agreement of the kind I am proposing, the single most difficult relationship is the one between wealthy and poor nations; there must be a careful balance between the burdens and obligations imposed on both groups of nations. If, for example, any single agreement has a greater impact on the poor nations, it may have to be balanced with a simultaneous agreement that has a greater impact on the wealthy nations. This approach is already developing naturally in some of the early discussions of global environmental problems. One instance is the implicit linkage between the negotiations to save the rain forests — which are found mostly in poor countries — and the negotiations to reduce greenhouse gas emissions — which is especially difficult for wealthy nations. If these negotiations are successful, the resulting agreements will become trade-offs for each other.

The design of a Global Marshall Plan must also recognize that many countries are in different stages of development, and each new agreement has to be sensitive to the gulf between the countries involved, not only in terms of their relative affluence but also their various stages of political, cultural, and economic development. This diversity is important both among those nations that would be on the receiving end of a global plan and among those expected to be on the giving end. Coordination and agreement among the donor countries might, for example, turn out to be the most difficult challenge. The two donor participants in the Marshall Plan, the United States and Great Britain, had established a remarkably close working relationship during the war, which was then used as a model for their postwar collaboration. Today, of course, the United States cannot conceivably be the principal financier for a global recovery program and cannot make the key decisions alone or with only one close ally. The financial resources must now come from Japan and Europe and from wealthy, oil-producing states.

The Western alliance has frequently been unwieldy and unproductive when large sums of money are at stake. Nevertheless, it has compiled an impressive record of military, economic, and political cooperation in the long struggle against communism, and the world may be able to draw upon that model just as the United States and Great Britain built upon their wartime cooperation in implementing the Marshall Plan. Ironically, the collapse of communism has deprived the alliance of its common enemy, but the potential freeing up of resources may create the ideal opportunity to choose a new grand purpose for working together.

Still, a number of serious obstacles face cooperation among even the great powers — the United States, Japan, and Europe — before a Global Marshall Plan could be considered. Japan, in spite of its enormous economic strength, has been reluctant to share the responsibility for world political leadership and thus far seems blind to the need for it to play such a role. And Europe will be absorbed for many years in the intricacies of becoming a unified entity — a challenge further complicated by the entreaties of the suddenly free nations in Eastern Europe that now want to join the EEC.

As a result, the responsibility for taking the initiative, for innovating, catalyzing, and leading such an effort, falls disproportionately on the United States. Yet in the early 1990s our instinct for world leadership often seems not nearly so bold as it was in the late 1940s. The bitter experience of the Vietnam War is partly responsible, and the sheer weariness of carrying the burden of world leadership has taken a toll. Furthermore, we are not nearly as dominant in the world economy as we were then, and that necessarily has implications for our willingness to shoulder large burdens. And our budget deficits are now so large as to stifle our willingness to consider even the most urgent of tasks. Charles Maier points out that the annual U.S. expenditures for the Marshall Plan between 1948 and 1951 were close to 2 percent of our GNP. A similar percentage today would be almost $100 billion a year (compared to our total nonmilitary foreign aid budget of about $15 billion a year).

Yet the Marshall Plan enjoyed strong bipartisan support in Congress. There was little doubt then that government intervention, far from harming the free enterprise system in Europe, was the most effective way to foster its healthy operation. But our present leaders seem to fear almost any form of intervention. Indeed, the deepest source of their reluctance to provide leadership in creating an effective environmental strategy seems to be their fear that if we do step forward, we will inevitably be forced to lead by example and actively pursue changes that might interfere with their preferred brand of laissez-faire, nonassertive economic policy.

Nor do our leaders seem willing to look as far into the future as did Truman and Marshall. In that heady postwar period, one of Marshall's former colleagues, General Omar Bradley, said, "It is time we steered by the stars, not by the lights of each passing ship." This certainly seems to be another time when that kind of navigation is needed, yet too many of those who are responsible for our future appear to be distracted by such "lights of passing ships" as overnight public opinion polls.

In any effort to conceive of a plan to heal the global environment, the essence of realism is recognizing that public attitudes are still changing — and that proposals which are today considered too bold to be politically feasible will soon be derided as woefully inadequate to the task at hand. Yet while public acceptance of the magnitude of the threat is indeed curving upward — and will eventually rise almost vertically as awareness of the awful truth suddenly makes the search for remedies an all-consuming passion — it is just as important to recognize that at the present time, we are still in a period when the curve is just starting to bend. Ironically, at this stage, the maximum that is politically feasible still falls short of the minimum that is truly effective. And to make matters worse, the curve of political feasibility in advanced countries may well look quite different than it does in developing countries, where the immediate threats to well-being and survival often make saving the environment seem to be an unaffordable luxury.

It seems to make sense, therefore, to put in place a policy framework that will be ready to accommodate the worldwide demands for action when the magnitude of the threat becomes clear. And it is also essential to offer strong measures that are politically feasible now — even before the expected large shift in public opinion about the global environment — and that can be quickly scaled up as awareness of the crisis grows and even stronger action becomes possible.

With the original Marshall Plan serving as both a model and an inspiration, we can now begin to chart a course of action. The world's effort to save the environment must be organized around strategic goals that simultaneously represent the most important changes and allow us to recognize, measure, and assess our progress toward making those changes. Each goal must be supported by a set of policies that will enable world civilization to reach it as quickly, efficiently, and justly as possible.

In my view, five strategic goals must direct and inform our efforts to save the global environment. Let me outline each of them briefly before considering each in depth.

The first strategic goal should be **the stabilizing of world population,** with policies designed to create in every nation of the world the conditions necessary for the so-called demographic transition

— the historic and well-documented change from a dynamic equilibrium of high birth rates and death rates to a stable equilibrium of low birth rates and death rates. This change has taken place in most of the industrial nations (which have low rates of infant mortality and high rates of literacy and education) and in virtually none of the developing nations (where the reverse is true).

The second strategic goal should be **the rapid creation and development of environmentally appropriate technologies** — especially in the fields of energy, transportation, agriculture, building construction, and manufacturing — capable of accommodating sustainable economic progress without the concurrent degradation of the environment. These new technologies must then be quickly transferred to all nations — especially those in the Third World, which should be allowed to pay for them by discharging the various obligations they incur as participants in the Global Marshall Plan.

The third strategic goal should be **a comprehensive and ubiquitous change in the economic "rules of the road" by which we measure the impact of our decisions on the environment.** We must establish — by global agreement — a system of economic accounting that assigns appropriate values to the ecological consequences of both routine choices in the marketplace by individuals and companies and larger, macroeconomic choices by nations.

The fourth strategic goal should be **the negotiation and approval of a new generation of international agreements** that will embody the regulatory frameworks, specific prohibitions, enforcement mechanisms, cooperative planning, sharing arrangements, incentives, penalties, and mutual obligations necessary to make the overall plan a success. These agreements must be especially sensitive to the vast differences of capability and need between developed and undeveloped nations.

The fifth strategic goal should be **the establishment of a cooperative plan for educating the world's citizens about our global environment** — first by the establishment of a comprehensive program for researching and monitoring the changes now under way in the environment in a manner that involves the people of all nations, especially students; and, second, through a massive effort to disseminate information about local, regional, and strategic threats to the environment. The ultimate goal of this effort would be to foster new patterns of thinking about the relationship of civilization to the global environment.

Each of these goals is closely related to all of the others, and all should be pursued simultaneously within the larger framework of the Global Marshall Plan. Finally, the plan should have as its more general, integrating goal **the establishment, especially in the developing world — of the social and political conditions most conducive to the emergence of sustainable societies —** such as social justice (including equitable patterns of land ownership); a commitment to human rights; adequate nutrition, health care, and shelter; high literacy rates; and greater political freedom, participation, and accountability. Of course, all specific policies should be chosen as part of serving the central organizing principle of saving the global environment.

Let us now consider each goal at some length. For each, I will offer a general discussion of why the goal is important, specific proposals for achieving the goal, and the role of the United States in doing so.

I. STABILIZING WORLD POPULATION

No goal is more crucial to healing the global environment than stabilizing human population. The rapid explosion in the number of people since the beginning of the scientific revolution — and especially during the latter half of this century — is the clearest single example of the dramatic change in the overall relationship between the human species and the earth's ecological system (see illustration, pages 32–33). Moreover, the speed with which this change has occurred has itself been a major cause of ecological disruption, as societies that learned over the course of hundreds of generations to eke out a living within fragile ecosystems are suddenly confronted — in a single generation — with the necessity of feeding, clothing, and sheltering two or three times as many individuals within those same ecosystems.

The raw numbers alone tell a dramatic story: as we saw in

Chapter 1, from the first recognizable humans more than 2 million years ago until the end of the last Ice Age, there were never more than a few million people. And 10,000 years later, there were still only about 2 billion. Yet in just the last forty-five years, the population has increased by more than that total — to almost 5.5 billion people. And in the next forty-five years, it will double yet again, pushing the world population to an estimated 9 billion people. Although until recently experts were predicting that population would stabilize at 10 billion sometime in the next century, they now say the total could reach 14 billion or even higher before leveling off. And what is in some ways even more remarkable is that an estimated 94 percent of the increases will occur in the developing world, where poverty and environmental degradation are already the most severe.

To put these numbers in a different perspective, consider that the world is adding the equivalent of one China's worth of people every ten years, one Mexico's worth every year, one New York City's worth every month, and one Chattanooga's worth every single day. If these increases continue at the current rate, the impact on the environment in the next century will be unimaginable. In thinking about ways to limit population growth, it is important to appreciate the powerful momentum toward continued increases that comes from the sheer size of our present population, especially the huge number of people now in or soon to be entering their childbearing years. Even if the entire world suddenly shifted to much lower growth rates, this momentum would still lead to continued increases in total numbers of people for many decades. It is also crucial to remember that the difference between ultimately stabilizing the population at 10 or 11 billion rather than 14 or 15 billion is profound in terms of our human impact on the environment — as well as the people of the earth.

Numbers aside, the way these masses of people live and the technologies they use are critical in determining their impact on the environment. Any child born into the hugely consumptionist way of life so common in the industrial world will have an impact on the environment that is, on average, many times more destructive than that of a child born in the developing world. For this reason, some Third World leaders resent the argument that the global environment is threatened principally by population growth in their countries.

But the absolute numbers are staggering. Consider the plight of several countries, as estimated by the "best case" scenarios projected by the United Nations Fund for Population Activities. Kenya, which now has 27 million people, will have within thirty years an estimated 50 million people. Egypt's population, 55 million people today, is increasing by an amount equal to the entire population of Israel every four years; within thirty years it will be at least 100 million people; Nigeria, which already has 100 million people, will have within thirty years at least 300 million people. All three countries are already putting great strains on their natural resources and threatening the integrity of their ecological systems, so it is truly frightening to imagine the impact of doubling or tripling their numbers — not to mention the pitiful quality of life these extra scores of millions can expect. Already new epidemics — from cholera to the Black Plague to AIDS — have emerged in societies knocked off balance by rapid population growth and the consequent disruption of their traditional patterns of living, and the degradation of their surrounding environments. Moreover, in some fast-growing areas like the Sahel, large-scale famines are no longer merely occasional but are now endemic.

The social and political tensions associated with growth rates like these threaten to cause the breakdown of social order in many of the fastest-growing countries, which in turn raises the prospect of wars being fought over scarce natural resources where expanding populations must share the same supplies. Take water, for example. Every one of the fourteen countries that depend on the Nile River is experiencing a population explosion, yet the Nile has no more water flowing between its banks today than it did in biblical times. Similarly, every nation depending upon the River Jordan has a rapidly expanding population, and conflict over that small stream is now beginning to add measurably to the political, social, and religious tensions that have long existed in the area. The Tigris and Euphrates present the same dilemma: limited supplies of water must be shared among several populations, all of whom are growing dramatically.

Consider the problem in another way. Imagine that someone

invented a miraculous technology that enabled human civilization to cut per capita emissions of greenhouse gases in half; imagine how much that might reduce our concern about global warming. (Actually, what may be even harder to imagine is that we will have to cut emissions even more than that.) But with world population doubling in less than a half century, all of the potential reductions in greenhouse gases coming from even extraordinary advances in technology would be completely wiped out — and soon greenhouse gases would be accumulating just as rapidly as they are today.

Consider also the impact on soil erosion of the current efforts to feed 5.5 billion people, and try to imagine the impact of trying to harvest twice as much food worldwide in only four decades. And what about well water and wood for heating and cooking? In many areas, women already walk several miles each day to scavenge for firewood and search for fresh water. Their horizons are denuded of trees and shrubs and the water tables are falling. As the number of these human scavengers doubles — and in some countries triples — the result is certain to be ecologically and socially catastrophic. In many regions, of course, it already is.

But there is ample cause for hope that the problem can be solved if the right solutions are properly pursued. Fortunately, population specialists now know with a high degree of confidence what factors dramatically reduce birth rates. It will take time and money, of course, but not that much of either when compared to the ingredients that have been most lacking: political resolve, imagination, leadership, and a willingness to address the problem on a truly global basis. Just as no problem better illustrates the dramatic change in humankind's impact on the global environment, none better illustrates the necessity of adopting a truly global solution and designing it in a strategic fashion.

Most of the developing world (with some important exceptions) have high birth and death rates — and rapidly growing populations. By contrast, the United States, Canada, Japan, Taiwan, South Korea, Hong Kong, Singapore, Australia, New Zealand, and every nation in Western Europe and Scandinavia now have low birth and death rates — and relatively stable populations. But the nations in the second category, including the United States, were all once in the first category. In fact, most of them did not make the demo-graphic transition until the 1930s, and in some cases even later. But in the developing world, death rates declined dramatically in the 1960s while birth rates did not. Why?

When we initially examine the changes industrial countries experienced as they began to achieve relatively stable rates of population growth, we are tempted to focus mainly on the dramatic increases in per capita income — and then conclude that rising incomes are the secret. Indeed, incomes in these countries did rise, but they contributed indirectly, not directly, to the changes in thinking that led to smaller families.

A more careful analysis suggests that rising per capita income is also associated with several of the basic causes of demographic transition. *High literacy rates and education levels* are important, especially for women; once they are empowered intellectually and socially, they make decisions about the number of children they wish to have. *Low infant mortality rates* give parents a high level of confidence that even with a small family, some of their children will grow to maturity, carry the family name and genes (and in the belief of some societies, the spirits of ancestors), and provide physical security for their parents when they are old. *Nearly ubiquitous access to a variety of affordable birth control techniques* gives parents the power to choose when and whether to have children.

These are the major factors, but there is a final secret to success. Experience has proven that the crisis can be solved only if it is approached holistically — that is, by addressing all of the critical dimensions simultaneously, with careful attention to how they relate to one another. In this sense, the problem challenges us to understand population growth as a complex system of causes and effects. And because all of the conditions necessary for stability must be present simultaneously and must be sustained — in some cases for several decades — before the transition to stability can begin, the population explosion also poses a challenge to our stamina and persistence. What's required, then, is foresight, maturity of commitment and philosophical cohesion — qualities that are much more likely to emerge if the challenge is confronted on a global basis.

Much confusion, disappointment, and despair have accompanied the effort to control population growth. Failure has usually occurred when politicians did not establish all of the necessary conditions to induce the desired change in the dynamics of the system. For example, a great deal of emphasis has been placed on the availability of birth control techniques and devices. But unless a number of other changes take place simultaneously, simply flooding a nation with condoms, pills, IUDs, and sterilization operations will make little difference in the birth rate. Yet most of the controversy over population policy today is about programs to make birth control more available. Little debate — and even less effort — is devoted to literacy and education levels. And even though infant mortality receives a great deal of attention, its connection to population growth is usually ignored.

Unfortunately, many supporters of strong programs for economic development in the Third World assume that the aggressive promotion of birth control and rising national income will eventually cause the population growth rates to stabilize. But too many of these programs increase national income by stripping away whatever natural resources can be quickly sold on the world market, further impoverishing the countryside. Tropical countries, for example, have frequently been encouraged to cut down their rain forests and sell the lumber as a strategy for development, but much of the cash ends up in the hands of a wealthy few (and in bank accounts in industrial countries), leaving the populace even worse off, stripped of their natural resources with little in return. The availability of birth control often has little effect under such circumstances; sometimes, in fact, population growth rates actually increase after this kind of development as the cycle of rural poverty and population increases spins even more rapidly out of control.

Money intended to develop the means for economic growth, and lead to a rise in per capita income, can also be siphoned away to provide subsistence to larger numbers of new people — and the cycle continues. Moreover, when the countryside is degraded, the migration to urban areas accelerates, and so does the consequent breakdown of traditional social patterns (some of which had served to restrain population growth). Ethiopia is an example of this cycle: although it has received a great deal of development assistance, its leaders have misused it; per capita income has not improved. Its literacy rate is extremely low, and its infant mortality rate is among the highest in the world. So too, of course, is its population growth rate — persistently.

But there are some stunning success stories that show what can happen with a strategic approach. One of the most interesting case studies of demographic transition in the Third World comes from the Kerala province of southwestern India, where the population growth has stabilized at zero even though per capita incomes are still extremely low. The provincial leaders, with assistance from international population funding, developed a plan that is keyed to Kerala's unique cultural, social, religious, and political characteristics and focuses on a few crucial factors. First, they have achieved an extremely high rate of literacy, especially among women. Second, through good health care and adequate nutrition, they have lowered their infant mortality rate dramatically. And third, they have made birth control readily and freely available. The consequences are little short of remarkable: in an area of the world characterized by uncontrollable population growth, Kerala's rate more nearly resembles that of Sweden than nearby Bombay.

The world's strategy for inducing a global demographic transition to lower growth rates should be based on the strategy used in Kerala and elsewhere. Specifically, the Global Marshall Plan should:

1. Allocate resources to fund carefully targeted functional literacy programs keyed to every society where the demographic transition has yet to occur. Although the emphasis should be on women, the programs should be directed toward men as well. Coupled with this program should be a plan for basic education, emphasizing simple techniques in sustainable agriculture, specific lessons on preventing soil erosion, planting trees, and safeguarding clean water supplies. Although literacy and education have always been seen as worthwhile goals, in the past they have been subordinated to the more general goal of economic development. This effort should now be given top priority.

2. Develop effective programs to reduce infant mortality and ensure the survival and excellent health of children. Several decades ago, the African leader Julius Nyerere said, "The most powerful contraceptive is the confidence by parents that their children will survive." In most societies, there is of course no such thing as "social security," and parents often rely on their adult children to care for them in their old age. If they believe there is a good possibility that their offspring will die young, then parents have a strong incentive to have many children in order to ensure that at least some survive into adulthood; besides, in a subsistence economy, children can help gather firewood and water, bring in the harvest, tend the garden and watch over the livestock. Again, programs for reducing infant mortality and improving child and maternal health have been established in the past, but they too have been seen as secondary to the general — if poorly defined — goal of development.

3. Ensure that birth control devices and techniques are made ubiquitously available along with culturally appropriate instruction. At the same time, scientists must be charged with stepping up research into improved and more easily accepted contraception techniques. Depending upon the culture, delayed marriages and birth spacing should also be emphasized, along with traditional practices such as breast feeding (which simultaneously improves the health of children and suppresses fertility).

The U.S. Role

It is time to act boldly to implement these three specific policies designed to enable the world to reach the strategic goal of the demographic transition. And it is time for the United States to take the leading role — because no one else can or will. But in the face of this clear challenge, the United States is — unbelievably — reducing its commitment to world population programs, essentially because President Bush depends upon a political coalition that includes a tiny minority within a minority who strongly oppose contraception and object to government funds being used to purchase any birth control technology.

Ironically, the vast majority of people in the anti-abortion movement do not object to birth control at all, but for the sake of their political coalition, they do not challenge the few who insist on opposing it. The movement as a whole acquiesces in the exaggerated claim that virtually any birth control program will almost inevitably lead to abortion. As a result, even when Congress adds language to proposed foreign aid legislation explicitly prohibiting the use of any government money for abortion, the anti-abortion movement still opposes it. At its urging, the United States has even gone so far as to prohibit its participation in any birth control program wherein some other participant sanctions abortion, using money from another source. While they try to show how our foreign aid might be used for abortions, in reality the anti-abortion advocates are primarily keeping peace in their own political family by opposing birth control.

It is particularly ironic that George Bush — of all the potential leaders in his generation of Republican politicians — would, as president, find it impossible to summon the courage to resist such an unreasoning demand by a tiny part of his electoral coalition. As a congressman, Bush became chairman of the Republican task force on the population issue in Congress and introduced appropriate legislation. Indeed, he was a leader on the issue. Both then and later, as President Nixon's representative to the United Nations, Bush made many eloquent speeches about the need for vigorous U.S. leadership in world family planning programs. He even contributed a foreword to a 1973 book on the population crisis, in which he described how his own determination to fight for birth control had come from his father, who had suffered unfair, demagogic attacks on the issue:

> My own first awareness of birth control as a public policy issue came with a jolt in 1950 when my father was running for the United States Senate in Connecticut. Drew Pearson, on the Sunday before Election Day, 'revealed' that my father was involved with Planned Parenthood. My father lost that election by a few hundred out of close to a million votes. Many political observers felt a sufficient number of voters were swayed by his alleged contacts with the birth controllers to cost him the election.

He was full of courage on the issue then, proudly proclaiming defiance of the political risks he said he knew all too well. But his courage has vanished — and I think it is because he is especially vulnerable when confronted with demands from one part of the coalition first put together by President Reagan, a coalition that Bush inherited and had to hold together at all costs to win the White House.

But again, it is a mistake to focus exclusively on birth control. The resulting oversimplification of what is an extremely complex problem is, in my view, one of the reasons for the strange lack of urgency many Americans seem to feel when it comes to the population crisis. There is also much more productive work to be done if we are ever to reach the strategic goal of the demographic transition.

Narrow-minded views of the problem also serve to alienate some natural allies. The Catholic church, for example, despite its opposition to contraception, is one of the most forceful and effective advocates for literacy and education programs and for measures to dramatically reduce infant mortality. Significantly, it has worked actively on these issues in many developing countries as part of coalitions in which other members are distributing birth control devices; furthermore, Catholic countries and non-Catholic countries with similar social conditions have identical rates of contraception use and population growth. Spokesmen for the Holy See have repeatedly signaled that although the Church's formal view is not likely to change, it will not block others who wish to promote contraception, and it is anxious to play a vigorous role in addressing the other factors that help to hasten the demographic transition. Isn't that good enough? Isn't it time to set the old arguments aside and instead find more ways to work together?

The argument over abortion, on the other hand, is not likely to end anytime soon. Personally, I favor the right of a woman to choose whether to conceive and have a child; I am deeply troubled by the reports from China of coerced abortions and the extension of totalitarianism to the workplace, where supervisors sometimes monitor each woman's menstrual cycle. I am also troubled by evidence that in some industrial countries where birth control is not readily available, abortion rates are astronomical. For example, in the Russian republic, the average woman has more than ten abortions in her lifetime. In my view, U.S. policy should not support or in any way encourage such practices. But isn't it obvious that a wider availability of birth control ultimately reduces the number of abortions? That is what the evidence shows.

The United States should restore full funding of its share of the cost of international population stabilization programs and increase the effort to make birth control available throughout the world — but it should do much more as well. It should also take the lead in organizing worldwide efforts to increase literacy and lower infant mortality rates; otherwise, the efforts to promote the use of birth control will be for naught.

Some theorists maintain that the demographic transition is a nearly inevitable process that will occur sooner or later in all countries as they develop economically. But they make two crucial errors. First, the process they describe can take centuries, assuming intervening events do not reverse its direction. Second, with populations now as large as they are, the momentum of further growth already built into the population is pushing many countries over the edge of an economic cliff as their resources are stripped away and the cycle of poverty and environmental destruction accelerates. Clearly, it is time for a global effort to create everywhere on earth the conditions conducive to stabilizing population.

II. DEVELOPING AND SHARING APPROPRIATE TECHNOLOGIES

The second strategic goal of the Global Marshall Plan should be a highly focused and well-financed program to accelerate the development of environmentally appropriate technologies that will abet sustainable economic progress and that can be substituted for the ecologically destructive technologies currently in place. These new technologies must be efficiently and quickly transferred to those nations incapable of developing or purchasing them on their own.

It is important, however, to remember that there is a great danger in seeing technology alone as the answer to the environmental crisis. In fact, the idea that new technology is the solution to all our problems is a central part of the faulty way of thinking that created the crisis in the first place.

Unless we come to a better understanding of both the potential and the danger of technology, the addition of more technological power simply ensures further degradation of the environment, and no matter what new technologies we discover, no matter how cleverly and efficiently we manage to get them into the hands of people throughout the world, the underlying crisis will worsen unless, at the same time, we redefine our relationship to the environment, stabilize human population, and use every possible means to bring the earth back into balance.

Still, the dissemination of new, appropriate technologies will likely be critical to our success in saving the environment. After all, once a technology — whether environmentally destructive or not — becomes well established, it acquires a staying power that makes it extremely difficult to dislodge. Individuals, corporations, social institutions, even entire cultures, adapt to the requirements of their technologies, in the process making such large investments of wealth, effort, time, and experience as to render any thought of change impractical or even unimaginable. And the elaborate web of economic incentives and disincentives that grow up around these technologies and their related activities serves as a further barrier.

New technologies should not be embraced too eagerly, either; careful study of their environmental impact is essential. CFCs offer a case in point. Originally developed as substitutes for an earlier generation of chemicals that were harmful on contact, CFCs were determined to be nontoxic before they were introduced. Ironically, they do not react chemically to human contact because of their stability as molecules, which also enables them to float steadily upward — unimpeded by any transforming reaction in the lower atmosphere — until they rise all the way to the stratosphere, where the sun's ultraviolet rays slice them apart and touch off the destruction they cause to the ozone layer. Although no amount of research can determine every possible impact of a technology, our experience with CFCs reminds us of the importance of caution when marveling over the magical powers of any new tool or technology.

The latest chapter in the story of CFCs teaches another important — and much happier — lesson: the search for new chemical compounds that can be quickly substituted for CFCs, institutionalized in the Montreal Protocol, the international treaty on CFCs that was adopted in 1987, can be seen as an important precedent for the much larger challenge confronting us. In encouraging the search for CFC replacements, the Montreal Protocol looked beyond governmental research agencies and aimed a number of measures at the private sector. The protocol included agreements on steadily tighter quotas on the amount of CFCs and related chemicals that firms in each country are allowed to produce in any given year, heavy taxes on those that are still being produced, and prospective bans just a few years hence on virtually all CFC production. Because of the rising demand for air conditioning, refrigeration, and all the other prominent uses for this family of chemicals, these measures mean that huge profits await any company that can quickly come up with acceptable substitutes for CFCs, which in turn means that huge amounts of money have been invested in the race to develop these substitutes.

During the debates over the Montreal Protocol, spokesmen for the CFC industry said that it was futile for the world to expect substitutes anytime soon; the good news, however, is that — for most uses — replacement chemicals are already being found, and they are being developed much faster than the naysayers predicted. Moreover, under the protocol, the substitutes will be made available to developing countries, which will ensure that this technology spreads as rapidly as possible.

So even though a great deal more needs to be done to rid the world of CFCs and related compounds, a success story is unfolding, and it should give us confidence that we can succeed in an even larger effort. The challenge is formidable, and the fundamental problem is how to find a mechanism that will effectively encourage a worldwide effort to quickly develop substitute technologies for the wide range of dangerous technologies now in common use around the world. Clearly, the world's nations need to develop a comprehensive and cooperative program that is strategic in scope and aggressive in its approach.

With this urgent need in mind, I propose the worldwide development of a **Strategic Environment Initiative** (SEI), a program that would discourage and phase out these older, inappropriate technologies and at the same time develop and disseminate a new generation of sophisticated and environmentally benign substitutes. As soon as possible, the SEI should be the subject of intensive international discussions, first among the industrial nations and then between them and the developing world. The initiative should include, at a minimum, the following:

1. Tax incentives for the new technologies and disincentives for the old.
2. Research and development funding for new technologies and prospective bans on the old ones.
3. Government purchasing programs for early marketable versions of the new.
4. The promise of large profits in a market certain to emerge as older technologies are phased out.
5. The establishment of rigorous and sophisticated technology assessment procedures, paying close attention to all of the costs and benefits — both monetary and ecological — of the new proposed substitute technologies.
6. The establishment of a network of training centers around the world, thus creating a core of environmentally educated planners and technicians and ensuring that the developing nations will be ready to accept environmentally attractive technologies and practices. We have a model for this initiative as well: during the Green Revolution, agricultural research centers of exactly this sort were set up throughout the world.
7. The imposition of export controls in developed countries that assess a technology's ecological effect, just as the Cold War technology control regime (known as COCOM) made careful and unusually accurate analyses of the potential military impact of technologies proposed for export.
8. Significant improvements in the current patchwork of laws, especially in those countries that now effectively fail to safeguard the rights of inventors and developers of new technology. This is no minor point; it is one of the principal ways to insure the viability of a major technology transfer program, and adequate protection for intellectual property rights is already a major bone of contention in global trade negotiations.
9. Better protection for patents and copyrights, improved licensing agreements, joint ventures, franchises, distributorships,

and a variety of similar legal concepts. All will be essential to unleash the creative genius on which we must draw.

I have chosen the phrase Strategic Environment Initiative purposely to imply an environmental equivalent of the Strategic Defense Initiative (SDI), the crash program to develop a series of technological breakthroughs focused on a common, if highly controversial, military objective. I have always opposed the deployment of a grand-scale SDI. Nevertheless, the research program has been remarkably successful in drawing together previously disconnected government programs, in stimulating the development of new technologies, and in forcing upon us a wave of intense new analysis of subjects previously thought to have been exhausted.

We need that same focus and intensity, and similar levels of funding, to deal comprehensively with the global environmental crisis. Just as SDI has led to well-focused programs aimed at such activities as target acquisition, instantaneous management of complex computer data streams, ultra-high-speed interception, and missile launch detection from orbit, the SEI must focus on the development of environmentally appropriate technologies. One caution, however: we should not make the mistake of equating technology with only "high" technology. Often, the most appropriate and environmentally benign approach involves "low" technology — an uncomplicated but clever approach or a passive rather than active approach. In the effort to develop the new technologies we need rapidly, all of the policy tools I have suggested here can and should be applied in a variety of fields, especially agriculture, forestry, energy production and use (in, for example, transportation and manufacturing), building technology, and waste reduction and recycling. Following is a brief discussion of how the SEI would address what's needed in each area.

Agriculture. Although the Green Revolution produced vast growth in Third World food production, it often relied on environmentally destructive techniques: heavily subsidized fertilizers and pesticides, the extravagant use of water in poorly designed irrigation schemes, the exploitation of the short-term productivity of soils (which sometimes led to massive soil erosion), monocultured crops (which drove out diverse indigenous strains), and accelerated overall mechanization, which often gave enormous advantages to rich farmers over poor ones. Now that so much more is known about the ecological consequences of some modern agricultural practices, we need a second Green Revolution that will focus on the needs of the Third World's poor, increase the productivity of small farms with low input agricultural methods, and promote environmentally sound policies and practices. The new Green Revolution, whose components should be not only scientific but financial, social, and political as well, may hold the key to satisfying the land hunger of tens of millions of poor and dispossessed people who are now being driven to activity that destroys their fragile environments. Worldwide recognition that the plight of these people has features that are substantially the same throughout the world — and that elements of a just solution, like land reform, are also the same in most countries — can lead to a powerful and effective global effort to link guarantees of justice for the dispossessed to the granting of financial assistance and technology transfers under the SEI.

Fortunately, many new, environmentally appropriate agricultural technologies are now available, and these would all be promoted under the SEI:

• New refinements in irrigation technology make it possible to simultaneously reduce water consumption, increase yields, and reclaim the productivity of overly salinated soils.

• New techniques for low-input crop management make it possible to drastically reduce soil erosion while maintaining yields and keeping costs down.

• New advances in plant genetics make it possible to introduce "natural" resistance to some crop diseases and predators without the heavy use of pesticides and herbicides.

• New approaches to crop rotation and multiple uses of land, including agroforestry, can provide alternatives to the common Third World practice of seasonally burning vast areas of land.

• New discoveries in aquaculture and fishing techniques promise alternatives to such grotesquely destructive practices as driftnet fishing.

• More sophisticated techniques of food distribution offer ways to sharply reduce the absurdly high losses during distribution in many less developed nations and offer energy savings as well.

Forestry. A strategic initiative to plant billions of trees throughout the world, especially on degraded lands, is one of the most easily understandable, potentially popular, and ecologically intelligent efforts on which the Global Marshall Plan should concentrate. The symbolism — and the substantive significance — of planting a tree has universal power in every culture and every society on earth, and it is a way for individual men, women, and children to participate in creating solutions for the environmental crisis. But for a tree planting program to be truly successful, two other tasks must be performed, one preceding the planting and the other following it. First, the seedling must be both genetically appropriate for its particular ecological niche and available in adequate numbers at the right times and in the right places. Second, whatever incentives are used to encourage the planting of trees must be keyed, not to the planting itself, but to the appropriate follow-up visits to ensure the survival of the seedling with adequate water and protection from grazing animals until it is sufficiently well established to grow on its own.

As for the first prerequisite, there is no doubt that an SEI, properly organized, could identify the tree types most appropriate for particular areas and then replicate many hundreds, thousands, or millions of the seedlings needed. In some areas, in fact, this is already being done on a small scale. But it is needed on a very large scale; as the National Research Council of the National Academy of Sciences reported in a lengthy 1991 study:

> At present, no adequate global strategy exists for systematically identifying, sampling, testing, and breeding trees with potential use. The development of improved varieties of tree species for use in industry, agroforestry, and the rehabilitation of degraded lands has been given little attention. . . . Sustained political support and expanded independent funding must be provided for long-term forest conservation operations, for professional and technical staff training, and for stabilization of institutions that address the needs of tree genetic resource conservation and management. . . . This is no longer the responsibility of a few nations, but will be achieved only through a global, cooperative effort.

As for the second prerequisite — incentives for nurturing seedlings — some useful models have begun emerging in the developing world. I visited some site projects of one of the most successful, the Kenya's Greenbelt Movement, led by Wangari Matthai, which combines tree planting with an educational program for women about birth control. Most of the seven million trees planted by the women in Matthai's movement have survived because a planter receives the small compensation for each seedling planted only after it has been sufficiently nurtured and protected to have an excellent chance of surviving on its own. The movement is now offering instruction on self-sufficiency in agriculture, and devoting some of the space in its nurseries to the development of seed stock for gardens and fields.

Another example of a tree planting movement that reclaims degraded land while serving related goals has been the effort throughout this century by Zionists to enlist the Jewish diaspora in planting millions of trees in Israel to create new forests. Indeed, the reclamation of the desert and degraded lands in Israel is one of the great ecological success stories, reversing centuries of land abuse and restoring productivity. (Unfortunately, more recent industrial approaches to agriculture have led to unsustainable water and soil depletion in some areas of Israel.)

Meanwhile, the Jewish National Fund's tree planting movement continues to serve as a model for what could be accomplished all over the world, both in degraded areas of the underdeveloped world and in industrial societies. Generations of Jewish children in the United States, for example, have raised money to plant entire forests in memory of a relative or in honor of a friend. In the process, these children have been given a valuable lesson in the dynamics of soil and water conservation — and more subtly, the importance of loving the land.

Millions of trees must be planted, but new forestry techniques must also be developed to improve timber harvesting methods. Though clear-cutting — the practice of cutting large tracts of forest to the ground — is alleged to be the most cost-efficient for the companies doing the harvesting, it often badly degrades the land and so imposes devastating long-term costs. In contrast, selective tree harvesting techniques have been pioneered in northern Europe and, with modifications, could improve harvesting practices in many areas of the world.

Energy. Energy is, of course, the lifeblood of economic progress. Unfortunately, the most common technologies for converting energy into usable forms of power happen to release enormous quantities of pollutants, including most prominently the growing concentrations of carbon dioxide (CO_2) now circling the earth. The energy component of an SEI should therefore focus on developing energy technologies that do not produce large amounts of CO_2 and other pollutants. In the near term, by far the most effective technologies for accomplishing this goal are those that improve both energy efficiency and conservation. For example, inexpensive but energy-efficient ovens and "cookers" distributed on an experimental basis in some Third World societies that depend on charcoal and firewood have dramatically decreased the amount of energy resources scavenged from the countryside.

Much larger energy savings, of course, and CO_2 reductions as well, can be accomplished when the industrial world develops more efficient internal combustion engines. And here the automobile deserves special attention.

Consider that the United States spends tens of billions of dollars on frenzied programs to upgrade and improve the technology of bombers and fighter planes to counter an increasingly remote threat to our national security, but we are content to see hundreds of millions of automobiles using an old technological approach not radically different from the one first used decades ago in the Model A Ford. We now know that their cumulative impact on the global environment is posing a mortal threat to the security of every nation that is more deadly than that of any military enemy we are ever again likely to confront. Though it is technically possible to build high-mileage cars and trucks, we are told that mandating a more rapid transition to more efficient vehicles will cause an unacceptable disruption in the current structure of the automobile industry. Industry officials contend that it is unfair to single out their industry while ignoring others that also contribute to the problem; I agree, but their point only illustrates further the need for a truly global, comprehensive, and strategic approach to the energy problem. I support new laws to mandate improvements in automobile fleet mileage, but much more is needed. Within the context of the SEI, it ought to be possible to establish a coordinated global program to accomplish the strategic goal of completely eliminating the internal combustion engine over, say, a twenty-five-year period.

Sixty years ago, Will Rogers noted the irony of a great nation in the midst of the depression becoming the first to "go to the poorhouse in an automobile." Today, we must recognize that our heavy reliance on cars as our primary means of transportation accounts for a large proportion of the CO_2 emitted into the atmosphere from the industrial world. Objectively, it makes little sense for each of us to burn up all the energy necessary to travel with several thousand pounds of metal wherever we go, but it is our failure to think strategically about transportation that has led to this absurd state of affairs.

In the early 1990s, two Japanese auto companies announced dramatic improvements in miles per gallon which they said they had achieved with no significant technological breakthroughs. The secret, it turned out, was in using higher mixtures of air to gasoline as the fuel was ignited. It had been known for a long time that such mixtures were more efficient, but in the past the technique had always proven too difficult to manage; frequently, too much air would be introduced and the engine would stop. However, better manufacturing tolerances and the use of microprocessors in controlling the flow of air and gasoline combined to make the technique suddenly feasible. In public policy, the trick is to mix intelligence with money; a higher ratio of intelligence is usually efficient and preferable, but all too often the entire apparatus comes to a halt when the mixture is too lean in money. The real challenge now is to improve our understanding of policy enough to sustain a higher ratio of intelligence to money.

We should be emphasizing attractive and efficient forms of mass transportation. To begin with, more money in the Highway Trust Fund should be made available to communities wishing to upgrade and extend subway, bus, and trolley lines. New and improved forms of mass transit, like the magnetically levitated trains in Japan and France, should be enthusiastically encouraged. We can also replace conventional commuting wherever possible, with what is now known as telecommuting. This technology is already in widespread use, as increasing numbers of people work at home but keep a direct connection to co-workers through a communications link between their computer work stations. As the capacity of computer networks increases, this trend is likely to accelerate. For a dozen years, I have been the principal author and advocate of a proposal to build a national network of "information superhighways" that would link supercomputers, work stations, and "digital libraries" to create "co-laboratories" and make it possible for people to work together despite being in different locations.

But telecommuting is not feasible in nations that lack elaborate electronic communication and power grids. And power grids are themselves no longer necessarily desirable: the economics of decentralized electricity generation are gradually becoming competitive with older technologies that generate massive amounts of electricity at a single large power plant and distribute it over transmission lines crossing the countryside. The most promising of these decentralized techniques is the generation of electrical current from the rays of the sun through photovoltaic cells, small flat panels of silicon or similar materials that are designed to produce currents of electricity. But this technology is still in its infancy, and what is required — as part of the proposed SEI — is a global effort to accelerate the development of cost-effective photovoltaic cells. The technical obstacles to developing them are becoming less important than the political and institutional barriers. And the SEI would have to address this. If, indeed, cost-effective forms of photovoltaic technology can be demonstrated, public demand may quickly sweep away the political and organizational roadblocks and in the process create the prospect of enormous profits for those entrepreneurs who quickly adapt photovoltaic technology to new uses.

Almost every discussion of substitutes for fossil fuels includes an argument over the role of nuclear power in our energy future. In fact, some opponents of positive action to save the environment try to cut short discussions of global warming with a dismissive reference to the political difficulties involved in building new nuclear reactors and expressions of exaggerated frustration with environmentalists, who, they imply, are the principal obstacles to adopting nuclear power as the obvious substitute for coal and oil.

Of course, uncertainties about future projections of energy demand and economic problems like cost overruns were the major causes of the cancellation of reactors by utilities, well before accidents like those at Three Mile Island and Chernobyl heightened public apprehension. Growing concern about our capacity to accept responsibility for the safety of storing nuclear waste products with extremely long lifetimes also adds to the resistance many feel to a dramatic increase in the use of nuclear power.

In my own view, the present generation of nuclear technology, light water-pressurized reactors, seems now rather obviously at a technological dead end. The research and development of alternative approaches should focus on discovering, first, how to build a passively safe design (whose safety does not depend upon the constant attention of bleary-eyed technicians) that eliminates the many risks of current reactors, and second, whether there is a scientifically and politically acceptable means for disposing of — in fact, isolating — nuclear waste.

In any event, the proportion of world energy use that could practically be derived from nuclear power is fairly small and is likely to remain so. It is a mistake, therefore, to argue that nuclear power holds the key to solving global warming. Nevertheless, research and development should continue vigorously, especially in technologies like fusion power, which offer the prospect, however distant, of somewhat safer and more abundant sources of electricity. Meanwhile, the emphasis in the short term should be on conservation and efficiency, and the SEI would encourage the aggressive exploration of a number of other options:

• Fuel switching can play an important role in reducing emissions of CO_2 and other pollutants. Natural gas, for example, can replace coal and oil for many uses and supply the same amounts of energy with only a tiny fraction of the unwanted by-products. The technology for recovering, transporting, and burning natural gas — an inherently more efficient and less polluting fuel — should receive particular attention to facilitate the effectiveness of our increased reliance on it as a fuel.

• Perhaps the single most sensible short-term step is to improve the efficiency of natural gas pipelines in Eastern Europe and the Soviet Union, which presently release enormous quantities of natural gas into the atmosphere, where it becomes a potent greenhouse gas. In fact, some have estimated that as much as 15 percent of all the methane released annually into the atmosphere leaks from these poorly designed pipelines. By transferring the technology of modern pipeline design to these countries, we can simultaneously reduce greenhouse gas emissions and use more of the fuel to substitute for dirtier coal and oil.

• Another need crying out for new technology is the recovery of methane currently leaking from landfills, which could also become a substitute for oil and coal instead of just another source of greenhouse gases.

• In general, the most valuable source of energy that might be substituted for our harmful technologies is all of the energy currently produced as a by-product of other activities and wasted in the process. Most industries, for example, generate enormous quantities of heat in the process of fabricating, assembling, transporting, or transforming the various materials that go into our factories and come out as finished products.

• Methods for recovering waste heat and using it to help generate power — whether by helping to generate electricity in steam turbines or by some other technique — is called cogeneration. By many estimates, an enormous amount of energy is available from the proper exploitation of new technologies for cogeneration. Unfortunately, many utilities discourage the use of cogeneration in a variety of ways, including the reluctance to purchase electricity from cogenerators as a source of power for other customers hooked into the same power grid. Laws encouraging and even requiring the efficient use of cogeneration technology have an important role to play in reducing the consumption of fossil fuels. Just the few cogeneration projects already installed are preventing as much as 80 million tons of CO_2 emissions each year in the United States alone, according to a study by the Gas Research Institute. Moreover, as with the planting of trees, the widespread adoption of cogeneration techniques also spreads a new way of thinking about the importance of energy conservation and the virtues of designing human activities with thought to how their various components relate to one another and the full consequences of any activity on the whole.

• New ways of thinking about the manufacturing process can lead to large savings in consumption not only of energy but of raw materials as well. Advanced manufacturing techniques that emphasize computer-assisted design and fabrication can lower costs dramatically and sharply reduce adverse impacts on the environment.

• The most advanced new manufacturing processes even include the concept of "electronic inventories" — the storage in digital form of products that can be quickly and precisely created in metal and plastic whenever the demand for a particular variety, style, or size is sensed by the factory's distribution network. The prospective savings from the elimination of inventories are staggering and illustrate the convergence of trends that may be able to transform our impact on the environment without damaging what we think of as our standard of living.

• A little-noticed but surprisingly competitive source of electrical power is wind-generated energy, using a new generation of windmills with advanced aerodynamic designs.

• New technologies for the storage and distribution of energy from current sources may offer savings almost as large as those that can come from new ways to generate power. This is especially true of electricity, which requires expensive facilities capable of generating the maximum amount of power that may be needed at any given moment (e.g., when everybody uses their air conditioners all at once, even if that only happens on one day of the year). More efficient ways to store electricity (e.g., Superconducting Magnetic Storage) can reclaim the amounts often wasted during "nonpeak hours." The inefficiency of current storage techniques is also the main reason that electric cars are still considered impractical.

• Similarly, the amount of energy lost in the transmission of electricity from one location to another is so large as to make long-distance transport extremely inefficient. New technologies like superconductivity may bring dramatic changes, making it possible to distribute energy over very long distances and manage peak loads innovatively. (Eventually, such advances may even make possible the visionary suggestion of Buckminster Fuller two decades ago that the Eastern and Western hemispheres be linked by underwater cable to assist each other in managing peak energy demand, since the high daytime use of one hemisphere occurs at precisely the time of low nighttime consumption by the other.)

• The consequences of the transition to these new technologies for those whose jobs are part of the older technologies — coal miners, for example — must be carefully heeded in designing an energy SEI. Retraining on a massive scale, financial assistance during the transition to new employment, and a continued search for technologies that might use the older sources of energy in an environmentally benign way must all be included. This is not only essential in order to sustain sufficient political support, it is also a matter of compassion and common sense.

• Probably the best contribution new technology can make will be found in a far more efficient relationship between our activities and our energy demand. Already, microprocessors are reducing energy consumption and managing energy flows within machines, causing some truly dramatic reductions in the amounts of energy required. Similarly, we can be much more intelligent about managing our use of energy in almost every activity — and these efficiencies of common sense can bring the largest savings of all.

Building Technology. The benefits of better design in reducing energy consumption are regularly demonstrated to everybody who pays utility bills in their homes and businesses. When energy prices suddenly escalated in 1973 and again in 1979, by far the most effective responses to the call for conservation came from homeowners, who insulated their walls and ceilings, installed storm windows, and performed a hundred other small and mundane tasks. Some discovered that so-called passive solar techniques were very effective in reducing heating bills. In less than two years, some utilities experienced a change from 7 percent annual increases in electricity demand to increases of less than 1 percent a year; in a few cases, overall energy demand actually declined. Of course, when energy prices stabilized and then declined in real terms, energy use began climbing again.

But the lesson of the experience remained: existing buildings can be modified to consume far less energy. Moreover, when new buildings are designed and constructed with a concern for energy use, the results can be quite startling. Decidedly low-tech measures, such as placing shade trees to decrease air conditioning needs; the use of the ground itself to insulate partially submerged walls; the strategic alignment of windows, doors, skylights, and the building itself with prevailing wind patterns and the path of the sun in the various seasons; and thicker, more effective insulation, can all help achieve substantial savings.

The redesign of devices that use energy inside the building can also have a dramatic effect. One of the most striking examples is the new generation of light bulbs — though they are still not widely used — which give the same amount of light as the older generation of bulbs while using a fraction of their electricity. As the energy experts Amory and Hunter Lovins have long argued, the widespread use of these new bulbs could by itself dramatically reduce energy consumption throughout the industrial world. The light bulbs now in common use are based on a pre–World War I design that sends the electrical current through a metal filament made mostly of tungsten; the filament glows and emits light, but the metal filament produces almost twenty times as much heat as light, meaning that most of the electricity is wasted — especially in the summer, when the extra heat often also leads to extra air conditioning to offset it. The new bulbs are based on an improved fluorescent design that sends the electrical current, not through metal, but through a gas that glows with light but loses very little energy in the form of heat. Unlike earlier fluorescent bulbs, the new ones fit regular sockets and fixtures and offer a quality of light every bit as pleasing as that of our incandescent bulbs and just as much of it. Yet they last more than ten times as long.

One may well ask why they are not being used. The answers are instructive. First, there is simple inertia. Consumers are generally not aware that the new bulb exists: few stores sell it, wholesale distributors don't keep it in stock, and the resulting low demand has limited the interest of manufacturers in high-volume production that would bring down the price. Moreover, the government provides no leadership whatsoever to encourage the transition. But there is something else at work as well: the initial cost of purchasing each new bulb is about $15, several times greater than that of a standard incandescent bulb. Over the lifetime of the bulb, the electricity savings far exceeds the total cost of the bulb, but most people — and governments — do not calculate costs and benefits that far into the future. This is a shame, because a single new energy-saving bulb, compared to a standard bulb, saves *a ton of coal* over its lifetime.

The SEI would improve our approach to building technology in two other ways:

• Household appliances can be redesigned to use less energy. Such efficiencies would also help reduce CO_2 emissions, hardly an insignificant point given that the Department of Energy has estimated that refrigerators, freezers, lighting, air conditioning, space heating, and water heating in the United States account for approximately 800 million tons of CO_2 emissions each year. Even though its secretary of energy recommended laws and programs to encourage the use of less wasteful appliances, the Bush administration has actively fought against them at the behest of appliance manufacturers, who argue that accelerating the transition to more efficient energy use interferes with market forces. (Some businessmen in an earlier generation felt the same way about five-day work weeks, minimum wages and child labor laws.)

• Building codes currently set certain standards for fire safety and structural integrity to minimize the chance of collapse. Given the extreme danger posed by our destruction of the environment, and the role that inefficiently designed buildings play in that process, why shouldn't building codes require environmentally friendly technologies, especially when they cost less? Builders are discouraged from incorporating these new technologies at present, because buyers are more interested in the initial price than the lifetime cost of the building. If improved building codes eliminated wasteful and inefficient designs, builders would no longer have to fear competitors willing to cut corners at the expense of responsible and energy-efficient design. The SEI should recommend new ground rules requiring the incorporation of these cost-effective, efficient designs; with new codes in place, the building industry could help us make this aspect of the transition to sensible energy use fairly quickly.

Waste Reduction and Recycling. Conservation and efficiency are not just techniques; they represent a way of thinking about human activity that is fundamentally different from the wasteful approach embodied by our current preoccupation with short-term results despite longer-term costs. As it happens, the same new technologies that make it possible to reduce energy consumption typically lead to reductions in the amount of waste produced. But this is not enough; the SEI would also focus on ways to improve waste management and the recovery and recycling of waste. Action should be taken on the following fronts:

• In some cases, entire product lines should be redesigned to make them easier to recycle. For example, some plastic beverage containers contain tiny amounts of metal that make recycling them impossible. Some newspaper supplements contain clay-based glossy coatings that make it impossible for recyclers to put the entire newspaper into their apparatus. Paying people to search through hundreds of thousands of newspapers each day to remove these supplements by hand can make the entire process uneconomical. There are many similar design flaws in products that might be easily recycled with minor changes in the overall process by which they are manufactured and distributed.

• On a large scale, this same "design flaw" is responsible for most of the raw sewage dumped into rivers throughout the industrial world. Rainwater flows into drainage sewers and is "recycled" into the rivers and ultimately oceans of the world. Sewage is funneled toward facilities where it is treated to minimize the consequences of dumping large, concentrated amounts into the environment. However, in almost all older cities, by thoughtless design, the two sewage collection systems — wastewater and storm runoff — flow into each other. As a result, whenever it rains, the sewage treatment plant is overwhelmed by the huge volumes of runoff and has to open its gates, dumping sewage and all straight into the rivers and lakes.

• Not only new designs for existing technology but also entirely new methods of waste treatment are greatly needed — especially for some of the particularly dangerous new forms of waste that have accompanied the chemical revolution. Also, the SEI should emphasize the immediate development of ways to more quickly and accurately assess the toxic potential of those new chemical substances produced as waste in manufacturing processes. In some cases, the problems of disposal may be so severe as to warrant a decision not to allow the process to begin in the first place.

In general, as discussed in Chapter 8, the goal should be to reuse everything that becomes a part of a new product and everything produced as a by-product of the process.

The U.S. Role

The development of an SEI to accelerate the development and dissemination of environmentally appropriate technologies may be the one facet of the Global Marshall Plan that other nations will be interested in leading if the United States chooses not to. Not coincidentally, it is the one element most likely to produce great economic benefits for the nation providing the most leadership.

Japan, which has excelled at developing profitable new technologies, has publicly identified this challenge as one it wishes to take on. On the subject of global warming, for instance, Japan has announced a breathtakingly comprehensive — if disappointingly slow — 100-Year Plan to develop successive waves of technology to help halt the process and to cope with the consequences of the changes we have already set in motion. But it doesn't stop with global warming: in almost every area of technology relevant to the environmental crisis, Japan is boldly taking the lead. What is maddening to many Americans who have advocated U.S. leadership in this area is that almost all of the key discoveries that led to these new technologies were made in the United States and then ignored by industry and government alike. For example, after making virtually every important breakthrough in solar energy technology, the United States is now a net importer of solar energy devices and systems, most of which come from Japan and it subcontractors elsewhere in Asia. Similarly, after General Motors invested money, time, and effort to develop the world's leading technology for electric cars, the effort died; now, by most estimates, Japanese automobile manufacturers will be the first to market commercially viable electric vehicles (even as they appear to be ahead in the race for very high mileage gasoline-powered cars). Once again, the Japanese auto companies have disproved the mercantilist chestnut "What's good for General Motors is good for America."

But all is not yet lost: what appears as still another example of a serious deficiency in America's ability to compete may actually be an ideal opportunity for the United States to address a pervasive and persistent structural problem in its approach to economic competition. The urgent need for environmentally appropriate technology raises a crucial question: How can we better translate our superior talent in research and development into better applied research and finally to commercially profitable products and processes?

This problem has sparked a divisive debate in recent years over the appropriate role for government in coordinating a national approach to technological development, sometimes called an industrial policy. Opponents of a coordinated approach — the Bush administration, for one — believe that government coordination would distort the marketplace and lead to inefficient decisions about allocation of effort, capital, and resources. It is interesting to note, however, that in another area involving our national interest, these same opponents of industrial policy are the most vigorous advocates of an aggressive role for government — namely, in matters concerning the Strategic Defense Initiative and other expensive programs to develop new military technologies.

The United States has almost always had an industrial policy for the military; in fact, the first contracts for what is now called mass production were let by the government to Eli Whitney for the manufacture of rifles with interchangeable parts. In each of our wars, especially those in this century, our government has aggressively pursued policies designed to stimulate industrial activity in areas directly relevant to the war effort. But coordinated government leadership of private industry has also been deployed to stimulate a few nonmilitary efforts with national security overtones. The crash program to land a man on the moon within ten years was such an effort, and, as with similar efforts in the two world wars, the Apollo Program led to important advances in U.S. leadership of a wide variety of technologies. Indeed, the entire modern computer industry was born as a spinoff of efforts first made in trying to land a man on the moon.

One of the problems with the ongoing debate over industrial policy is that, unlike in the past, the policy has no clear focal point. Instead, we hear little more than broad assertions about the need to compete more effectively or improve our productivity. The debate seems sterile, an argument about means but not ends, so it is not surprising that many Americans conclude that, in principle, it is better to limit government's role in directing or distorting the activities of private firms. But as soon as a worthy goal becomes the focus of a national effort that cries out for coordinated national leadership, the terms of the debate shift dramatically; the debate becomes a discussion of ends as well as means, and the natural American "can do" instinct to reach the stated objective begins to take over.

The fundamental purpose of the SEI is to enable us to make dramatic progress in the effort to heal the global environment; in my opinion, that goal will eventually become so compelling that America will demand the kind of determined effort that made the Apollo Program so productive and inspiring. The new program could reinvigorate our ability to excel at applied as well as basic research, spur gains in productivity, lead to innovations, break-throughs, and spinoffs in other fields of inquiry, and reestablish the United States as the world's leader in applied technology.

III. A NEW GLOBAL ECO-NOMICS

The third strategic goal of the Global Marshall Plan should be a categorical change in the economic rules of the road by which our civilization — or at least the large and growing majority of it that is committed to market economics — determines the value of our choices. Just as firmly entrenched technologies are difficult to dis-lodge even when their harmful consequences become painfully obvious, so too with established methods of calculating costs and benefits: once they are accepted as holy writ they are virtually impossible to change without a major determined effort.

The stunning victory of free market economics over communism in the global war of ideas has brought with it a new obligation to change those features of our economic philosophy that we know are flawed in light of the ecological destruction they legitimize, and even encourage. For example, as we saw in Chapter 10, the current method for calculating gross national product (GNP) completely excludes any measurement for depletion of natural resources. Everything in nature is simply assumed to be limitless and free. A developing nation that clear-cuts its rain forest may add the money from the sale of the lumber to its income, but is not required to place a value on the depreciation of its natural resources or in any way reflect in its calculation of GNP the fact that next year it will not be able to sell its rain forest because it is gone.

But recognizing the problem is not enough; the world community, led by the United States, should move to change this widely used formulation — and others like it that badly mislead decision-makers who might otherwise place more appropriate economic values on the protection of the global environment. There is no excuse for not changing the definition of GNP.

As a member of the Joint Economic Committee, I have attempted in a series of hearings to catalogue all of the formulas in our current version of economic theory that need to be changed in order to eliminate the serious distortions in the way free markets calculate the value of the environment. But for each misleading formula, we need to substitute an appropriate method of valuing the ecological consequences of market decisions.

Though this task may seem theoretical, I am convinced that it is among the most important and far-reaching changes we can hope to accomplish. For every large decision by a national authority, there are billions of small choices by individuals that add up to an aggregate force completely dwarfing most policy decisions by governments. It follows, then, that influencing the criteria and values used to inform and guide these billions of everyday choices represents the real key to changing the direction of human civilization.

There is already a specific mechanism for changing the way we calculate GNP — the United Nations System of National Accounts — but thus far no one has developed replacements for other faulty economic formulas. Take, for example, the formula we currently use to measure productivity. By excluding most environmental costs and benefits from our methods of assessing the productive potential of changes in policy, we severely distort our assessments. To remedy this ecological blindness, we should work with the appropriate professional communities (e.g., accountants, actuaries, auditors, corporate counsels, statisticians, economists of all stripes, city planners, investment bankers, and so on) and encourage them to change their formulations. This task may seem only a little less difficult than reducing CO_2 emissions smokestack by smokestack, but the ideas and ways of thinking embodied in these faulty economic formulas led to the emissions from the smokestacks in the first place, and it is important to try to change them.

We must also change our current use of discount rates, the device by which we systematically undervalue the future consequences of our decisions. Our technologically enhanced power has dramatically changed our ability to modify the world around us in ways that have important consequences. Yet we still calculate the effects of our actions in essentially the same way we did at the beginning of the industrial revolution: we still assume that whatever we do now will have little impact on the future. If it was ever valid, that assumption is now patently dishonest, and the formulas that embody it must be changed. But again, the actual work of changing them requires a strategic plan and a systematic program.

To accomplish the transition to a new economics of sustainability, we must begin to quantify the effects of our decisions on the future generations who will live with them. In this, we have much to learn from the Iroquois nation, which requires its tribal councils to formally consider the impact of their decisions on the seventh generation into the future, approximately 150 years later. Of course, it is sometimes genuinely difficult to project the future, but even where it is not, we have obstinately refused even to consider it. That must change — again, not just in theory but in practice, with the sustainability of economic choices factored into decision-making at every level of commerce.

A number of specific steps can be taken to accelerate the shift toward economic rules that promote sustainability. The first and most obvious changes involve the elimination of those public expenditures — both national and international — that encourage and subsidize environmentally destructive economic activity. For example, the World Bank should halt the funds that subsidize the building of roads through the Amazon rain forest as long as there are no credible safeguards to stop what has been until now the primary use of such roads: providing direct access to the heart of the forest for chain saws and torches.

These unnatural government subsidies ought to be the easiest mistakes to correct, and ultimately, as our awareness of the ecological consequences grows, they will be. But right now, they are extremely difficult to change, both at the international and at the national level. In the developed world, especially in the EEC, Australia, Canada, and the United States, the forces of supply and demand are distorted by agricultural subsidies that encourage the repeated plowing of marginal land until it becomes highly vulnerable to unsustainable losses from soil erosion.

Similarly, the cutting down of many ancient forests might never take place without the massive taxpayer subsidies that allow the construction of logging roads into the heart of the areas with the oldest trees. And in Florida, the destruction of the Everglades is being actively subsidized by taxpayers and consumers through artificial price supports for sugarcane — a crop that otherwise would never be grown in that area. In fact, I myself have supported sugar price supports and — until now — have always voted for them without appreciating the full consequences of my vote.

Perhaps a brief description of my motives in these votes would be instructive because, looking back, I recognize in myself many of the same habits of thought and action that I am belatedly attempting to help change in others. Several of my colleagues have tried for years to persuade me to drop my support for the sugar subsidies, but since other considerations always seemed more important, I never seriously considered what they had to say. As a member of the Southern "farm bloc" in Congress, I have followed the general rule that I will vote for the established farm programs of others in farm states — especially the ones important to my region — in return for their votes on behalf of the ones important to my state. That principle of mutuality is certainly not bad in and of itself; it is part of what helps hold our country together. But when it is allowed to become an overriding consideration, it can lead to the kind of willful blindness that makes objective consideration of an issue impossible. I found this issue doubly difficult to consider objectively because I was also lobbied by the owners and employees of a company in my state which makes a substitute for sugar — corn syrup — which is priced according to the price of sugar. Theoretically, if the price of sugar comes down once the subsidy is removed, jobs might be lost, not only on the plantations that are speeding the destruction of the Everglades, but also in the Tennessee plant that makes corn syrup. Thus are established patterns of political calculation difficult to change. (But change is possible: I, for one, have decided as I write this book that I can no longer vote in favor of sugarcane subsidies; looking beyond this particular case, I want to shift the burden of proof to the advocates of subsidies, to show that ecological problems will not occur as a result of distorting the market.) Consider how much more difficult it is to change destructive patterns that have even deeper roots in our society — such as the constant and extravagant burning of fossil fuel.

As our awareness of the extent of the ecological crisis grows generally, it will be increasingly important to ensure that information is available in the marketplace about the environmental consequences of our choices. One of the most effective ways to encourage market forces to work in environmentally benign ways is to give concerned citizens a better way to take the environment into account when they purchase goods or make other economic decisions. But consumers must have confidence in the information they are given, and unfortunately, a number of companies try to mislead consumers with unwarranted claims of environmental responsibility. In response, some environmentalists like Denis Hayes, one of the founders of Earth Day, are trying to develop a universally recognizable "green label" to identify products that meet sound ecological standards, including recyclability. The government can also help with this problem by ensuring that information about such things as automobile mileage and the energy efficiency of appliances is as accurate as possible. In addition, I believe the government should establish legal standards for green labels. The effort to do it entirely in the private sector has a lot of appeal — but almost certainly cannot work without the force of law.

Full information about who is responsible for environmental damage will also be an increasingly important way to make market forces work *for* the environment instead of against it. For example, some of the large corporations that used to engage in driftnet fishing have stopped because of the damage to their corporate reputation. This is a promising start. But at a recent hearing on the appearance of so-called pirate fleets of driftnet fishermen, I heard testimony that some of the ships are still secretly owned by a few of the large corporations that allegedly got out of the business; in one case, a ship still unloads its catch at facilities owned by a subsidiary of one of the same companies, Mitsubishi. If that information is made public when more and more people care about it, then the products of the parent corporation may suffer in the marketplace.

Public concern can prod even the largest corporations to take action, and some companies have found that in the process of addressing their environmental problems they have been able to improve productivity and profitability at the same time. For example, 3M, in its Pollution Prevention Pays program, has reported significant profit improvements as a direct result of its increased attention to shutting off all the causes of pollution it could find. What some of our best companies have come to realize is that as their way of thinking about the environmental consequences of the industrial process changes, their way of thinking about other consequences of the process changes too. An effective quality control program that reduces the number of defects, for example, requires a level of attention to detail and to the interaction of all facets of the production process, and this is precisely the approach required to identify the best ways to eliminate pollution. That is yet another reason why some companies have begun to feel that an emphasis on environmental responsibility makes good business sense. Some, like du Pont, have even begun to calculate executive compensation and bonuses partly on the basis of performance in environmental stewardship.

Some companies that have not changed their approach are beginning to face stockholder challenges from institutional investors. In fact, a large, highly organized program — the Coalition for Environmentally Responsible Economies — now works full time to focus the attention of investors and corporate boards on the environmental performance of corporations with publicly traded stock, and an increasing number of large pension funds, universities, and churches are basing their investment decisions on the Valdez Principles, which embody criteria for assessing the environmental performance of corporations.

In order to protect against distorted priorities and promote sound decision-making by corporations, we should also attempt to pass a new generation of environmental antitrust laws that insist on carefully scrutinizing the ways in which vertical integration, for

example, can work to the disadvantage of the environment. Under current applications of the antitrust laws, it is illegal for railroads to own trucking companies because the combination can stifle beneficial competition between the two modes of transportation. But what about large paper users buying forestland and thus reducing to zero whatever incentive they may have had to use recycled paper over virgin forest? Similarly, what about chemical companies that produce pesticides and fertilizers buying up seed companies and selecting and breeding seeds that maximize use of their chemical products, neglecting other varieties that might feature a greater degree of natural resistance to pests? In neither case should there be an automatic prohibition against cross-ownership, but there ought to be a requirement to consider the potential for harmful consequences to the environment and, if necessary, the right to prevent such mergers.

It will also be increasingly important to incorporate standards of environmental responsibility in the laws and treaties dealing with international trade. Just as government subsidies of a particular industry are sometimes considered unfair under the trade laws, weak and ineffectual enforcement of pollution control measures should also be included in the definition of unfair trading practices.

Especially as the United States attempts to expand free market principles and encourage freer and fairer trade throughout Latin America, environmental standards must be included among the criteria for deciding when to liberalize trading arrangements with each country. The mixture of environmental protection with trade negotiations is volatile, but so is the mixture of any other consideration with trade talks. Furthermore, valuable precedents are being set even now — witness Germany's aggressive effort to impose stricter environmental standards within the EEC and the decision by the General Agreement on Trade and Tariffs to activate a panel to review the relationship between trade and the environment.

Returning to the difficult issue of foreign economic development, I have reluctantly concluded that several of the international financing institutions established for the worthy purpose of "developing" the Third World are — by ignoring the ecological consequences of large-scale projects — often doing more harm than good. While they have made some progress in integrating environmental concerns into their loan criteria, they are still falling far short of their responsibility. As a result, several of us in the Senate have begun looking for ways to hold their feet to the fire. One key may be greater orchestration among all lenders to a given country, to create a "broad spectrum" package. However, some environmentalists are so impatient that they are actively considering efforts to completely eliminate U.S. funding for any international institution that fails to thoroughly reform the mechanisms by which it considers environmentally sound practices. These institutions simply must become part of the solution instead of part of the problem.

And as noted above, there are similar problems now confronting the international trading system, whose rules were not designed with the environment in mind and must now be quickly adapted to foster environmental protection.

At the same time, financial institutions ought to reconsider their approach to currency definitions. The harsh distinction between hard currency (the money of an industrial nation, which is acceptable in payment of international debts) and soft currency (the typically inflated and unstable money of a debtor nation, which often can be used only for domestic purchases) distorts land use patterns and resource allocations in developing countries. For example, a great deal of land traditionally used to grow indigenous food crops is routinely plowed up for other crops that can be sold into the export market; the latter brings hard currency and the former brings soft currency. In a sad irony, the hard currency is often used to buy nonindigenous food from importers to feed populations who can no longer grow their own food. The entire arrangement makes very little sense.

Although it is a tall order, we must also seek financial reforms in the developing countries receiving assistance and new technology under the Global Marshall Plan. One of the most serious but least recognized problems is capital flight, the process by which wealthy elites in the developing countries siphon large sums of money out of their national economies and into private bank accounts in the West. In fact, in many Third World nations, the amount of capital flight goes up and down in almost direct proportion to the amount of foreign aid. A more equitable distribution of political power, wealth, and land is a prerequisite in many of these countries for any successful effort to rescue their environments and societies.

One of the best development ideas of the last ten years is one first proposed by the biologist Tom Lovejoy of the Smithsonian, the so-called debt-for-nature swaps. Under this plan, a version of which was finally agreed to by Brazil in the summer of 1991, debts owed by developing countries to the industrial nations are forgiven in return for enforceable agreements to protect vulnerable parts of the environment in the debtor nation. Since most of the debts are unlikely to be paid in any event, and since the protection of these environments is in the best interest of the debtor country as well as of the rest of the world, everybody wins. Moreover, a massive reduction of the crippling burden of debt now carried by the developing world is absolutely essential to creating the prospect for sustainable societies — and eventually a global economy that functions with these nations as integral and healthy participants.

The insanity of our current bizarre financial arrangements with the Third World is even more apparent when one realizes that fully half of all Third World debt has been accumulated in order to purchase weapons with which to wage wars among themselves — with consequent murder and mayhem sometimes unleashed on entire societies and often with horrendous environmental destruction in the process. This is what occurred in Iraq's 1991 invasion of Kuwait. Stopping these wars (partly by choking off the obscene flow of advanced weaponry from the industrial world) is one of the single most important steps toward environmental protection the world can take.

In addition to debt-for-nature swaps, another new idea for using market-based mechanisms to help the world deal with the global environmental crisis is the establishment of a market for CO_2 emission "credits," not only here at home but internationally as well. I favor an international treaty limiting the amounts of CO_2 individual nations are entitled to produce each year, and it should incorporate a mechanism for establishing these emission credits. Once the treaty is complete, those nations that have more success in reducing their emissions could then sell their emission rights to others who need more time to make the adjustment. In practice, this would become a way to rationalize investments in the most efficient alternatives for all CO_2-generating activities, whether substituting renewable forms of energy for fossil fuels, developing new techniques for efficiency and conservation, or designing entirely new approaches to large activities that are now taken for granted. Of course, agreeing on how to parcel out the emission rights will not be easy, nor will determining the feasibility of reducing the overall limits each year. But once enough nations come to recognize the grave threat posed by CO_2 emissions, working out a treaty may not be an insurmountable challenge.

Clearly, then, if sustainable development is to become feasible, our approach to economic policy must be transformed. At the earliest opportunity, world leaders and their economic ministers should convene a global summit to discuss new approaches to this challenge. Their agenda should include the immediate adoption of a new set of economic rules of the road. Following is a summary of those I have proposed:

1. The definition of GNP should be changed to include environmental costs and benefits.
2. The definition of productivity should be changed to reflect calculations of environmental improvement or decline.
3. Governments should agree to eliminate the use of inappropriate discount rates and adopt better ways to quantify the effects of our decisions on future generations.
4. Governments should eliminate public expenditures that subsidize and encourage environmentally destructive activities.
5. Governments should improve the amount and accuracy of information on the environmental impacts of products and provide it to consumers.
6. Governments should adopt measures to encourage full disclosure of companies' responsibility for environmental damage.
7. Governments should adopt programs to assist companies in the study of the costs and benefits of environmental efficiency.
8. Nations should revise their antitrust laws to encompass environmental harm.
9. Governments should require the incorporation of standards

to protect the environment in treaties and international agreements, including trade agreements.

10. Environmental concerns should be integrated into the criteria used by international finance institutions for the evaluation of all proposed grants of development funds.

11. Governments should make accelerated use of debt-for-nature swaps to encourage environmental stewardship in return for debt relief.

12. Governments should develop an international treaty establishing limits on CO_2 emissions by country and a market for the trading of emission rights among countries that need more and countries that have an excess amount.

The U.S. Role

As the world's leading exemplar of free market economics, the United States has a special obligation to discover effective ways of using the power of market forces to help save the global environment. Yet even as we correctly point out the dismal failures of communism, and even as we push the underdeveloped world — appropriately, in my view — to adopt a market-based approach to economics, we have been reluctant to admit our failure to bring environmental values into our economic decisions. Further, the Bush administration has shown little interest in changing the government policies that presently distort the principles of market economics in ways that encourage the destruction of the environment.

Many U.S. policy-makers seem content to leave the environmental consequences of our economic choices in the large waste-basket of economic theory labeled externalities. As I stated in Chapter 10, anything that economists wish to forget about is called an externality and then banished from serious thought. For example, consider this analysis by President Bush's Council of Economic Advisers of the impact of global warming on agriculture: "The costs of today's agricultural policies are estimated to be more important in economic terms than even pessimistic estimates of the effects of global warming, largely because the former must be borne in the present and the latter may occur, if at all, in the relatively distant future."

That's it. As far as the council is concerned, global warming need be given no further thought. Since it has been discounted into insignificance, they figure we can just forget about it. And meanwhile we go on preaching to the rest of the world that our brand of market economics takes everything into account in the most efficient way imaginable.

What if we took a broader view and began incorporating factors affecting the environment into our economic system? How would we do it? Well, here are some specific proposals.

None is likely to be more effective than finding ways to put a price on the environmental consequences of our choices, a price that would then be reflected in the marketplace. For example, if we were to tax the pollution dumped by factories into the air and water, we would get less of it. And we might well notice a sudden increase in the amount of interest companies show in improving the efficiency of their processes in order to reduce the pollution they cause.

To most of us, the principle sounds unassailable: let the polluter pay. But what about when it applies to each of us instead of to a nameless, faceless, corporation? For example, rather than require homeowners to pay higher property taxes to cover the cost of garbage collection, why not lower property taxes and then charge for garbage collection directly — by the pound? Those responsible for creating more garbage would pay more; those who found ways to cut down would pay less. The interest in recycling might rise dramatically. And when choosing between products at the store, people might even start avoiding unnecessary and bulky packaging if they knew it was going to end up in their garbage. There is an economic rule of thumb: whatever we tax, we tend to get less of; whatever we subsidize, we tend to get more of. Currently, we tax work and we subsidize the depletion of natural resources — and both policies have contributed to high unemployment and the waste of natural resources. What if we lowered the tax on work and simultaneously raised it on the burning of fossil fuels? It is entirely possible to change the tax code in a way that keeps the total amount of taxes at the same level, avoids unfairness and "regressivity," but discourages the constant creation of massive amounts of pollution.

Accordingly, I propose:

1. That we create an Environmental Security Trust Fund, with payments into the Fund based on the amount of CO_2 put into the atmosphere. Production of gasoline, heating oil and other oil-based fuels, coal, natural gas, and electricity generated from fossil fuels would trigger incremental payments of the CO_2 tax according to the carbon content of the fuels produced. These payments would be reserved in a trust fund, which would be used to subsidize the purchase by consumers of environmentally benign technologies — such as low-energy light bulbs or high-mileage automobiles. A corresponding reduction in the amount of taxes paid on incomes and payrolls in the same year would ensure that the trust fund plan does not raise taxes but leaves them as they are — while having sufficient flexibility to ensure progressivity and to deal equitably with special hardships encountered in the transition to renewable energy sources (such as those faced by someone with no immediate alternative to the purchase of large quantities of heating oil, gasoline, or the like). I am convinced that a CO_2 tax which is completely offset by decreases in other taxes is rapidly becoming politically feasible.

But CO_2 taxes will not be enough to stop the profligate waste of virtually all other natural resources, and as a result I also propose:

2. That a Virgin Materials Fee be imposed on products at the point of manufacture or importation based on the quantity of nonrenewable, virgin materials built into the product. For example, paper mills would be charged a materials fee on the basis of the percentage of their paper made from freshly cut trees as opposed to recycled pulp and paper. The manufacturers and processors paying the tax would then qualify for tax credits to subsidize the purchase of equipment necessary for recycling and for the efficient collection and use of recycled materials, assuming confirmation of a net environmental benefit.

In addition to these two broad proposals, I also recommend several other specific changes in U.S. policy that would rewrite our economic rules of the road for the sake of the environment:

3. The government should adopt a policy of purchasing environmentally appropriate substitutes wherever they are competitive — taking into account full life-cycle costs — with older, less responsible technology. For example, it should replace every light bulb (when it wears out) with one of the new long-life bulbs that consume only a fraction as much electricity to produce the same amount of light. The government should also be required to purchase recycled paper in amounts that each year represent a larger percentage of its total needs until virtually all the paper it uses is recycled. If the government — with its substantial needs — can lead by example, it can make a tremendous difference to the ability of the new product manufacturer to achieve economies of scale and become viable enough to break into the market.

4. The government must establish higher mileage requirements for all cars and trucks sold in the United States. Although the carbon taxes and corresponding subsidies of the Environmental Security Trust Fund will be far more effective in accelerating the transition to more efficient vehicles, mandated improvements in the average fuel economy can be an important supplementary measure. With two of the largest automobile manufacturing plants in the world located in my own state, I have faced an awkward political problem in supporting tougher mileage requirements, and I do recognize some of the very real practical difficulties with the proposed legislation. Nevertheless, the amount of CO_2 emissions in the United States is so great that I think even imperfect measures to force remedial action are better than none at all.

5. Efficiency standards throughout the economy — for buildings, for industrial motors and engines, and for appliances — must also be strengthened. With tougher standards, there is less temptation for manufacturers to compete by cutting corners. The Bush administration — inexplicably — has fought hard against such standards.

6. Utility rate reform must encourage full use of conservation and efficiency measures. At present, some state governments are moving aggressively on this front, but federal policy has lagged far behind. All utilities should encourage conservation rather than build new generating capacity; helping to finance conservation has proved remarkably effective. Finally, utilities should be encour-

aged to plant forests to offset some portion of the CO_2 they generate.

7. Tree planting programs — with carefully selected seedlings appropriate to the areas being planted and careful follow-up to ensure tree survival — should be part of workfare programs in communities where work requirements are attached to welfare payments. Similarly, tree planting projects should be given higher priority in summer jobs programs for teenagers.

8. Accelerated phaseout of all ozone-destroying chemicals. We should also subsidize the development of truly benign substitutes.

There are, of course, many other problems that must be attended to if the government is going to play the key, constructive role necessary for it to lead the world environmental movement. We need to pay careful attention to the deep social and attitudinal causes of America's relative economic decline, some of which also contribute to the environmental crisis:

• The neglect of our human resources and falling levels of proficiency in literacy, numeracy, geography, and basic reasoning skills.

• Our unwillingness to make decisions with an eye to their long-term effects, coupled with our insistence on basing strategy on short-term time horizons: for example, the practice of rewarding business leaders on the basis of quarterly earnings, the willingness of investors to allocate capital on the basis of short-term profits instead of the quality of the goods produced, jobs provided, and long-term market share gained; the tendency of political leaders to base important decisions on the perception of their impact on the next election or even on the next public opinion polls.

• Our complacent pursuit of outdated strategies that used to work in the postwar markets, when we were the only strong economy left in the free world, but that have long since been surpassed by more streamlined and effective strategies.

• Our tolerance of government and industry working at cross purposes and failing to plan together or find ways to resolve persistent conflicts, not according to the Japanese model, but according to an original and innovative American model of a kind that existed in past national efforts such as the Apollo Program.

• Our inability to translate new discoveries in the laboratory into new advantages for American companies and workers.

All of these problems are deeply interrelated and all, I believe, can be solved with the same shift in thinking and a focused national effort represented by the Strategic Environment Initiative and the Global Marshall Plan.

IV. A NEW GENERATION OF TREATIES AND AGREEMENTS

The fourth strategic goal of the Global Marshall Plan should be the successful negotiation and resolution of an entirely new generation of international treaties and agreements aimed at protecting the environment. Just as democracy and market economics are important to a successful effort to restore the earth's ecological balance, so is a further extension of the rule of law.

The discussion earlier in this chapter about the trade-offs essential for the successful design of a Global Marshall Plan — the balance between the industrial nations and the developing nations — bears directly on virtually all of this new generation of treaties and agreements, and there will be a great many of them.

The prototype of this new kind of agreement was the Montreal Protocol, which was global in scope and called for a worldwide phasing out of the chemicals (like CFCs) that are destroying the ozone layer and featured a cost-sharing arrangement between the industrial and the impoverished nations. In addition, it anticipated the need for regular reviews as new information about the atmosphere became available. Indeed, the provisions of the original agreement have already been toughened considerably with the attachment of the London amendments, and further toughening is expected in the wake of the latest evidence — which continues to point toward an ever more serious threat.

Much attention is now focused on the efforts by the world community to conclude a counterpart to the Montreal Protocol that deals with greenhouse gases, especially CO_2. And many of the innovations in the Montreal agreement will be directly applicable to the new agreement. But the new treaty will be infinitely more difficult to conclude than its predecessor, and as a result, nations now in the process of negotiating it will be tempted to look for some way to escape the necessity of actually agreeing to sharp reductions in the gases that are causing global warming and will try to substitute symbolic action.

As co-chairman of the Senate Observer Group on the climate change negotiations, I have paid careful attention to the twists and turns in these talks. For example, I witnessed an interesting twist in the early stages of the negotiation in 1991, when Japan formally proposed an approach called pledge and review: it stipulated that each nation would merely pledge to take action on its own and that the world community would later review the record to see how it did. The Bush administration, of course, was thrilled because the proposal offered a chance to appear to take action before the 1992 election without having to actually do anything at all. But the problem, of course, is that the politicians in power who make the pledge may not be the same ones in power when the performance is reviewed. And politicians are often tempted to make a promise that is not binding and hope for some unexpectedly easy way to keep the promise, though some, like Bush, are perfectly prepared to break promises if it turns out that there is no easy and painless way to keep them. The Japanese later called for tougher action.

The U.S. Role

The diversity and complexity of this new generation of global agreements are remarkable. There are, for example, two separate treaties and a third "statement of principles" now being negotiated in preparation for the so-called Earth Summit in Rio de Janeiro in June 1992: in addition to the "framework" treaty on climate change, negotiators are working on a treaty to protect biodiversity and a statement of principles to preserve the forests of the world. As these documents were under negotiation, another treaty was actually concluded and signed in 1991: a treaty to protect Antarctica from oil drilling and coal mining. But the way in which it was finally resolved caused many observers to be apprehensive about whether the United States would allow the kind of progress in the other negotiations that will be necessary if agreements are to be reached in time for the Brazil summit meeting. Even after every other nation that was a party to the negotiations had agreed to the language — which had been negotiated over a period of years — and even after our own negotiators had given tentative approval pending a final sign-off by the White House, President Bush insisted that the proposed language was too restrictive — even though no American mining or drilling companies had expressed any interest whatsoever in prospecting for mineral deposits in Antarctica. Ironically, President Bush had one year earlier signed legislation making it illegal for a U.S. company to engage in such activities in Antarctica — pending completion of the treaty.

When the White House's refusal was announced, the other nations involved were outraged and demanded that President Bush review the matter. Eventually, after the United States had suffered an embarrassing barrage of criticism, the president quietly caved in and did what his negotiators had recommended all along.

The problem with this kind of approach is not the embarrassment or the criticism. The problem is that such behavior demonstrates the very opposite of leadership. And if the world is to have any chance at all of successfully negotiating the excruciatingly difficult treaties that are now necessary, the United States will simply have to take a leadership role. After the Antarctica fiasco, there seemed little chance that the Bush administration was prepared to do so. Yet it also seemed clear that this administration will change if it senses a change in the political winds that is significant enough to force it to reassess its policy.

So with time running out, the real source of hope still lies in the prospect of a change in the way people at the grass roots think about the global environment.

V. A NEW GLOBAL ENVIRONMENTAL CONSENSUS

The fifth major goal of the Global Marshall Plan should be to seek fundamental changes in how we gather information about what is happening to the environment and to organize a worldwide education program to promote a more complete understanding of the crisis. In the process, we should actively search for ways to promote a new way of thinking about the current relationship between human civilization and the earth.

This is perhaps the most difficult and yet the most important challenge we face. If a new way of thinking about the natural world emerges, all of the other necessary actions will become instantly more feasible — just as the emergence of a new way of thinking about communism in Eastern Europe made feasible all of the steps toward democracy that had been "unthinkable" only a few months earlier. And indeed, the model of change we use in designing and implementing our strategy should be based on the assumptions that there is a threshold we must cross and that not very much change will be apparent and obvious until we reach that threshold, but when it finally is reached, the changes will be sudden and dramatic.

Central to any strategy for changing the way people think about the earth must be a concerted effort to convince them that the global environment is part of their "backyard" — as it really is. I have always been struck by the way a proposal for an incinerator or a landfill mobilizes a lot of people who do not want the offending entity near them. In the midst of such a controversy, no one seems to care much about the economy or unemployment rate; the only thing that matters is protecting their backyard. The famous "not in my backyard" syndrome, NIMBY, has been much maligned but is often on target and is an undeniably powerful political force. How might its energy be focused against threats to the environment? Is that possible? The key lies with the definition of "backyard," and in truth, our backyards are threatened by problems like global warming and ozone depletion.

An important step in the right direction would be to take a new approach to the collection of information about what precisely is happening to the global environment. As chairman of the Space Subcommittee in the Senate, I have strongly urged the establishment of the new program that NASA calls Mission to Planet Earth. Sally Ride, the first American woman in space, coined the phrase, and it is meant to be taken ironically. As she points out, we have undertaken highly sophisticated planetary studies by sending spacecraft into orbit around Mars and Venus, and we have used that unique perspective to study other, more distant planets. Yet we have not used the same techniques to improve our understanding of our own planet just when we desperately need to understand much more about the changes that are taking place.

Even more important than gathering new information, though, we must start to take action now — and the information collection system should enhance that goal. This conclusion carries with it two implications: first, the information should be collected as quickly as possible; and second, it should — wherever feasible — be collected in a manner that facilitates public education and fosters a greater understanding of what the new information means within the larger context of rapid global change.

In other words, the Mission to Planet Earth should be a Mission by the people of Planet Earth. Specifically, I propose a program involving as many countries as possible that will use schoolteachers and their students to monitor the entire earth daily, or at least those portions of the land area that can be covered by the participating nations. Even relatively simple measurements — surface temperature, wind speed and direction, relative humidity, barometric pressures, and rainfall — could, if routinely available on a more nearly global basis, produce dramatic improvements in our understanding of climate patterns. Slightly more sophisticated measurements of such things as air and water pollutants and concentrations of CO_2 and methane would be even more valuable. But the first step is collecting the kind of rudimentary information necessary to monitor the environment closely, just as hospital emergency rooms monitor the vital signs of patients receiving intensive care.

The mass production of uniform instruments for this program could bring the unit costs down to trivial levels, and the instruments themselves could be designed to facilitate daily electronic "polling" or data collection. By deploying relatively cheap low-orbit satellites capable of rapidly redistributing the information gathered from the many scattered monitoring stations, the data could be fed into regional, national, and global collation and analysis centers, where they could be studied and incorporated into computer models on a regular basis. As the schools gained experience and confidence, the range of activities in the program could be expanded to include, for example, soil sampling (to map soil types, monitor soil erosion rates, and measure residues of pesticides and salt) and an annual tree census, using sampling techniques that monitor deforestation and desertification.

If the program worked as planned, those involved might eventually be persuaded to go even further and actually plant trees and establish nurseries for trees and crops indigenous to their individual areas. And a different sort of seed might be planted in the process: for example, the world's leading scientist on the problem of ozone depletion, Dr. Sherwood Rowland, first became interested in the atmospheric sciences as a youngster when he was asked to look after a backyard weather station by a neighbor who went on vacation for several weeks. The virtue of involving children from all over the world in a truly global Mission to Planet Earth is, then, threefold. First, the information is greatly needed (and the quality of the data could be assured by regular sampling). Second, the goals of environmental education could hardly be better served than by actually involving students in the process of collecting the data. And, third, the program might build a commitment to rescue the global environment among the young people involved.

There are now efforts to improve the Mission to Planet Earth, which NASA first organized along lines that resemble sprawling Defense Department weapons procurement programs; most of the money was budgeted for large pieces of hardware that will take ten to fifteen years to build and then deploy in space. We need the information faster and cheaper, if it is at all possible — and I am convinced it is. Toward that end, Senator Barbara Mikulski and I have been working together to force changes in the NASA program, with some success. Even as NASA is proposing new space platforms built by defense contractors to collect more data, the Bush administration is refusing to spend tiny sums of money to safeguard the valuable information already collected — by the Landsat system, for example, a series of satellites that have made a unique photographic record of the earth's surface for twenty years. The administration has allowed the data collected to go to waste and is now proposing to stop the launch of the next Landsat satellite, thus eliminating the chance to assemble new portraits of our planet and provide a rare and invaluable perspective on the changes we are causing to the earth's surface.

Another difficulty with the current design of the Mission to

Planet Earth is that no one yet knows how to cope with the enormous volume of data that will be routinely beamed down from orbit. Nothing even approaching this amount of data has ever been dreamed about. In order to help organize it — and interpret it — I have proposed something called the Digital Earth program, which is designed to build a new global climate model capable of receiving data from several different sources that are not considered compatible by today's definitions; furthermore, Digital Earth would be designed to actually learn from its mistakes, when predictions based on information from the known climate record are run on the models of environmental change so that the results can be compared with what actually happened. Even though the global climate models all have serious limitations, they still give us the best information available about what is likely to happen to climate in the future, and I believe this new approach can substantially improve the quality and usefulness of the models.

Because of the unprecedented volume of data, it may also be necessary to disperse the means of storing and processing it much more widely. Most experts in the United States and Japan now believe in the inherent advantages of a computer architecture or system design known as massive parallelism, and massively parallel computers will undoubtedly play a key role in Mission to Planet Earth. These computers are valuable in another way too, for they provide a metaphor that I think is particularly useful in figuring out how to best cope with the task of collecting and processing the enormous quantity of data and how best, in the process, to change minds and hearts all over the world on the subject of the environment.

The power of massively parallel computers comes from their ability to process information, not in one central processing unit, but in tiny, less powerful units throughout the computer's memory field in locations immediately next to the spot where the information itself is stored. For many applications, the inherent advantage of this design is crucial: the computer wastes less time and energy in retrieving raw data from the memory field, bringing it to the powerful central processor, waiting for processing, then taking the processed data back to the memory field to be stored again. By locating each small portion of the data with enough processing

capacity to handle it, more data can be processed simultaneously, then transported only once, not twice, between the memory field and the center.

When you stop to think about this approach in generic terms, it seems obvious that both democracy, as a political system, and capitalism, as an economic system, work on the same principle and have the same inherent "design advantage" because of the way they process information. Under capitalism, for example, people free to buy and sell products or services according to their individual calculations of the costs and benefits of each choice are actually processing a relatively limited amount of information — but doing it quickly. And when millions process information simultaneously, the result is incredibly efficient decisions about supply and demand for the economy as a whole. Communism, in contrast, attempted to bring all of the information about supply and demand to a large and powerful central processor. Forced to deal with ever more complex information, the system's inherent inefficiencies led to its collapse and the collapse of the idea on which it was based.

Similarly, representative democracy operates on the still revolutionary assumption that the best way for a nation to make political decisions about its future is to empower all of its citizens to process the political information relevant to their lives and express their conclusions in free speech designed to persuade others and in votes — which are then combined with the votes of millions of others to produce aggregate guidance for the system as a whole. Other governments with centralized decision-making have failed in large part because they literally do not "know" what they or their citizens are doing.

Unfortunately, we are now on the verge of ignoring this powerful truth in designing the Mission to Planet Earth. The current plan is to bring all the data to a few large centers where they will be processed; somehow the results will then be translated into policy changes that are in turn shared around the world. The hope is that this mission will eventually help change thinking and behavior worldwide to the extent necessary to save the global environment.

The alternative approach — or architecture — that I am recommending here is to distribute the information collecting and processing capability in a "massively parallel" way throughout the world by involving students and teachers in every nation. This way, some of the essential work may well be accomplished much faster and much more efficiently — and we can then work to upgrade and improve the information handling capacity in each location. Furthermore, we ought to be establishing environmental training centers and technology assessment centers throughout those areas of the world (especially the Third World) where major environmental remediation efforts are needed and where major technology transfers from the West are expected.

In discussing information and its value, it is also worth remembering that some self-interested cynics are seeking to cloud the underlying issue of the environment with disinformation. The coal industry, for one, has been raising money in order to mount a nationwide television, radio, and magazine advertising campaign aimed at convincing Americans that global warming is not a problem. Documents leaked from the National Coal Association to my office reveal the depth of the cynicism involved in the campaign. For example, the strategy memorandum notes their "target groups" as follows: "People who respond most favorably to such statements are older, less-educated males from larger households, who are not typically active information-seekers . . . another possible target is younger, lower-income women [who are] likely to soften their support for federal legislation after hearing new information on global warming. These women are good targets for magazine advertisements."

In order to counter entrenched interests like this one, we will have to rely on the ability of an educated citizenry to recognize propaganda for what it is. And the economic and political stakes in this battle are so high, there will be a relentless onslaught of propaganda.

The key, again, will be a new public awareness of how serious is the threat to the global environment. Those who have a vested interest in the status quo will probably continue to be able to stifle any meaningful change until enough citizens who are concerned about the ecological system are willing to speak out and urge their leaders to bring the earth back into balance.

Conclusion

Life is always motion and change. Fueled by the fruits of sun and soil, water and air, we are constantly growing and creating, destroying and dying, nurturing and organizing. And as we change, the world changes with us. The human community grows ever larger and more complex, and in doing so demands ever more from the natural world. Every day, we reach deeper into the storehouse of the earth's resources, put more of these resources to use, and generate more waste of every kind in the process. Change begets change, then feeds on its own momentum until finally the entire globe seems to be accelerating toward some kind of profound transformation.

Much earlier, I described two kinds of changes: the slow and gradual change that is typical of our daily lives and the rapid, systemic change that occurs when a pattern shifts from one state of equilibrium to another, a shift that comes as a surprise. But there is yet a third kind of change, which combines elements of the first two; one version of it is described in a new theory called self-organized criticality put forward by Per Bak and Kan Chen, physicists at Brookhaven National Laboratory. It may at first sound a bit complicated, but I think it sheds much light on the dynamics of change — both in our lives and in the world at large.

Bak and Chen begin by studying something profoundly simple: sandpiles. They watch very carefully as sand is poured — grain by grain — on a table, first to form a pile and then to build it higher. With slow-motion videotapes and computer simulations, they count exactly how many grains of sand are dislodged as each new grain falls on top of the pile. Sometimes, as the pile grows, a single grain of sand triggers a little avalanche. Less frequently, bigger avalanches occur — again, they are triggered by a single grain of sand. But the potential for each avalanche, regardless of its size, is built up slowly as a result of the accumulated impacts of all the grains of sand. Small changes reconfigure the sandpile and ultimately render it vulnerable to larger changes.

As common sense would have it, most of the falling grains of sand dislodge only a few other grains and have little apparent impact on the pile as a whole. Yet even the grains in this large majority have a profound influence on what happens later. Indeed, they create the potential for future changes, both small and large. Amazingly, there is a precise mathematical relationship between the number of grains of sand dislodged by each new grain and the frequency with which sand avalanches of various magnitudes occur.

It's important to note, however, that this predictable response of the sandpile to each falling grain cannot occur until the pile reaches what is known as a critical state, in which every single grain is in direct or indirect physical contact with the rest of the sandpile. (These sandpiles never reach equilibrium.) But once enough sand is poured to form a single sandpile, and once there is physical contact between all the grains of sand, each new grain sends "force echoes" of its impact cascading — however faintly — down through the pile, in effect communicating its impact to the rest of the sandpile, causing some grains to shift in position and in the process shifting or reconfiguring the entire sandpile. In that sense, the sandpile "remembers" the impact of each grain that is dropped and stores that memory holistically (or holographically) in the physical position of all the grains in relation to one another and in the full three-dimensional shape of the pile itself.

The sandpile theory — self-organized criticality — is irresistible as a metaphor; one can begin by applying it to the developmental stages of a human life. The formation of identity is akin to the formation of the sandpile, with each person being unique and thus affected by events differently. A personality reaches the critical state once the basic contours of its distinctive shape are revealed; then the impact of each new experience reverberates throughout the whole person, both directly, at the time it occurs, and indirectly, by setting the stage for future change. Having reached this mature configuration, a person continues to pile up grains of experience, building on the existing base. But sometimes, at midlife, the grains start to stack up as if the entire pile is still pushing upward, still searching for its mature shape. The unstable configuration that results makes one vulnerable to a cascade of change. In psychological terms, this phenomenon is sometimes called a midlife change, an emotional avalanche releasing the combined force of many small and subtle changes accumulated over time. When it comes — and it can be touched off by a single traumatic event — this large change can cause a consolidation of the personality, leaving its mature configuration essentially unchanged but with thicker sides and a larger mass.

In describing their sandpiles, Bak and Chen use different terms: what I have called the "formative" stage is their "subcritical" state; what I have referred to as the "mature configuration" is their "critical" state; and what I have described as the building up of unstable configurations is their "supercritical" state. With that terminology in mind, consider one of their conclusions:

> A subcritical pile will grow until it reaches the critical state. If the slope is greater than the critical value — the supercritical state — then the avalanches will be much larger than those generated by the critical state. A supercritical pile will collapse until it attains the critical state. Both subcritical and supercritical piles are naturally attracted to the critical state.

One reason I am drawn to this theory is that it has helped me understand change in my own life. Most important, it has helped me come to terms with my son's terrible accident and its aftermath. After his near death, and after several other accumulated changes immediately before the accident, I felt as if my life had grown, in Bak and Chen's term, supercritical; a number of painful experiences had piled on top of one another. But change came cascading down the slopes of my life, and I settled back into what had felt like maturity before but was now fuller and deeper. I now look forward to the future with both a clearer sense of myself and of the work I hope to do in the world.

The legendary psychologist Erik Erikson was the first to document and describe the developmental stages of life that all of us experience. He also noted the successive and predictable crises that confront us as we grow from one stage to the next, explaining that these crises are sometimes necessary so that we do not get stuck in an unresolved conflict that prevents further growth. I had the privilege of studying under Professor Erikson when I was at that awkward stage of life when the challenge of discovering and defining one's "identity" is the primary psychological task. Now, at midlife, I have reached the stage when, in Erikson's terms, "generativity" is the central focus. It is the time of life when most people, according to Erikson and his followers, are ready to move beyond the work of achieving communion and full mutual trust with another. The successful resolution of this struggle brings the ability to care for many others and to establish and guide the next generation. Generativity, then, emerges during the most productive and fertile stage of one's life: one focuses on being fruitful for the future.

Can these two metaphors help us understand the current stage of humankind's relationship to the earth? Perhaps it can be said that civilization has passed the subcritical or formative stage and has recently reached a mature configuration, a worldwide community or global village. But is our species now on the verge of a kind of midlife crisis? Increasingly, people feel anxious about the accumulation of dramatic changes that portend ever-larger "avalanches" cascading down the slopes of culture and society, uprooting institutions like the family and burying values like those that have always nurtured our concern for the future. The actions of any isolated group now reverberate throughout the entire world, but we seem unable to bridge the chasms that divide us from one another. Is our civilization stuck in conflict between isolated nations, religions, tribes, and political systems — divided by gender and race and language? And now that we have developed the capacity to affect the environment on a global scale, can we also be mature enough to care for the earth as a whole? Or are we still like adolescents with new powers who don't know their own strength and aren't capable of deferring instant gratification? Are we instead on the verge of a new era of generativity in civilization, one in which we will focus on the future of all generations to come? The current debate about sustainable development is, after all, a debate about generativity. But are we really ready to shift our short-term thinking to long-term thinking?

Answering these questions is difficult if not impossible, both because the changes now under way have been building over a long period and because what is occurring to civilization and to the relationship between humankind and the environment is now truly global in nature. Returning to the metaphor of the sandpile, consider this phenomenon (again reported by Bak and Chen), which complicates the task of predicting or even understanding very large changes in a critical system:

> An observer who studies a specific area of a pile can easily identify the mechanisms that cause sand to fall, and he or she can even predict whether avalanches will occur in the near future. To a local observer, large avalanches would remain unpredictable, however, because they are a consequence of the total history of the entire pile. No matter what the local dynamics are, the avalanches would mercilessly persist at a relative frequency that cannot be altered. The criticality is a global property of the sandpile.

The ozone hole is a case in point, since it represents an unpredictable consequence of a global pattern by which civilization has accumulated dangerous chemical gases in the atmosphere. The general phenomenon of ozone depletion was anticipated, but the sudden "avalanche" of nearly total depletion above Antarctica came as a complete surprise. Since we are continuing to pile up larger quantities of the same gases, more such changes are certain to take place, although we won't necessarily be able to predict when. Of course, the same pattern is likely to hold for the larger and more serious problem of global warming: as we send increasing quantities of greenhouse gases into the atmosphere, it will become more and more difficult to believe that the only consequence is the well-understood phenomenon of warming. "Avalanches" of change in climate patterns are certain to occur and persist if we keep making this sandpile steeper and larger; moreover, the combination of several significant changes occurring almost simultaneously increases the risk of catastrophe significantly.

Apart from our growing threat to the integrity of the global ecological system, the dramatic changes now taking place within civilization are also likely to pose serious threats of their own to the integrity and stability of civilization itself. The accumulation of another billion people every ten years is creating a whole range of difficult problems, and all by itself the exploding population is liable to push world civilization into a supercritical state, leaving it vulnerable to very large "avalanches" of unpredictable change. To cope with this dangerous turn of events, we must somehow find a way to accelerate our movement to a new stage of development, one that embraces a mature understanding of our ability to shape our own future. As Erikson once wrote, "The possibility of species-wide destruction creates for the first time the necessity of a species-wide ethic."

When considering a problem as large as the degradation of the global environment, it is easy to feel overwhelmed, utterly helpless to effect any change whatsoever. But we must resist that response, because this crisis will be resolved only if individuals take some responsibility for it. By educating ourselves and others, by doing our part to minimize our use and waste of resources, by becoming more active politically and demanding change — in these ways and many others, each one of us can make a difference. Perhaps most important, we each need to assess our own relationship to the natural world and renew, at the deepest level of personal integrity, a connection to it. And that can only happen if we renew what is authentic and true in every aspect of our lives.

The twentieth century has not been kind to the constant human striving for a sense of purpose in life. Two world wars, the Holocaust, the invention of nuclear weapons, and now the global environmental crisis have led many of us to wonder if survival — much less enlightened, joyous, and hopeful living — is possible. We retreat into the seductive tools and technologies of industrial civilization, but that only creates new problems as we become increasingly isolated from one another and disconnected from our roots. Concern with self — narrowly defined as completely separate from others and from the rest of the world — is further reinforced as the primary motivation behind all social interactions and civilization as a whole. We begin to value powerful images instead of tested truths. We begin to believe that in the face of possible destruction, only those images that reflect and enlarge the self matter. But that response cannot last, and ultimately it gives way to a sense that what is real and right in our lives is slipping away from us. To me, this response has become so pervasive that it suggests a kind of collective identity crisis. I have for several years now been engaged in an intensive search for truths about myself and my life; many other people I know are doing the same. More people than ever before are asking, "Who are we? What is our purpose?" The resurgence of fundamentalism in every world religion, from Islam to Judaism to Hinduism to Christianity; the proliferation of new spiritual movements, ideologies, and cults of all shapes and descriptions; the popularity of New Age doctrines and the current fascination with explanatory myths and stories from cultures the world over — all serve as evidence for the conclusion that there is indeed a spiritual crisis in modern civilization that seems to be based on an emptiness at its center and the absence of a larger spiritual purpose.

Perhaps because I have ended up searching simultaneously for a better understanding of my own life and of what can be done to rescue the global environment, I have come to believe in the value of a kind of inner ecology that relies on the same principles of balance and holism that characterize a healthy environment. For example, too much of a focus within seems to lead to a certain isolation from the world that deprives us of the spiritual nourishment that can be found in relating to others; at the same time, too much attention to others — to the exclusion of what is best understood quietly within one's heart — seems to make people strangers to themselves. The key is indeed balance — balance between contemplation and action, individual concerns and commitment to the community, love for the natural world and love for our wondrous civilization. This is the balance I seek in my own life. I hope and trust we will all find a way to resist the accumulated momentum of all the habits, patterns, and distractions that divert us from what is true and honest, spinning us first this way, then that, whirling us like a carnival ride until our very souls are dizzy and confused.

If it is possible to steer one's own course — and I do believe it is — then I am convinced that the place to start is with faith, which for me is akin to a kind of spiritual gyroscope that spins in its own circumference in a stabilizing harmony with what is inside and what is out. Of course, faith is just a word unless it is invested with personal meaning; my own faith is rooted in the unshakeable belief in God as creator and sustainer, a deeply personal interpretation of and relationship with Christ, and an awareness of a constant and holy spiritual presence in all people, all life, and all things. But I also want to affirm what people of faith from long ago apparently knew and that our civilization has obscured: that there is revelatory power in the world. This is the essence of faith: to make a surrendering decision to invest belief in a spiritual reality larger than ourselves. And I believe that faith is the primary force that enables us to choose meaning and direction and then hold to it despite all the buffeting chaos in life.

I believe also that — for all of us — there is an often poorly understood link between ethical choices that seem quite small in scale and those whose apparent consequences are very large, and that a conscious effort to adhere to just principles in all our choices — however small — is a choice in favor of justice in the world. By the same token, a willingness to succumb to distraction, and in the process fail to notice the consequences of a small choice made carelessly or unethically, makes one more likely to do the same when confronted with a large choice. Both in our personal lives and in our political decisions, we have an ethical duty to pay attention, resist distraction, be honest with one another and accept responsibility for what we do — whether as individuals or together. It's the same gyroscope; either it provides balance or it doesn't. In the words of Aristotle: "Virtue is *one* thing."

For civilization as a whole, the faith that is so essential to restore the balance now missing in our relationship to the earth is the faith that we do have a future. We can believe in that future and work to achieve it and preserve it, or we can whirl blindly on, behaving as if one day there will be no children to inherit our legacy. The choice is ours; the earth is in the balance.

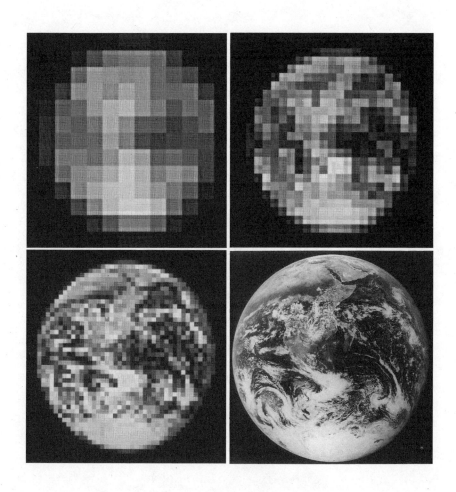

Acknowledgments

The individual to whom I owe the greatest debt for making this book possible is my wife, Tipper Aitcheson Gore. She was the only person to read every word in every draft — a task for which she volunteered with unflagging good cheer and enthusiasm, and invariably had excellent suggestions for improving the flow of ideas and sharpening the way they were expressed. She encouraged and sustained me throughout and arranged our family's activities to accommodate my concentrated attention on my writing. For this latter gift of encouragement, love, and enough peace and quiet to compose my thoughts, I also owe thanks to my children, Karenna, 18, Kristin, 14, Sarah, 13, and Albert, 9. I really wrote this book for them.

I began the writing process in April 1989, after several weeks in my son's hospital room at Johns Hopkins. Now that the book is finished, I have come to think of the whole undertaking as part of the process of healing and recovery my family and I went through after the accident described in the introduction. For that reason, before I acknowledge the individuals who made the book itself possible, I want to thank the men and women initially responsible for that healing process: first, two off-duty nurses from Hopkins, Victoria Costin-Siegel and Esther O'Campo, who had taken their medical kits with them to the baseball game and by chance passed by the scene of the accident soon after it occurred. The emergency room team was ably led by Dr. David Dudgeon, who performed surgery as soon as the ambulance arrived and then again three days later, when internal hemorrhaging posed another threat to Albert's life. Dr. Paul Griffin took care of all the broken bones; Dr. John Gearhart and Dr. William Zinkham, the internal injuries; Dr. Walter Tunnessen, the skin; Dr. Dave Cornblath, the nerve damage. Several months later, when microsurgery on his nerves was needed to restore the use of Albert's right arm, Dr. David Kline, of LSU, and Dr. Alan Hudson, of Toronto Children's Hospital, operated jointly at the Ochsner Clinic in New Orleans. Physical therapists Amy Kest, Keith Scott, Terri Pomeroy, and John Cummings contributed their substantial skill and patience over a long period. To each of them, I will always be grateful. And throughout our family's ordeal, we were lifted up spiritually and sustained emotionally by thousands of people — most of whom we had never met — who contacted us and prayed for Albert's recovery. It soon became apparent that those who had suffered the most in their own lives seemed to have the most comfort and understanding to give us. That sharing was an experience that moved me as deeply as any I have ever had; in a sense, it gave me permission to fully realize my grief and go, in Robert Bly's words, "into the ashes."

It is that experience of personal healing, in turn, which made it possible for me to write this book and which convinced me that the healing of the global environment depends initially upon our ability to grieve for the deep tragedy that our collision with the earth's ecological system is causing. Yet if we do so, I have no doubt whatsoever that the human spirit is capable of the transformation that that healing and recovery will entail.

In addition to Tipper, three friends were instrumental in encouraging me to go forward with the book: Gary Allison, Geoff Haines-Stiles, and Peter Knight. I talked with each of them during the final weeks at Johns Hopkins; then, in the early summer of 1989, all three of us met for intensive discussions. They pushed me to be more specific and to articulate more precisely the linkages between my ideas and concepts. Each of them has been a continual source of help and encouragement.

By December I had made enough progress to take the next step, and with Peter's help hired Mort Janklow as my agent. I quickly learned that his reputation for savvy is well earned; he was instrumental in matching me with the one editor who was ideal for this book, John Sterling of Houghton Mifflin. I say "this book" as if it would have been essentially the same with another editor but I cannot now imagine having worked with anyone else, and I want to acknowledge how much intelligence, skill, and patience Sterling brought to the book. I had worked for seven years as a newspaper reporter with some excellent editors (especially John Seigenthaler, who taught me about writing), but I had no idea how different a book is from the daily journalism I knew. It was Sterling who guided me through the various steps involved and gave invaluable advice whenever I reached an impasse.

From the end of 1990 to the early spring of 1991, when I was putting much of the first draft on paper, Yehudah Mirsky was an extremely able and erudite research assistant and a source of sound advice and encouragement. Moreover, at a critical time for me in the creative process, his dry humor, patience, and even disposition were much appreciated.

The next most important person to this project at Houghton Mifflin was Luise Erdmann, my manuscript editor, who also taught me a lot and was fun to work with. Rebecca Saikia-Wilson and Chris Coffin were both superb at turning the manuscript into a bound book, and Irene Williams is the best in the business of book promotion.

Among the many people who responded to requests for help with research were Charles Crawford at Memphis State University, Martha Cooper of the Smithsonian Institution, Julie Fisher of Fisher-Peck Associates in New Haven, Mahnaz Ispahani, Nati Krivatsky of the Folger Library, Cheryl McNab of the Ashoka Foundation, Kevin O'Rourke of Columbia University, John Tuxill of Cultural Survival, and Leon Wieseltier, who alerted me to some important books. In addition, I am grateful to the many talented people at the Library of Congress who were helpful in steering me toward the material I needed.

A number of scientists have been generous with their time in reading my penultimate draft and helping me avoid mistakes; in particular, Michael McElroy, chairman of the Department of Earth and Planetary Sciences at Harvard, spent many hours going over the manuscript, and I am extremely grateful for his contributions of fact and nuance. Sherwood Rowland, of the University of California, Irvine, gave the atmosphere chapter his expert scrutiny and enabled me to improve it significantly. Wally Broecker at Columbia reviewed the water chapter, with the assistance of Jim Simpson, Peter Schlosser, and Stephanie Pfirmann, and I am grateful to each of them for their suggestions. Among the other scientists who have been helpful in giving me advice during the writing of this book are Per Bak, Lester Brown, Jacques Cousteau, Richard Leakey, Thomas Lovejoy, Norman Myers, Rajendra K. Pachauri, Carl Sagan, Robert Watson, and Alexei Yablokov. Experts in other fields who helped by reading or advising on specific chapters include Robert Costanza, Herman Daly, Amy Fox, Paul Gorman, Lance Laurence, Charles Maier, Jerry Mande, Jim Morton, Michael Novak, Henry Peskin, Robert Repetto, Stephen Viederman, and Jim Wall. None of them is responsible for any remaining mistakes in the text. Those who read the six-hundred-page manuscript and returned it with comments — all in an absurdly short time — include several friends, Rick Adcock, Gary Allison, Tom Grumbly, Geoff Haines-Stiles, Nancy Hoit, Reed Hundt, Ward Hussey, Peter Knight, Jim Kohlmoos, Marty and Anne Peretz, Jack Robinson, Sr., my brother-in-law, Frank Hunger, and my parents, Albert and Pauline Gore.

Those to whom I owe thanks for illustrations used here include Tom Van Sant, a true visionary who has produced many unique images of the earth, one of which is the cover picture; Todd Gipstein and Patricia Corley of Gipstein Multi-Media, who tracked down and helped produce a number of the illustrations; Nancy Hoit, who made valuable suggestions; Gilbert Grosvenor, president of the National Geographic Society, and his able staff, including Karen Harshberger, Al Royce, and Barbara Shattuck, for their helpfulness in finding several of the pictures; Adele Medina O'Dowd for graphic artwork; Michael Kapetan for finding the fresco of Plato and Aristotle; Lorne Michaels for the Yard-a-Pult; William J. Kaufman III, for the Black Hole; Tom Boden of the Carbon Dioxide Information Analysis Center in Oak Ridge for the CO_2 graph; Robert G. Rossi, Christopher J. "Ragging" Waters, Wrisney Tan, and Brad Haynes for a quick turn-around on the lengthy transcript of the sessions in Carthage; and Bruce Reed for his patience, good humor, and help during the long night when I wrote the proposal.

And finally I want to thank Liza McClenaghan, who has been particularly helpful in locating materials, solving problems, and doing everything from fixing my word processor to catching the Federal Express man before the door closed for the night. In fact, it's that time right now.

Notes

Introduction. Much of the discussion in the early part of the introduction is based on congressional investigations in which I took part as a member of the House of Representatives. For example, the material on Agent Orange comes from hearings before the House Commerce Committee and the testimony of witnesses there.

The calculation of topsoil erosion and the amounts floating past Memphis in the Mississippi River is based on my conversations with the U.S. Army Corps of Engineers and the Iowa Department of Agriculture and Land Stewardship. Approximately 260,000 tons of topsoil per day floated past Memphis in 1991.

As a freshman congressman, I was assigned to the Oversight and Investigations Subcommittee of the House Commerce Committee and convinced its chairman, John Moss of California, to allow me to begin an investigation of hazardous chemical waste dumping. The first hearing examined the problems of Christine and Woodrow Sterling and their neighbors in Toone, Tennessee, and the problems of Lois Gibbs and her neighbors in the area near Buffalo, New York, called Love Canal. This series of hearings continued when the chairmanship of the committee passed to Bob Eckhardt of Texas. As the ranking Democrat under both chairmen, I was allowed to pursue the investigation through several dozen hearings over a period of a few years along with the subcommittee staff (principally Dick Frandsen, Pat McLean, Thomas Greene, Mark Raabe, Ben Smethurst, and Lester O. Brown). The Superfund Law, which was a direct result of these hearings, passed during the lame duck session of December 1980 and fell into the clutches of the new Reagan appointees: Rita Lavelle (later convicted for perjury), Ann Gorsuch Burford, and James Watt.

The hearings on global warming came before the Investigation and Oversight Subcommittee of the House Science and Technology Committee, which I chaired. The series of hearings was staffed by Tom Grumbly and Jim Jensen. After the first hearing with Roger Revelle, I telephoned Professor Carl Sagan at Cornell and asked him to be the first witness in the next series of hearings and to help popularize the issue of global warming. It was that second hearing that first brought the issue into the national news media and to public attention.

I'm especially indebted to Professor Revelle for awakening me as a young student to the dramatic alteration of the relationship between the human species and the ecological system of the earth. He headed the Center for Population Studies at Harvard and had worked at the Scripps Institution in La Jolla, California, before his death in 1991. He contributed insights in many different and wide-ranging fields of science.

The Interparliamentary Conference on the Global Environment deserves more than the brief mention I have given it in the text. It was the first meeting of its kind and produced three impressive days of discussion and agreement among participants from forty-two countries. But what I remember about it most was the extraordinary collegial approach taken by the bipartisan group of senators who joined me in hosting the conference, especially the ranking Republican member of the group, John Chafee of Rhode Island. The staff director of this enterprise was Frank Potter and the person most responsible for its success was Carol Browner, who was then my legislative assistant and is now Florida's secretary of the environment.

The quote from William Hutchinson Murray originally appeared in a 1978 *Forbes Magazine* article, but in tracking it down I found that Murray was one of those fascinating individuals who deserves a wider audience and, in fact, had one early in this century. A Scottish mountaineer, he wrote extensively about climbing in Scotland, Tibet, and Nepal. His work *The Story of Everest* has been translated into nine languages.

Chapter 1. For my discussion of the Aral Sea, I have relied on analysis by Soviet scientists in Uzbekistan and Moscow to whom I was introduced by the vice chairman of the Supreme Soviet Committee on the Environment, Alexei Yablokov. I also learned a great deal from the leading U.S. specialist, Dr. Philip Mecklin, of Western Michigan University in Kalamazoo. During my visit to the Aral Sea, I was deeply affected by the suffering of the people in Karakalpak, the region in Uzbekistan that borders the Aral on the south and that has borne the brunt of the ecological tragedy there.

The ice core experiment I visited in Antarctica was managed by the University of New Hampshire; the best-known such experiment is near the geographic center of the continent at the Russian Vostok research station. Other ice cores are now being drilled even further back in time near the center of the ice dome covering Greenland.

My discussion of the Arctic Basin is based on two trips under the ice there with the navy in 1990 and 1991. I am particularly indebted to Admiral Bruce DeMars, who, along with his staff, has been extremely responsive to the requests by the scientific community to facilitate research that would be impossible without the active cooperation of the nuclear submarine navy.

For my discussion on loss of species, I have relied heavily on the work of Tom Lovejoy, who was my guide and teacher during a visit to the Amazon rain forest in 1988, and on numerous meetings with Brazilian scientists since that time. Among the latter group I am especially indebted to Dr. Eneas Salati of Brazil, the world's leading expert on the hydrology of the Amazon Basin. I have also relied on the work of Professor E. O. Wilson, who was kind enough to discuss material on species loss from his forthcoming book, and on Norman Myers, the English biologist and political activist.

For my discussion of East Africa, I am indebted to Dr. Richard Leakey, the anthropologist who agreed to manage his native Kenya's efforts at conservation and who has done a spectacular job in unbelievably difficult circumstances.

For my discussion of coral bleaching, I have relied on a number of scientists, including Thomas Goreau, Raymond Hayes, Walter C. Jaap, Robert L. Wicklund, and Dr. Ernest Williams, who testified at a hearing on October 11, 1990. At the same hearing, experts from NASA who have compiled a temperature record from satellite measurements correlated the dates of major bleaching episodes with the dates of temperature peaks in the latitudes of the reefs in question.

A word here is in order about the number of children under the age of five who die in our world each day. Many people find the number — 37,000 — to be startling. It is, however, accurate according to the World Health Organization and other groups that deal with the horrendous tragedy these preventable deaths represent.

For the calculation of a 600 percent increase in the number of chlorine atoms in the atmosphere, I have relied upon Dr. Sherwood Rowland of the University of California, Irvine, whom most regard as the leading expert in the world on these topics. The exact numbers are 0.6 parts per billion (ppb) in 1950 and 3.9 ppb in 1992. I first made the emotional connection between these changes in the global atmosphere and the composition of each of my own individual breaths when I read *The End of Nature,* by Bill McKibben.

The reference to a 25 percent increase in heat-absorbing molecules requires some elaboration. It does not include increased water vapor, which is of course the principal greenhouse gas, but increases in it result from warming that is initially triggered by other gases added to the atmosphere by human activity. Of these, CO_2 has increased by 13 percent since the measurements began in 1958 and by an extra amount during the unmonitored years before then; methane (CH_4) has increased from about 1,050 ppb in 1945 to more than 1,700 ppb in 1991 for an increase of almost 60 percent. The next largest concentration is of CFCs, which, as already noted, have increased more than 600 percent. After weighing these molecules for their relative volumes and absorptive capacity, Rick Adcock, at Tufts University, arrived at the calculation of 23 percent, but adds that it would be more scientifically correct to say that the range is between 20 and 25 percent.

My discussion of population relies on figures calculated by the United

Nations Population Fund and the Population Research Council as well as on work by Paul and Anne Ehrlich.

Chapter 2. The extensive discussion of clouds and water vapor is based on a science roundtable discussion held on October 7, 1991 (the transcript is available from the Senate Commerce Committee) and the work of the twelve scientists who participated. Similarly, the discussion of the West Antarctic ice sheet and sea level rise is based on a hearing I chaired on May 13, 1991, on the role of ice in global climate change.

Professor Richard Lindzen's quote is from a letter printed in the *New York Times* on February 19, 1991.

The White House memorandum with the advice to "raise the many uncertainties" about global warming, instead of debating specific facts, was reported widely in the press the day after Earth Day and was allegedly leaked by a disgruntled executive branch employee who gagged on the depth of its cynicism.

Although I have mentioned the well-known ordeal of Galileo only briefly, I recommend the full text of his trial transcript, which I had never read and found utterly fascinating. Similarly, although I've mentioned the theory of continental drift only in passing, I found it worthwhile to read the transcripts of scientific society meetings in which proponents of the then radical theory were subjected to remarkable public scorn by their peers, who found the entire notion too ridiculous to bear. I do not remember the name of my sixth-grade classmate, whom I will always regard as the codiscoverer of continental drift.

The discussion of those who believed the earth was flat is worth a short comment since this book is being published in the five hundredth anniversary year of Columbus's voyage. Actually, as Carl Sagan describes entertainingly in *Cosmos,* the roundness of the earth — and even its nearly exact circumference — was established in the third century B.C. by Eratosthenes, an astronomer working in the Egyptian city of Alexandria. The courage later shown by Columbus was not in challenging a prevailing notion that the earth was flat but rather in challenging the accuracy of the calculations of the earth's circumference; mistakenly believing that the earth was one third smaller than it is, Columbus convinced his sponsors that his ships could carry enough rations to reach India. He was, of course, saved from starvation by the serendipitous discovery of what later became known as the New World, but which he believed was India.

Although the phrase I have quoted from Ivan Illich has appeared elsewhere, I first encountered it in an interview he gave to the *New Perspectives Quarterly* in the spring of 1989.

The discussion of Chaos Theory and its application to the global environment is based in part on discussions I had with climatologists at the University of East Anglia in Great Britain, although I have since encountered similar work elsewhere. A good discussion appears in James Gleick's best-selling *Chaos*.

The discussion of positive feedback loops has become quite common with scientists concerned with the environmental crisis. I have based most of my discussion on the results of a hearing I chaired on May 8, 1989. There are two areas of frequent misunderstanding that deserve some mention here. First, the common description of rain forests as the "lungs of the world" obscures a more complicated set of truths. Mature forests (what biologists call climax forests) typically have a neutral CO_2 balance; that is, they consume roughly the same amount of CO_2 that they produce. Forests that are still growing toward maturity, on the other hand, are large net consumers of CO_2, and all forests release large quantities of CO_2 when they are burned. The implications are: (1) halting the burning of forests is an important strategy for reducing the rate of CO_2 increase, and (2) massive tree planting programs represent a good way to pull carbon out of the atmosphere.

The second area of misunderstanding involves the oceans. I have heard on numerous occasions the assertion that a great deal of carbon is fixed in the oceans by the production of shells, which then sink to the bottom. However, as Professor Michael McElroy of Harvard has shown, the production of shells is actually associated with the release of CO_2 and a change in the alkalinity of the ocean through the net removal of calcium. Additionally, while the warming of the oceans interferes with its ability to absorb CO_2, this effect may well be limited to the topmost layers, and warming could further isolate the deep ocean (which contains most of the CO_2) from the atmosphere.

The temperature increase in Siberia was recorded by two NASA scientists in Huntsville, Roy Spenser and John Christy. While the large increase I have cited is only for March 1990, the trend toward disproportionate warming in Siberia is reflected in smaller but consistent increases throughout the rest of the record.

The case for positive benefits from the thawing of tundra has been made most persistently by a Soviet scientist, Mikhail I. Budyko, but most Soviet scientists take strong exception to his conclusions.

The quote from Robert McNamara is one I have heard him deliver personally on many occasions; the first time was at an Aspen Institute seminar chaired by Ambassador Richard Gardner.

Chapter 3. For my treatment of the "the year without a summer," I have drawn upon the work of John Dexter Post in *The Last Great Subsistence in the Western World,* on research done by Professor William Crawford at Memphis State University, on the voluminous work of Emmanuel Le Roy Ladurie, and on the work of Hubert Lamb.

The correspondence between volcanic eruptions and short-lived changes in the global environment was first brought to my attention in 1981 by Stephen Schneider, who testified at several of the early global warming hearings — including the first one — and has been one of the most active scientists in bringing the issue to public attention. I found especially valuable and have relied upon original research by Kevin D. Pang of the CalTech Jet Propulsion Lab, which he reported in "The Legacies of Eruption," a 1991 article in *The Sciences*. Le Roy Ladurie, who has produced the classic work in climate history, also has been a source in my discussion of volcanoes.

Pang's references to Benjamin Franklin's shrewd observations led me to a fascinating excursion into Franklin's scientific work that relates to the global environment. Most of what I learned was ultimately beyond the scope of this book but truly fascinating: for example, the Gulf Stream was discovered by Franklin.

In my discussion of migrations and prehistoric climate change, I have relied on the work of Randall White and Brian M. Fagan and Elisabeth S. Verba, Frederick E. Grine, Richard G. Klein, and David Pilbeam.

I constructed my own calendar on which I correlated the events, both climatic and sociopolitical — and couldn't resist coming to some of my own conclusions based on the concurrence of events that seemed to me to betray something other than simple coincidence. For example, the speculation that the same warming trend that allowed Leif Eriksson to travel to Vinland might be implicated as a causal factor in the sudden demise of the Mayan civilization grows out of this effort to match climate and historic events in different parts of the world. Similarly, the speculation about the link between the emigration from northern Europe caused by the sudden return of Ice Age conditions 11,000 years ago and the appearance soon thereafter of the first known cities in southern Turkey and Mesopotamia is based on this technique. In both cases, however, I have carefully searched the existing literature for what is known about the events. These texts appear in the bibliography.

For my discussion of the great famine, I have relied on the work of Cormac O'Grada, Cecil Woodham-Smith, and Emmanuel Le Roy Ladurie.

My treatment of the Dust Bowl is based on the work of Paul Bonnifield, Vernon Gill Carter, and Tom Dale.

Chapter 4. My discussion of air pollution in Eastern Europe, Asia, Mexico, and Latin America is based on conversations with parliamentarians and environmental leaders from the countries mentioned, on personal observations, and on contemporary news accounts. Of special value were the reports by *Time, Newsweek, U.S. News and World Report, National Geographic,* the *New York Times,* and the *Washington Post.*

The calculation of the width of the earth's atmosphere should be qualified: technically, since a gas expands to fill a vacuum, the earth's atmosphere cannot be described as having a definite line of demarcation between it and outer space, but most of the molecules that make up the atmosphere are found within the first few miles above the surface. If the density of air remained constant at every altitude at a value equal to its density at sea level, the entire atmosphere would extend upward to an altitude of only about seven kilometers.

The discussion of diminished oxidation of the atmosphere is based on extensive discussions with Sherwood Rowland, Michael McElroy, Robert Watson, and a number of other scientists. The cleansing or oxidation takes place mainly in the tropics, where ultraviolet radiation penetrates the deepest and where water vapor is most available. For this reason, the burning of vast areas of forest in the tropics is particularly dangerous because it produces such large quantities of carbon monoxide, which monopolizes the hydroxyl (the "detergent").

The effect of enhanced levels of ultraviolet radiation on the immune system is now a major area of research. Dr. Margaret Kripke of Texas has been one of the pioneers in this field, and it was her early work that alerted me to the issue.

The discussion of polar stratospheric clouds in the Arctic and Antarctic omits any effort to describe the complicated chemical actions involved: the ice crystals are actually mixtures of nitric acid and water called nitric acid tri-hydizate (NAT). These frozen particles are only formed in temperatures below –80 degrees Centigrade, conditions found only in three places: both polar regions and, ironically, high in the atmosphere of the tropics at the very top of convective columns of air that rise much higher there than elsewhere. In addition, although the discussion of ozone depletion is often dominated by chlorofluorocarbons, the role of bromines and halons should receive much more emphasis than they usually do because of the important roles they play in the chemistry of destruction.

A word about the amount of stratospheric ozone that has been lost as of 1992: the popular press usually dwells on calculations of cumulative loss since the beginning of careful measurement in 1978; this is misleading, however, because the destruction began well before the regular monitoring. A simple rule of thumb, according to Sherwood Rowland, is to double the loss calculated since 1978 to obtain the total loss since human beings changed the chemistry of the global atmosphere so profoundly in the years after World War II.

The reference to Professor Lindzen's public withdrawal of his hypothesis on the role of water vapor is to a statement he made in October 1991 during an Earth Science Roundtable. Lindzen graciously acknowledged that the credit for pointing out his error properly goes to Dr. Alan Betts, who works out of his home in Middlebury, Vermont. However, Lindzen continues to disagree with most other scientists and feels that water vapor is likely to be a negative feedback, and lead to cooling.

The three scientists referred to for the Marshall Institute study are Dr. Robert Jastrow, Dr. William Nierenberg, and Dr. Frederick Seitz. The same three authors had collaborated previously on a study purporting to prove that the comprehensive space-based star wars system proposed by

President Reagan was scientifically feasible. This second study, which was apparently intended to support yet another of Reagan's political goals, was apparently written mainly by Jastrow, now at Dartmouth. It was not peer-reviewed and was widely discredited in the scientific community. Subsequent analysis by experts in solar radiation revealed a large mathematical error in the authors' calculations, which changed an expected warming cycle in sunspot activity to one that they mistakenly described as a cooling period. The report received prominent attention nonetheless, because President Bush's chief of staff, John Sununu, referred to it frequently as a basis for his deep skepticism about global warming. Ironically, although the Marshall Institute study is still regarded as an unfortunate politicization of science, the relationship between sunspots and temperature fluctuations has received new attention because of work done on short-term climate variations at the Danish Meteorological Institute. Even if this new work is validated, as I expect, the rapid and huge increases in CO_2 are expected to completely overpower this sunspot effect.

The discussion of the earth's climate system as an engine for redistributing heat and cold was first explained to me in a comprehensible way at the University of East Anglia by Phil Jones, but it has been described by numerous scientists such as Stephen Schneider, Ralph Cicerone, and John Firor. It should be noted here that there are two factors that mitigate the accelerated warming at the poles. In Antarctica, the magnifying impact of melting at the ice edge is tempered by the mixing of deeper cold water with the surface meltwater; and at both poles, but especially in the Arctic, the increased evaporation produces clouds that complicate what might otherwise be straightforward calculations.

Chapter 5. The calculation of "the recipe for life" is derived information provided by the National Institutes of Health.

The ocean conveyor belt has been described at length by Wallace Broecker and Peter Schlosser and others. After reading their research, I invited Broecker to testify at a hearing on the role of oceans in climate change and subsequently visited him and his entire team at the Lamont-Doherty Geological Observatory of Columbia University. Much is still unknown about the operation of the conveyor: for example, the volume of warm water traveling westward from the Pacific to the Indian Ocean has not yet been measured. New research has focused on the movement of warm water from the Indian Ocean around the Horn of Africa to the South Atlantic, and still other scientists have proposed the existence of a shallower water conveyor in the Pacific, linked to convective activity in the tropics.

The discussion of California water patterns and the effect of warming on the snow pack comes in part from a hearing that Senator Tim Wirth and I had in Los Angeles in 1989.

The discussion of sea level rise is based on analyses by Roger Revelle,

Stephen Schneider, Lynn Edgerton, Michael Oppenheimer, and James Hansen.

The discussion of the West Antarctic ice sheet is based on a number of discussions with researchers during my visit there in 1988 and, as noted in the text, on the work of Robert Bindschadler.

The work by Lonnie and Ellen Thompson on glaciers had not been published when this book went to press, but it had been accepted and reviewed. Their work appears to be unprecedented in its scope and should be regarded as a major contribution to the temperature record.

My discussion on the effects of sea level rise on freshwater aquifers relies on the work of Lester Brown and his colleagues at the WorldWatch Institute.

The discussion of how forests attract rain deserves elaboration. Although small amounts of dimethylsulfide (DMS) are produced in forests, their principal role is in the oceans, where they form the nuclei of cloud droplets. Far more significant in the formation of cloud droplets above forests are terpenes, which are transformed by oxidation into a sulfate aerosol. Recent research has shown that industrially produced sulfates can have similar impacts; studies have linked sulfates in air pollution to increased cloud formation above the continental United States.

As noted earlier, I am indebted to Professor Eneas Salati of Brazil for the material in the discussion of the hydrology of the Amazon rain forest.

I have relied on official reports of the World Environment Programme and World Health Organization for many of the facts in the discussion of water-borne diseases. The work of Sandra Postal and others at the World-Watch Institute has been helpful in some of the discussion of problems with irrigation.

Chapter 6. In the discussion of rain forest destruction I have relied heavily on Dr. Thomas Lovejoy's work at the Smithsonian Institution and the discussions with Brazilian scientists that he arranged during my visit to the rain forest in 1988.

I have also relied on the books cited in the bibliography on the rain forest and research compiled by groups such as the Rain Forest Action Network.

The quotation from José Lutzenberger comes from his keynote address to the Interparliamentary Conference on the Global Environment, April 29, 1990.

I have relied here on numerous official reports of the United Nations Environment Programme. When I visited its headquarters in Nairobi in 1991, I was most impressed with the extensive work that scientists there have done on virtually all the problems mentioned in this chapter.

The figure quoted — 40 percent photosynthetic energy being consumed by human beings — comes from an often-cited study by Peter M. Vitousek and others. Although their methodology is laid out in painstaking detail, their conclusions have not gone unquestioned and remain controversial in the minds of some scientists.

The advance of the Sahara is worth some discussion. Specialists in this field have chafed over the years at the easy assumption by laypersons that deserts are steadily advancing, because field checks have often found green areas where the desert was assumed to have laid its claim. More recent studies, incorporating satellite observations, have shown that in fact the Sahara undulates irregularly forward and back — sometimes in the same direction for several years at a time. But the overall trend, just as clearly, has been much more of an advance than a retreat over the past two decades.

The study by Mamadou was delivered personally in a paper presented to the Planeterre Conference in Paris in 1989.

Chapter 7. I have relied for many of the ideas in this chapter on a series of books noted in the bibliography. I recommend especially the California Lands Project for its overall balance. In addition, I have relied on private conversations with Norman Myers, a specialist in the field, and, as noted in the text, an excellent article by *National Geographic.* I have also relied upon studies by the National Academy of Scientists and the transcripts of expert discussions organized by the Keystone Center.

Chapter 8. Much of the material here is based upon congressional hearings I have held for thirteen years on waste disposal practices in the United States and around the world. During those investigations I have developed a high respect for the work of the Congressional Research Service, and I have relied on a number of studies that the CRS has done in this field over the years. In addition, I have drawn on the strong work of the General Accounting Office, which also has a record of expertise in this subject. I have drawn upon an unusually good series of investigative stories by *Newsday* that was later published as a book and an excellent investigative series on backhauling by the *Seattle Post-Intelligencer.* I have also relied upon official reports from the Environmental Protection Agency and the United Nations Environment Programme.

The quotation from the homeless eight-year-old boy appeared in the *New York Times* in October 1990.

Chapter 9. The figures on the amount of borrowing came from the Congressional Budget Office. The opinion polls cited as evidence of growing distaste for politics have appeared in numerous places, most recently in the *Washington Post*/ABC poll of November 1991.

The discussion of coordinated strategy between the United States and Saudi Arabia is based on eyewitness reports from numerous participants at the preparatory meetings of the Climate Change negotiators during 1991.

The accusations against the minister of the environment in Sarawak were leveled during election campaigns in Malaysia in the last two years.

Chapter 10. My discussion of environmental economics is based on a series of hearings I chaired before the Joint Economic Committee to examine this issue in detail. Before these hearings, in 1990, I conducted a series of informal roundtable discussions with leading experts in the field, including Conn Nugent, Dr. Mohan Munasinghe, Dr. Salah El Serafy, Dr. Henry Peskin, Dr. Carol Carson, Barbara Brambl, and Dr. Peter Bartelmus.

I am particularly indebted to Dr. Robert Reppeto, Dr. Herman Daly, Dr. Robert Costanza, and Stephen Viederman. There is a difference between the extensive work some business economists have done on the micro-economics of this issue and the relative lack of attention paid to the macro-economics of the problem.

Chapter 11. My discussion of technology and its impact on perception and thinking owes a great deal to work of two thinkers, Marshall McLuhan and Maurice Merleau-Ponty. The first I studied as an undergraduate, the second, as a graduate student in divinity school.

The problem of migrating cultures in Kenya is based on personal study and discussions with the conservationist Richard Leakey in Kenya. The thought from Octavio Paz appeared in *The New Yorker*. The famous Erikson experiment with children and blocks was reported in *Childhood and Society*. The quotation from Father Thomas Berry is from a personal conversation, though I understand it will appear in his forthcoming book, *The Universe Story*.

The bizarre suggestions for altering the global climate with tin foil strips and iron fertilization come, believe it or not, from a subpanel report to the National Academy of Sciences.

Chapter 12. My treatment of Deep Ecology is based on a number of discussions with its exponents and on accounts such as *Green Rage*, by Christopher Manes.

I have based my discussion of the human brain in significant part on analyses by Carl Sagan in *The Dragons of Eden* and *Broca's Brain*.

My discussion of addiction theory and dysfunctional family theory is based on the books cited in the text and bibliography.

Chapter 13. My treatment of spirituality and the environment is based in part on a series of dialogues that I have organized, along with Dean James Morton of the Cathedral of Saint John the Divine in New York City and Carl Sagan, with active help from my colleague Senator Tim Wirth. These dialogues, between scientists and religious leaders, have been aimed at exploring the common ground between these two worlds. I have also relied on the books cited in the bibliography — and of course on the Bible.

The discussion of Greek philosophy is based on my own reading of Plato and Aristotle and on analyses of Renaissance philosophical developments such as the works of Paul O. Kristeller. I have also learned a great deal from my discussions with the theologian Michael Novak and from my old friend Jim Wall of Chicago, the editor of the *Christian Century,* and from the many religious leaders who have taken part in the dialogues to which I referred earlier.

The material on Arno Penzias comes from personal conversations with him.

The treatment of other religions is based on their sacred texts, which I encountered in my dialogues mentioned above. I am especially indebted to Paul Gorman and Amy Fox of the Cathedral of Saint John the Divine for helping me to find much of this material.

The phrase at the end of the chapter, "bright shining as the sun," is taken from one of the most famous Christian hymns, "Amazing Grace."

Chapter 14. The discussion of the siege of Leningrad is based on material reported in Shattering, by Fowler and Mooney, and an account by Steven Witt.

The stories of the other resistance fighters are based in large part on personal conversations with them except for that about Mechai Viravayda, which is based on Ruth Caplan's description in *Our Earth, Ourselves.* Although I could not speak with Mendes, I did talk with his widow and his close associates in the Amazon.

Chapter 15. My discussion of the Marshall Plan is based on the work of Charles Maier and Stanley Hoffman, both professors at Harvard, who organized an impressive review of the Marshall Plan a few years ago. I'm especially indebted to Professor Maier, who went out of his way to help me understand the material.

The passage written in 1973 by George Bush, by then the U.N. ambassador, comes from Phyllis Piotrow's *World Population Crisis: The United States Response.*

Conclusion. The discussion of sandpiles is based on personal conversations with Per Bak and on the paper he wrote with Kan Chen. Others have helped them develop sandpile theory, including Kurt A. Wiesenfeld of Georgia Tech, Chao Tang at the Institute for Theoretical Physics in Santa Barbara, and Glen A. Held at the IBM Thomas J. Watson Research Center.

Bibliography

About Stewardship of the Environment. South Deerfield, Mich.: Channing L. Bete Co., 1991.

Ackerman, Nathan. *The Psychodynamics of Family Life*. New York: Basic Books, 1958.

Anderson, Bruce N., ed. *Ecologue: The Environmental Catalogue and Consumer's Guide for a Safe Earth*. New York: Prentice Hall Press, 1990.

Ausubel, Jesse H., and Hedy E. Sladovich, eds. *Technology and Environment*. National Academy of Engineering, Washington, D.C.: National Academy Press, 1989.

Barraclough, Geoffrey, ed. *The Times Atlas of World History*. Maplewood, N.J.: Hammond, 1982.

Bates, Albert K. *Climate in Crisis*. Summertown, Tenn.: The Book Publishing Co., 1990.

Battan, Louis J. *Weather*. Englewood Cliffs, N.J.: Prentice-Hall, 1985.

Becker, Ernest. *The Denial of Death*. New York: The Free Press, 1973.

Belk, K. E., N. O. Huerta-Leidenz, and H. R. Cross. "Factors Involved in the Deforestation of Tropical Forests." College Station, Tex.: Texas A&M University, Department of Animal Science. n.d.

Benedick, Richard Elliott. *Ozone Diplomacy: New Directions in Safeguarding the Planet*. Cambridge, Mass.: Harvard University Press, 1991.

Benedick, Richard Elliot, et al. *Greenhouse Warming: Negotiating a Global Regime*. Washington, D.C.: World Resources Institute, 1991.

Berry, Thomas. *The Dream of the Earth*. San Francisco: Sierra Club Books, 1988.

Bonnifield, Paul. *Dust Bowl: Men, Dirt and Depression*. Albuquerque: University of New Mexico Press, 1979.

Bowen, Murray. *Family Therapy in Clinical Practice*. New York: J. Aronson, 1978.

Bradley, R. S., et al. "Precipitation Fluctuations over Northern Hemisphere Land Areas Since the Mid-Nineteenth Century," *Science,* vol. 237, 10 July 1987, pp. 171–75.

Bradshaw, John. *The Family: A Revolutionary Way of Self-Discovery.* Deerfield Beach, Fla.: Health Communications, 1988.

———. *Homecoming: Reclaiming and Championing Your Inner Child.* New York: Bantam Books, 1990.

Brahn, Paul G., and Jean Vertut. *Images of the Ice Age*. New York: Facts on File, 1988.

Broecker, Wallace S., and T.-H. Peng. *Tracers in the Sea*. Palisades, N.Y.: Lamont-Doherty Geological Observatory, 1982.

Brown, Lester. *The Changing World Food Prospect: The Nineties and Beyond*, Washington, D.C.: WorldWatch Paper, 1988.

Brown, Lester, et al. *State of the World*. New York, W. W. Norton, 1984–91.

Bullard, Fred M. *Volcanoes of the Earth,* 2nd ed. Austin: University of Texas Press, 1984.

Burkitt, Denis P., and S. Boyd Eaton. "Putting the Wrong Fuel in the Tank." *Nutrition*, vol. 5 (3), May/June 1989, pp. 189–91.

Cannadine, David. *Blood, Toil, Tears and Sweat: The Speeches of Winston Churchill*. Boston: Houghton Mifflin, 1989.

Caplan, Ruth, et al. *Our Earth, Ourselves*. New York: Bantam, 1990.

Capra, Fritjof. *The Turning Point*. New York: Bantam, 1982.

Carson, Rachel. *Silent Spring*. Boston: Houghton Mifflin, 1962.

Carter, Vernon Gill, and Tom Dale. *Topsoil and Civilization,* rev. ed. Norman: University of Oklahoma Press, 1974.

Cohen, Michael J. *A Field Guide to Connecting with Nature*. Eugene, Ore.: World Peace University, 1989.

Commission for Racial Justice. *Toxic Wastes and Race in the United States: A National Report on the Racial and Socio-Economic Characteristics of Communities with Hazardous Waste Sites*. New York: United Church of Christ, 1987.

Committee on Earth Sciences. "Our Changing Planet: The FY 1991 U.S. Global Change Research Program." Reston, Va.: U.S. Geological Survey, 1991.

Culbert, T. Patrick, ed. *The Classic Maya Collapse*. Albuquerque: University of New Mexico Press, 1973.

Daly, Herman E., and John B. Cobb, Jr. *For the Common Good: Redirecting the Economy Toward Community, the Environment, and a Sustainable Future*. Boston: Beacon Press, 1989.

Delphos, William A. *Environment Money: The International Business Executive's Guide to Government Resources*. Washington, D.C.: Venture Publishing, 1990.

Dickinson, Robert E., ed. *The Geophysiology of Amazonia: Vegetation and Climate Interactions* New York: John Wiley, 1987.

Donaldson, Peter J. *Nature Against Us: The U.S. and the World Population Crisis, 1965–1980*. Chapel Hill, N.C.: University of North Carolina Press, 1990.

———, and Amy Og Tsui. "The International Family Planning Movement." *Population Bulletin*, vol. 45 (3), November 1990.

Doyle, Jack. *Altered Harvest: The Fate of the World's Food Supply*. New York: Viking, 1985.

Dubos, René. *Man, Medicine, and Environment.* New York: Praeger, 1968.

Eaton, S. Boyd. "Primitive Health." *Journal of MAG,* vol. 80, March 1991, pp. 137–40.

——, and Melvin Konner. "Paleolithic Nutrition." *New England Journal of Medicine,* January 31, 1985, pp. 283–89.

Edgerton, Lynne. *The Rising Tide: Global Warming and World Sea Levels.* Washington, D.C.: Island Press, 1991.

Ehrlich, Paul R., and Anne H. Ehrlich. *The Population Explosion.* New York: Simon & Schuster, 1990.

Eisler, Riane. *The Chalice and the Blade: Our History Our Future.* San Francisco: Harper & Row, 1987.

Erikson, Erik H. *Childhood and Society.* New York: W. W. Norton, 1950.

——. *Insight and Responsibility.* New York: W. W. Norton, 1964.

Fagan, Brian M. *The Journey from Eden: Peopling Our World.* New York: Thames & Hudson, 1990.

Falk, Richard A. *This Endangered Planet: Prospects and Proposals for Human Survival.* New York: Vintage Books, 1971.

Feliks, Yehuda. *Nature and Man in the Bible.* London: Soncino Press, 1981.

"Fertility Behavior in the Context of Development: Evidence from the World Fertility Survey." Population Studies No. 100, United Nations, New York, 1987.

Firor, John. *The Changing Atmosphere: A Global Challenge.* New Haven: Yale University Press, 1990.

Fisher, Ron, et al. *The Emerald Realm: Earth's Precious Rain Forests.* Washington, D.C.: National Geographic Society, 1990.

Flavin, Christopher. *Slowing Global Warming: A Worldwide Strategy.* Washington, D.C.: Worldwatch Institute, 1989.

Fletcher, Susan. "Briefing Book: Selected Major International Environmental Issues." CRS, March 22, 1991.

——. "International Environmental Issues: Overview." CRS Issue Brief, June 3, 1991.

Fowler, Cary, and Pat Mooney. *Shattering: Food, Politics, and the Loss of Genetic Diversity.* Tucson: University of Arizona Press, 1990.

Gershon, David, and Robert Gilman. *Household Ecoteam Workbook.* Olivebridge, N.Y.: Global Action Plan for the Earth, 1990.

Gimbutas, Marija. *The Language of the Goddess.* San Francisco: Harper & Row, 1989.

Gleick, James. *Chaos: Making a New Science.* New York: Viking, 1987.

Gordon, Anita, and David Suzuki. *It's a Matter of Survival.* Cambridge, Mass.: Harvard University Press, 1991.

Gribbin, John. *The Hole in the Sky.* New York: Bantam, 1988.

Goldstein, Eric A., and Mark A. Izeman. *The New York Environmental Book.* Washington, D.C.: Island Press, 1990.

Halberstam, David. *The Next Century.* New York: Morrow, 1991.

Harmon, Leon D. "The Recognition of Faces." *Scientific American,* November 1973, vol. 229 (5), pp. 70–82.

Hoffman, Stanley, and Charles Maier, eds. *The Marshall Plan: A Retrospective.* Boulder, Colo.: Westview Press, 1984.

Hong, Evelyne. *Natives of Sarawak: Survival in Borneo's Vanishing Forests.* Malaysia: Institut Masyarakat, 1987.

Hughes, J. Donald. *Ecology in Ancient Civilizations.* Albuquerque: University of New Mexico Press, 1975.

Hulteen, Bob, and Brian Jaudon. "With Heart and Hands." *Sojourners,* February/March 1990, pp. 26–29.

Human Exposure Assessment for Airborne Pollutants. *Advances and Opportunities.* Washington, D.C.: National Academy Press. 1991.

Interparliamentary Conference on the Global Environment. Final Proceedings. April 29–May 2, 1990, Washington, D.C.

John Paul II. "The Ecological Crisis a Common Responsibility." Message of His Holiness for the Celebration of the World Day of Peace, January 1, 1990.

Johnson, Lawrence E. *A Morally Deep World.* Cambridge: Cambridge University Press, 1991.

Kates, Robert W., et al. *The Hunger Report: 11988.* Providence: Alan Shawn Feinstein Hunger Program, Brown University, 1988.

Kelly, Brian, and Mark London. *Amazon.* New York: Holt, Rinehart & Winston, 1983.

Korten, David C. *Getting to the 21st Century: Voluntary Action and the Global Agenda.* West Hartford, Conn.: Kumarian Press, 1990.

Kristeller, Paul Oskar. *Renaissance Concepts of Man and Other Essays.* New York: Harper Torchbooks, 1972.

———. *Renaissance Philosophy and the Medieval Tradition.* Latrobe, Pa.: Archabbey Press, 1966.

———. *Renaissance Thought and Its Sources.* New York: Columbia University Press, 1979.

———, and Philip Wiener. *Renaissance Essays from the Journal of the History of Ideas.* New York: Harper Torchbooks, 1968.

Laing, R. D. *The Politics of the Family and Other Essays.* New York: Vintage Books, 1972.

Lamb, Hubert H. *Climate, History and the Modern World.* New York: Methuen, 1982.

———. *Weather, Climate and Human Affairs: A Book of Essays and Other Papers.* London: Routledge, 1988.

Lee, Charles. "The Integity of Justice." *Sojourners,* February/March 1990, pp. 22–25.

Le Roy Ladurie, Emmanuel. *Times of Feast, Times of Famine: A History of Climate since the Year 1000.* Garden City, N.Y.: Doubleday, 1971.

Lipske, Michael. "Who Runs America's Forests?" *National Wildlife,* October/November 1990, pp. 24–28.

Ludlum, David M. *The Weather Factor.* Boston: Houghton Mifflin, 1984.

Lyman, Francesca, et al. *The Greenhouse Trap.* Boston: Beacon Press, 1990.

McCarthy, James E. "Hazardous Waste Fact Book." CRS, January 30, 1987.

———. "Hazardous Waste Management: RCRA Oversight in the 101st Congress." October 12, 1990.

———. "Solid and Hazardous Waste Management." CRS Issue Brief, March 5, 1991.

McCarthy, James E., et al. "Interstate Shipment of Municipal Solid Waste." CRS, August 8, 1990.

MacIntyre, Alasdair. *Three Rival Versions of Moral Enquiry: Encyclopedia, Genealogy and Tradition.* Notre Dame, Ind.: University of Notre Dame Press, 1990.

McKibben, John. *End of Nature.* New York: Random House, 1989.

Managing Global Genetic Resources. *Forest Trees.* Washington, D.C.: National Academy Press, 1991.

———. *The U.S. National Plant Germplasm System.* Washington, D.C.: National Academy Press, 1991.

"Managing Planet Earth." *Scientific American* Special Issue, September 1989.

Manes, Christopher. *Green Rage: Radical Environmentalism and the Unmaking of Civilization.* Boston: Little, Brown, 1990.

Matthews, Jessica Tuchman, ed. *Preserving the Global Environment: The Challenge of Shared Leadership.* New York: W. W. Norton, 1991.

Merleau-Ponty, M. *Phenomenology of Perception.* London: Routledge & Kegan Paul, 1962.

Merrick, Thomas W. "World Population in Transition." *Population Bulletin,* vol. 41 (2). Population Reference Bureau, April 1986.

Miller, Alice. *The Drama of the Gifted Child: The Search for the True Self.* New York: Basic Books, 1981.

Mokyr, Joel. *The Lever of Riches: Technological Creativity and Economic Progress.* New York: Oxford University Press, 1990.

Montagu, Ashley. *Human Heredity.* Cleveland: World, 1959.

Myers, Norman. *The Gaia Atlas of Future Worlds: Challenge and Opportunity in a Time of Change.* New York: Doubleday, 1990.

———. *A Wealth of Wild Species: Storehouse for Human Welfare.* Boulder, Colo.: Westview Press, 1983.

Naar, John. *Design for a Livable Planet: How You Can Clean Up the Environment.* New York: Harper & Row, 1990.

Nasr, Seyyed Hossein. *The Encounter of Man and Nature: The Spiritual Crisis of Modern Man.* London: George Allen and Unwin, 1968.

Newsday staff. *Rush to Burn: Solving America's Garbage Crisis?* Washington, D.C.: Island Press, 1989.

Norse, Elliott A. *Ancient Forests of the Pacific Northwest.* Washington, D.C.: Island Press, 1990.

Novak, Michael. *The Experience of Nothingness.* New York: Harper & Row, 1970.

Oelschlaeger, Max. *The Idea of Wilderness: From Prehistory to the Age of Ecology.* New Haven: Yale University Press, 1991.

O'Grada, Cormac. *The Great Irish Famine.* Dublin: Gill & Macmillan, 1989.

Oppenheimer, Michael, and Robert H. Boyle. *Dead Heat: The Race Against the Greenhouse Effect.* New York: Basic Books, 1990.

Ornstein, Robert, and Paul Ehrlich. *New World, New Mind: Moving Toward Conscious Evolution.* New York: Doubleday, 1989.

Palais, Julie M. "Polar Ice Cores." *Oceanus* 29 (4), Winter 1986–87, pp. 55–63.

Pang, Kevin D. "The Legacies of Eruption." *The Sciences,* vol. 31 (1), January 1991, pp. 30–35.

Parry, Martin. *Climate Change and World Agriculture.* London: Earthscan Publications, 1990.

Piotrow, Phyllis Tilson. *World Population Crisis: The United States Response.* New York: Praeger, 1973.

"Policies for Fertility Reduction." Asia-Pacific Population & Policy, Population Institute East-West Center, Honolulu, June 1989.

Policy Implications of Greenhouse Warming. Washington, D.C.: National Academy Press, 1991.

Post, John Dexter. *The Last Great Subsistence Crisis in the Western World.* Baltimore: Johns Hopkins University Press, 1977.

Pyne, Stephen J. *The Ice: A Journey to Antarctica.* New York: Ballatine Books, 1986.

Redford, Kent. "The Ecologically Noble Savage." *Cultural Survival Quarterly,* vol. 15 (1), 1991, pp. 46–48.

Reisner, Marc. *Cadillac Desert: The American West and Its Disappearing Water.* New York: Viking, 1986.

Repetto, Robert, and Malcolm Gillis. *Public Policies and the Misuse of Forest Resources.* Cambridge: Cambridge University Press, 1988.

Repetto, Robert, William Magrath, et al. *Wasting Assets: Natural Resources in the National Income Accounts.* Washington, D.C.: World Resources Institute, 1989.

Revkin, Andrew. *The Burning Season: The Murder of Chico Mendes and the Fight for the Amazon Rain Forest.* Boston: Houghton Mifflin, 1990.

Rhoades, Robert E. "The World's Food Supply at Risk." *National Geographic,* April 1991, pp. 74–105.

Roan, Sharon L. *Ozone Crisis: The 5-Year Evolution of a Sudden Global Emergency.* New York: John Wiley, 1989.

"Russia's Greens." *The Economist,* November 4, 1989, pp. 23–26.

Sagan, Carl. *Broca's Brain: Reflections on the Romance of Science.* New York: Random House, 1974.

————. *The Dragons of Eden: Speculations on the Evolution of Human Intelligence.* New York: Random House, 1977.

Sarna, Nahum M. *Exploring Exodus.* New York: Schocken Books, 1986.

Satir, Virginia. *The New Peoplemaking.* Mountain View, Calif.: Science and Behavior Books, 1988.

Schaef, Anne Wilson. *When Society Becomes an Addict.* San Francisco: Harper & Row, 1987.

Scheffer, Victor B. *The Shaping of Enviornmentialism in America.* Seattle: University of Washington Press, 1991.

Schindler, Craig, and Gary Lapid. *The Great Turning.* Santa Fe, N. Mex.: Bear & Company, 1989.

Schneider, Stephen H. *Global Warming: Are We Entering the Greenhouse Century?* San Francisco: Sierra Club Books, 1989.

————, and Randi Londer. *The Coevolution of Climate and Life.* San Francisco: Sierra Club Books, 1984.

————, and Lynne E. Mesirow. *The Genesis Strategy: Climate and Global Survival.* New York: Plenum, 1976.

Schumacher, E. F. *Small Is Beautiful.* New York: Harper & Row, 1973.

Sheldrake, Rupert. *The Rebirth of Nature.* New York: Bantam, 1991.

Shoumatoff, Alex. *The World Is Burning: Murder in the Rain Forest.* Boston: Little, Brown, 1990.

Smith, W. Eugene. *Minamata.* New York: Holt, Rinehart & Winston, 1975.

Socio-Economic Development and Fertility Decline: A Review of Some Theoretical Approaches. New York: United Nations, 1990.

Solkoff, Joel. *The Politics of Food.* San Francisco: Sierra Club Books, 1985.

Teilhard de Chardin, Pierre. *The Phenomenon of Man.* New York: Harper & Brothers, 1959.

Tickell, Crispin. "Environmental Refugees: The Human Impact of Global Climate Change." Unpublished lecture at the Royal Society, June 5, 1989.

United Nations Environment Programme. *The African Elephant.* Nairobi, Kenya: UNEP/GEMS, 1989.

————. *The Greenhouse Gases.* Nairobi, Kenya: UNEP/GEMS, 1987.

————. *The Ozone Layer.* Nairobi, Kenya: UNEP/GEMS, 1987.

————. *Profile.* Nairobi, Kenya: UNEP/GEMS, 1987.

United Nations Environment Programme Environment Brief No. 4. *Hazardous Chemicals.* Nairobi, Kenya: UNEP/GEMS 1987.

van Andel, Tjeerd H., and Curtis Runnels. *Beyond the Acropolis: A Rural Greek Past.* Stanford: Stanford University Press, 1987.

Vitousek, Peter M., et al. "Human Appropriation of Products of Photosynthesis," *BioScience,* vol. 36 (6), June 1986, pp. 368–73.

Wann, David. *Biologic: Environmental Protection by Design.* Boulder, Colo.: Johnson Books, 1990.

Weisman, Steven B. "Where Births Are Kept Down and Women Aren't." *New York Times,* January 29, 1988.

Westbroek, Peter. *Life as a Geological Force: Dynamics of the Earth.* New York: W. W. Norton, 1991.

White, Randall. *Dark Caves, Bright Visions: Life in Ice Age Europe.* New York: American Museum of Natural History with W. W. Norton, 1986.

Wigley, T. M. L., M. J. Ingram, and G. Farmer, eds. *Climate and History: Studies in Past Climates and Their Impact on Man.* Cambridge: Cambridge University Press, 1981.

Wilson, E. O., ed. *Biodiversity.* Washington, D.C.: National Academy of Sciences, 1988.

Witt, Steven C. *BriefBook: Biotechnology and Genetic Diversity.* San Francisco: California Agricultural Lands Project, 1985.

Woodham-Smith, Cecil. *The Great Hunger: Ireland 1845–49.* London: Hamish Hamilton, 1962.

World Meteorological Organization. *Scientific Assessment of Climate Change.* Geneva: Intergovernmental Panel on Climate Change, 1990.

World Rainforest Movement and Sahabat Alam Malaysia. *The Battle for Sarawak's Forests.* Malaysia: Jutaprint, 1989.

World Resources Institute. *World Resources 1988–1989: An Assessment of the Resource Base That Supports the Global Economy.* n.d.

Worster, Donald. *Nature's Economy: A History of Ecological Ideas.* Cambridge: Cambridge University Press, 1977.

Young, Louise B. *Sowing the Wind: Reflections on the Earth's Atmosphere.* New York: Prentice Hall Press, 1990.

Television Programs

Burke, James. "After the Warming." PBS, November 21, 1990.

Moyers, Bill. "Spirit and Nature." PBS, June 5, 1991.

Index

Credits

The author is so grateful for permission to reprint the illustrations and photographs on the following pages.

Page 5: C.D. Keeling, R.B. Bacastow, A.E. Carter, S.C. Piper, T.P. Whorf, M. Heimann, W. G. Mook, and H. Roeloffzen, "A Three-Dimensional Model of Atmospheric CO_2 Transport Based on Observed Winds: Observational Data and Preliminary Analysis" Appendix A of *Aspects of Climate Variability in Pacific and the Western Americas*, Geophysical Monograph, American Geophysical Union, vol. 55, 1989 (Nov.). *Page 20:* David C. Turnley/Corbis. *Page 24*: Global Tomorrow Coalition, *The Global Ecology Handbook*. Copyright © 1990 by the Global Tomorrow Coalition. Reprinted by permission of Beacon Press, Boston. *Pages 32–33*. Population figures were based on historical estimates and data provided by the United Nations Population Fund and the Population Reference bureau. *Page 45:* Computer mosaic by Todd Gipstein, Gipstein Multi-Media Productions, from an 1865 photograph by Alexander Gardner. *Page 48:* Reprinted with permission from William J. Kaufman, *Black Holes and Warped Spacetime*. Copyright © 1979 by W. H. Freeman & Company. *Page 76*: R. S. Bradley, "Precipitation Fluctuations over Northern Hemisphere Land Areas Since the Mid-Nineteenth Century." From *Science*, Vol. 237, p. 171, July 10, 1987. Copyright © 1987by the American Association for the Advancement of Science. *Page 94*: J. M. Barnola, D. Raynaud, C. Lorius, and Y.S. Korotkevich. 1991. Atmospheric CO_2—Atmospheric CO_2 from Ice Cores, Vostok, pp. 4–7. In T. A. Boden, R. J. Sepanski, and F. W. Stoss, eds., *Trends '91: A Compendium of Data on Global Change*, ORNL/CDIAC-46. Carbon Dioxide Information Analysis Center, Oak Ridge National Laboratory, Oak Ridge, Tennessee. *Page 96:* P.D. Jones and T. M. L. Wigley. 1991. Temperature, Global and Hemispheric Anomalies, *pp. 512–17.* In T. A. Boden, R. J. Speanski, and F. W. Stoss, eds., *Trends '91: A Compendium of Data on Global Change*, ORNL/CDIAC-46. Carbon Dioxide Information Analysis Center, Oak Ridge National Laboratory, Oak Ridge, Tennessee. *Page 118:* James P. Blair. Copyright © National Geographic Society. *Page 157:* Reprinted with permission of National Broadcasting Company, Inc. Photograph by Todd Gipsein, Gipstein Multi-Media Productions. *Page 179:* James Natchwey/VII Photo. *Page 193:* Global Stewardship: A Statement of the Context and Challenges Facing the White House Conference on Science and Economics Research Related to Global Change. *Page 198:* Courtesy of Culver Pictures. *Page 235:* Steve Raymer. Copyright © National Geographic Society. *Page 251:* Alinari/Art Resource, New York. I. (And 1104) Raphael, *The School of Athens*: detail of Aristotle and Plato. Vatican, Stanza della Segnatura. *Page 287:* Copyright © 1988 by Miranda Smith Productions, Inc. *Page 299*: NASA. *Pages 17, 165, 267, 369*: computer mosaics by Todd Gipstein, Gipstein Multi-Media Productions. Photo: NASA. Original graphs and images appearing on *pages 5, 20, 24, 32–33, 94, 96, 102, 118, 193, and 299* in the 1992 edition were updated by mgmt.design. *Jacket photo:* Satellite Composite View of Earth by Tom Van Sant and the GeoSphere Project. All rights reserved by Tom Van Sant, Inc., 146 Entrada Drive, Santa Monica, California; with assistance from NOAA, NASA, EYES ON EARTH; technical direction by Lloyd Van Warren; source data derived from NOAA/TIROS-N Series Satellites, completed April 15, 1990.